Evolution of a Taboo

Evolution of a Taboo

Pigs and People in the Ancient Near East

MAX D. PRICE

OXFORD
UNIVERSITY PRESS

Oxford University Press is a department of the University of Oxford. It furthers the University's objective of excellence in research, scholarship, and education by publishing worldwide. Oxford is a registered trade mark of Oxford University Press in the UK and certain other countries.

Published in the United States of America by Oxford University Press
198 Madison Avenue, New York, NY 10016, United States of America.

First issued as an Oxford University Press paperback, 2023

Library of Congress Cataloging-in-Publication Data
Names: Price, Max D., author.
Title: Evolution of a taboo : pigs and people in the Ancient Near East /
Max D. Price.
Other titles: Pigs and people in the Ancient Near East
Description: New York, NY : Oxford University Press, 2020. |
Includes bibliographical references and index.
Identifiers: LCCN 2020023549 (print) | LCCN 2020023550 (ebook) |
ISBN 9780197543276 (hardback) | ISBN 9780197682647 (paperback) |
ISBN 9780197543290 (epub) |ISBN 9780197543306
Subjects: LCSH: Food habits—Middle East . | Swine—Middle East—History—To 1500. |
Swine—Religious aspects—Judaism. | Swine—Religious aspects—Islam. | Taboo—Middle East. |
Mammal remains (Archaeology)—Middle East. | Middle East—Antiquities.
Classification: LCC GT2853.M628 P75 2020 (print) | LCC GT2853.M628 (ebook) |
DDC 394.1/20956—dc23
LC record available at https://lccn.loc.gov/2020023549
LC ebook record available at https://lccn.loc.gov/2020023550

Paperback printed by Integrated Books International, United States of America

To my parents, for encouraging me even during my "pig phases."
And to Jess, for everything else.

Contents

List of Figures xi
Acknowledgments xv

1. The Power of Pigs 1
   Cairo's Pig Problem 1
   The Power of Zooarchaeology 3
   Pigs as a Lens into the Past 5

2. Animals in a Landscape 10
   The Setting: Geography of the Near East 10
   Dramatis Personae: Pigs and Wild Boar 12
   Pig Domestication 16
   Raising Pigs 20
   "Pig Principles" and Types of Data 24

3. From Paleolithic Wild Boar to Neolithic Pigs 27
   Reluctant Hunters of the Lower and Middle Paleolithic 28
   The Upper Paleolithic and Epipaleolithic 30
   The Road to Domestication: The Pre-Pottery Neolithic A 32
      The Intensification of Wild Boar Hunting at Hallan Çemi 35
      Taking Boars on Boats to Cyprus 37
      Wild Boars in Rituals 37
      Domestication by Two Pathways 41
   First Domestic Pigs: The Early and Middle PPNB 42

4. Out of the Cradle 48
   The (Delayed) Adoption of Pig Husbandry in the Near East 49
   The First European Pigs 51
   New Approaches in the Late Neolithic 52
      Penning Pigs 52
      Feasting on Pork 54
      Pigs in Other Rituals 55
   Pigs in the Chalcolithic: The Great Transformation 56

5. Urban Swine and Ritual Pigs in the Bronze Age 62
   Plows, Wool, Warhorses (and Wardonkeys), and Wealth 64
   Regional Diversity in Pig Husbandry in the Early and Middle Bronze Age 67

The (Informal) City Pigs of Early and Middle Bronze Age Mesopotamia 68
The Inheritance of Tradition in Western Anatolia, Egypt, and Iran 71
The Erosion of Pig Husbandry in the Levant and Western Syria 73
Pigs and the Changing Environment 76
Pigs in Texts 77
Pork Consumption in the Late Bronze Age 80
Pigs and the Gods 83

6. Theorizing the Taboo 92
The Anthropology of Taboo 92
Theories of the Pig Taboo 96
Biblical Explanations 96
Classical Writers on Pig Taboos 98
Health-Related Explanations 99
Religious Explanations 101
Douglas's Physiological Explanation 103
Ecological Explanations 104
Political-Economic Explanations 106
Ethnic-Political Explanations 107
The Chicken Explanation 110
Making Sense of the Pig Taboo 111
The Pig Taboo as an Evolving Cultural Element 114

7. The Coming of the Taboo: Pigs in the Iron Age 116
The Iron Age and Israelite Ethnogenesis 116
The Writing of the Torah and the Pork Taboo 118
The Archaeology of the Israelites and the First Jews 119
Pig Husbandry in the Iron Age 124
Pig Husbandry and Avoidance in the Southern Levant 124
The Evolution of the Israelite Pig Taboo 128
Writing the Taboo 130
Pig Taboos in Other Parts of the Near East 133
The Genetic Turnover 137

8. Clash of Cultures in the Classical Period 142
Pigs in Greek and Roman Cultures 144
Economic Roles 145
Ritual and Cultural Significance 148
Zooarchaeology of the Greco-Roman Near East 151
Raising Pigs in the Classical Period 152
Judaism and the Levant in the Classical Period 154
Unholy of Unholies: Pigs and the Clash of Cultures 159
Between Judaism and Christianity 164
Pigs in Christian Thought 169

9. Islam and the Modern Period 172
Islam: Orthopraxy, Food Laws, and the Pig Taboo 173
Zooarchaeological and Historical Data on Pigs 176
Raising Domestic Pigs in the Near East Today 179
Informal Economics: Pig Husbandry in Egypt 179
Formal Economics: "White Steak" and Pig Husbandry in Israel 181
Wild Boar in the Near East Today 183
Swine, Bigotry, and Intolerance 184
Transgression 189

10. The Complexity of Swine 195
Swine in Retrospect: A Window onto Complexity 195
Domestication 197
The Ritual Significance of Swine 199
The Unpredictable Evolution of the Pig Taboo 200
The Uniqueness of the Pig Taboo 203
Tradition and Fate 204

Appendix 207
Notes 213
References 247
Index 309

Figures

2.1 Topographic map of Middle East. Modified from image downloaded from Wikipedia Commons, https://de.wikipedia.org/wiki/Datei:Middle_East_topographic_map-blank.svg. 11

2.2 Physiological and behavioral changes in domestic pigs. Animal silhouettes drawn by Michel Coutureau in collaboration with Vianney Forest (INRAP), ©1996 ArcheoZoo.org. 18

3.1 Stone pillar from Göbekli Tepe (Enclosure C) decorated with bas-relief image of a male wild boar and other animals. Photograph D. Johannes. © Deutsches Archäologisches Institut, Berlin. 40

3.2 The gradual decline of pig dental size over time at Çayönü and nearby sites. Numbers inside boxplot squares indicate sample size. This figure uses the log-size index of molar breadth measurements. This method compares observed values against those of a standard, in this case the mean values of the modern wild boar presented in the figure. Data from Payne and Bull 1988 (modern wild boar reference); Monahan 2000; Price 2016; Price and Hongo, in press. 45

4.1 Steady and continual reduction in the size of pigs. This figure was drawn with log-size index values, using average modern wild boar as a standard. Data from Payne and Bull 1988 (modern wild boar reference); Price 2016; Price and Hongo, in press. 53

4.2 4th millennium BC cylinder seal impressions containing scenes of wild boar hunting. Images kindly provided by Robert Englund (see Englund 1995). 57

4.3 Proto-Elamite vessel of a pig. Southwestern Iran, ca. 3100–2900 BC. Purchase, Rogers Fund and Anonymous Gift, 1979. Accession number 1979.71. The Metropolitan Museum of Art. 58

5.1 Number of administrative texts mentioning pigs and other livestock species. Cuneiform terms for species in parentheses. Texts refer to live animals or animal products. Data from Cuneiform Digital Library Initiative, https://cdli.ucla.edu. 70

5.2 Relative abundances of pigs in southern Levant. Numbers inside pig silhouettes show percentage of pigs. Numbers below indicate site. *Key to sites*: 1, Pella; 2, Marj Rabba; 3, Tel Teo; 4, Meser; 5, Tel Aviv Jabotinsky St.; 6, Teleilat al Ghassul; 7, Abu Hamid; 8, Tel esh Shuna; 9, Tel Ali; 10, Bir es Safadi; 11, Horvat Beter; 12, Abu Hamid; 13, Shiqmim; 14, Gilat; 15, Grar; 16, Ai et Tell; 17, Tel Halif; 18, Megiddo; 19, Tell Abu al Kharaz; 20, Tel Yaqush; 21, Tel es Sakan; 22, Yiftahel; 23, Ashkelon; 24, Tel Hartuv; 25, Tel Bet Yerah; 26, Qiryat Ata I III; 27, Tel Dalit; 28, Tell Madaba; 29, Tel Lod; 30, Tel Yarmouth; 31, Tel Arad; 32, Tel Erani; 33,

Khirbet al Minsahlat; 34, Kh. ez Zeraqon; 35, Tel es Safi; 36, Tell Handaquq; 37, Tell al 'Umayri; 38, Tell Abu en Niaj; 39, Tel Dan; 40, Refaim Valley; 41, Tell el Hayyat; 42, Shiloh; 43, Tel Aphek; 44, Tel Yoqne'am; 45, Tel Haror; 46, Tell Jemmeh; 47, Tel Kabri; 48, Tel Hazor; 49, Tel Nagila; 50, Jericho; 51, Tel Harasim; 52, Beth Shean; 53, Lachish; 54, Tel Dor; 55, Beth Shemesh; 56, Tel Rehov; 57, Tel Kinrot; 58, Miqne Ekron; 59, Nahariya. Data from Allentuck 2013; Allentuck and Rosen 2019; Fall et al. 1998; Hellwing and Gophna 1984; Horwitz 1997, 2003, 2009; Horwitz and Lernau 2005; Kansa 2004; Lev Tov 2010; Marom et al. 2014; Marom and Zuckerman 2012; Price et al. 2013, 2018; Sapir Hen et al. 2013, 2016; Van den Brink et al. 2015; Vila and Dalix 2004; Wapnish and Fulton 2018; Wapnish and Hesse 1988. 74

5.3 Pig sties in Building 400 at Amarna, Egypt. Photo courtesy of The Egypt Exploration Society and the University of Oxford Imaging Papyri Project. Plan courtesy of Barry Kemp. 81

5.4 Amulet or clay plaque from Nippur showing a boar mounting a nursing sow (probably Old Babylonian period, early 2nd millennium BC). 10.5 × 7 cm. Image courtesy of the Penn Museum, Image #296755. 85

5.5 Obsidian amulet of Lamashtu with dog and pig. Early 1st millennium BC. 5.7 × 4.7 × 0.9 cm. Purchase, James N. Spear Gift, 1984. Accession number 1984.348. The Metropolitan Museum of Art. 88

7.1 The genetic replacement of pig haplogroups across the Near East. Four regions compared: Anatolia, Levant, N. Mesopotamia/S. Anatolia, and Iran/Caucasus. Points indicate locations of sites. Data from Meiri et al. 2017; Ottoni et al. 2012. 138

8.1 Marble funerary stela for a pig killed in an accident en route to a Dionysia festival. Pella, Greece, 2nd–3rd century AD. The inscription reads: "I, the Pig, beloved of all, a four footed youngster, am buried here. I left the land of Dalmatia, when I was given as a gift. I stormed Dyrrachion and yearned for Apollonia, and I crossed every land on foot, alone and invincible. But now I have departed the light on account of the violence of the wheel, longing to see Emathia and the wagon of the phallic procession. Now here I am buried in this spot, without having reached the time to pay my tribute to death." Translation by Onassis Cultural Center, New York, "A World of Emotions." The credits on the depicted monument belong to the Ministry of Culture and Sports of the Hellenic Republic. The monument belongs to the Ephorate of Antiquities of Pella (object number AKA 1674). Ministry of Culture and Sports / Archaeological Resources Fund. 144

8.2 Terracotta statuette of Eros, god of love and sex, astride a pig. Southern Italy, 3rd century BC. Height 11.1 cm. Rogers Fund, 1919. Accession number 19.192.75. The Metropolitan Museum of Art. 149

8.3 Herakles delivering the Erymanthean Boar alive to Eurystheus. Athens, ca. 510 BC. 43 × 28.2 cm. Object number 86.AE.83. The J. Paul Getty Museum. 150

8.4 Intensification of pig husbandry in the Classical period at Gordion (central Anatolia). 153
Data reflect ages at death reconstructed from dental eruption and wear patterns in Roman and Hellenistic occupations. Lines represent the declining probabilities that a piglet born in either phase will reach successive age classes. Data from Çakırlar and Marston 2019.

8.5 *St. Anthony*, 1564. Engraving by Hieronymus (Jerome) Wierix (ca. 1553–1619). Bequest of Phyllis Massar, 2011. Accession number 2102.136.699, The Metropolitan Museum of Art. 170

9.1 Woodcut entitled *Das grosse Judenschwein* (The Jews' Big Pig). Germany. 15th century. See Fuchs 1921. Image downloaded from Wikipedia Commons, https://commons.wikimedia.org/wiki/File:Judensau_Blockbuch.jpg. 187

Acknowledgments

What is it about the pig? An animal familiar and foreign, inhuman and eerily all too human. We use its name to offend those we dislike, while its fulsomest qualities encapsulate the deepest fears we harbor about ourselves. We rarely admit what we know to be true: swine are our uncanny reflections.

A sense of the uncanny has lingered around my work on pigs. My singular focus on their history in the ancient Near East has been an exercise in exploring an unusual animal within cultural contexts that seem recognizable but are ultimately unfamiliar. Rather than telling a good story, as so many others have done, my goal has been to develop a more accurate picture of the pig and the development of the taboo. I argue against single-theory explanations for pigs' historical trajectory and instead explore the evolution of their cultural significance.

This book also represents a moment in a long and personal struggle with a Jewish identity that I have often held at arm's length—not only because, from an early age, I embraced atheism, and later Marxism, but also because, as the product of a "mixed marriage," I was often rejected by Judaism. Perhaps it is in an effort to understand this mutual estrangement that I have followed the rabbinic commandment and sought refuge in study—but a study of that which the rabbis have utterly rejected.

There are many people who helped this book along to publication. Some did so through their mentorship: Richard Meadow, Gil Stein, Kate Grossman, Cheryl Makarewicz, Heather Lechtman, Jason Ur, and Rowan Flad. Others contributed by reading drafts and exchanging ideas, or just chatting about pigs, Israelites, taboos, domestication, and animal husbandry: Josh Walton, Hitomi Hongo, Deirdre Fulton, Tate Paulette, Yorke Rowan, Morag Kersel, Lee Perry-Gal, Janling Fu, Canan Çakırlar, Nimrod Marom, Yitzchak Jaffe, Noa Corcoran-Tadd, Ari Caramanica, Bridget Alex, Bastien Varoutsikos, Jeff Dobereiner, Bridget Alex, Nat Erb-Satullo, Roz Gillis, Louise Bertini, Ben Arbuckle, Salima Ikram, Allowen Evin, Abra Spiciarich, Keith Dobney, Brian Lander, Mike Fisher, Thomas Cucchi, Ajita Patel, Mindy Zeder, Siavash Samei, Umberto Albarella, Lidar Sapir-Hen, Greger Larson, Laurent Frantz, Kathy Twiss, Jesse Wolfhagen, Naomi Sykes, Sadie Weber, Francesca Slim,

Taylor Hermes, Sarah Pleuger, and Kathryn O'Neil Weber. Thanks for your kindness and wisdom. Many thanks also to Stefan Vranka, my editor at Oxford University Press. I was fortunate to have two academic postings while writing this book—at MIT as a lecturer in the Department of Materials Science and Engineering and at the University of Kiel on a fellowship supported by the Alexander von Humboldt Stiftung.

And to my family, for all their patience with me throughout this project and their unwavering support, thank you.

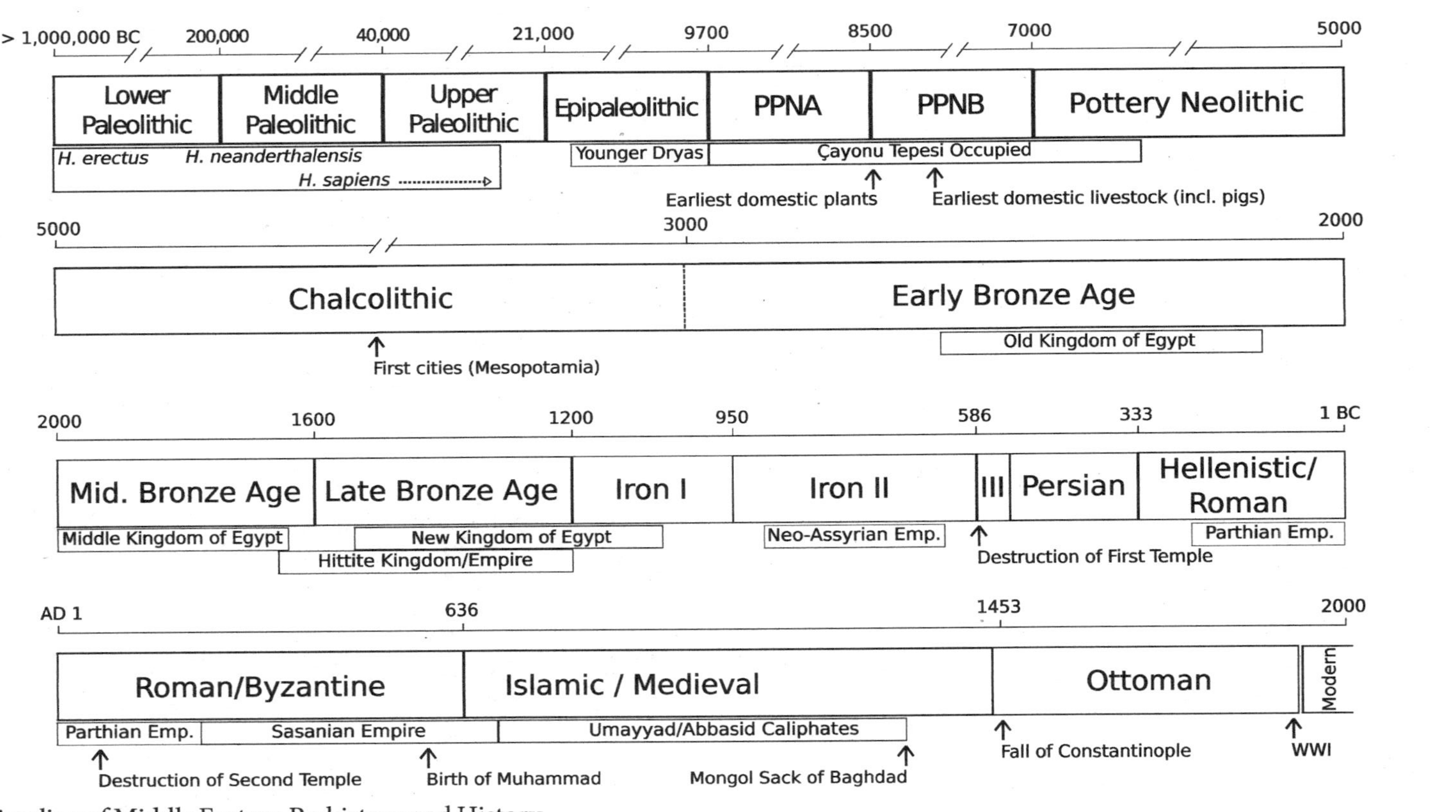

Timeline of Middle Eastern Prehistory and History

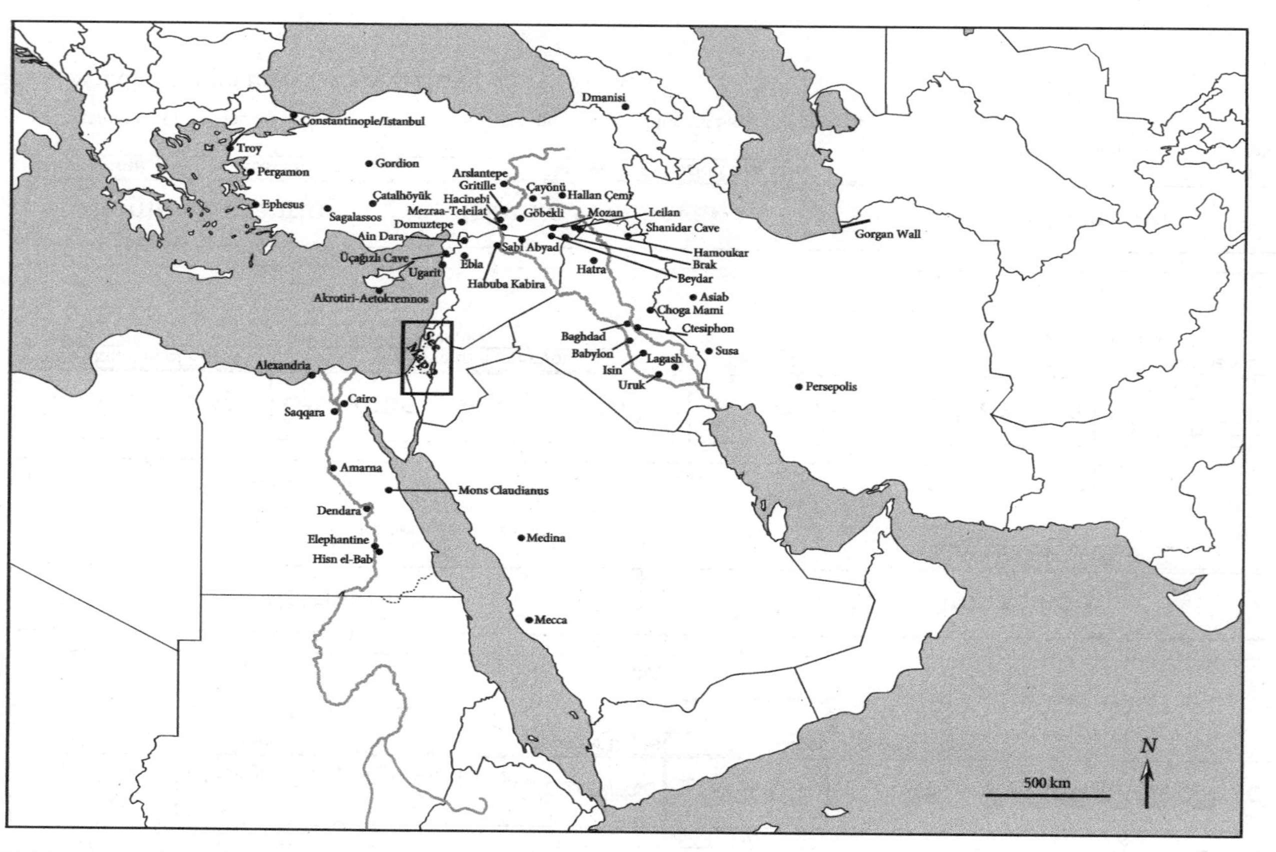

MAP 1 Map of Middle East with Sites Mentioned in Text

MAP 2 Map of Southern Levant with Sites Mentioned in Text

Evolution of a Taboo

1
The Power of Pigs

Cairo's Pig Problem

In the spring of 2009, the Near East stood poised on the threshold of a regionwide crisis. In the Levant, Israel had just finished a campaign against Gaza that had cost the lives of 1,500 Palestinians, mostly civilians, and Hamas was still launching rockets into southern Israel, explicitly targeting civilians. In Mesopotamia, as the American occupation of Iraq approached its seventh year and the insurgency against it faded, its most vicious opponents began reorganizing themselves into a group that would, four years later, declare itself the Islamic State in Iraq and Syria (ISIS). In Iran, anti-government demonstrations erupted across the country to protest the reelection of President Mahmoud Ahmadinejad. And in Syria, a country held together since 1971 by a brutal Baathist regime, a multi-year drought that was likely intensified by global warming had forced hundreds of thousands of Syrian farmers from their homes.[1] They gathered in the cities, desperately searching for income in the shadows of a global economy shattered by Wall Street's recklessness.

To add to the growing calamity, a major health crisis was developing. In what would become a dress rehearsal for the much more devastating COVID-19 pandemic a decade later, the H1N1 "swine flu" rapidly turned into a worldwide epidemic. Governments around the globe scrambled to prevent outbreaks. In Egypt, President Hosni Mubarak's government placed medical personnel at airports and began a campaign to vaccinate anyone traveling for the Hajj, the annual pilgrimage to Mecca.

In April, the government also decided to slaughter all the pigs in Cairo and its suburbs.

The decision would come at a high price. A Christian minority, referred to as the Zabaleen, had raised pigs in Cairo and its outskirts for generations. Their livelihood leaned heavily on collecting refuse from the city's streets and feeding its organic components to their pigs, the meat of which they would sell to supplement their often meager incomes.[2] But the Zabaleen and

Evolution of a Taboo. Max D. Price, Oxford University Press (2020). © Oxford University Press.
DOI: 10.1093/oso/9780197543276.001.0001.

their pigs played a much larger role in the Egyptian economy than simply subsisting off its urban waste. They stood at the heart of an informal waste management and recycling system on which the 20 million residents of metropolitan Cairo depended to keep their city clean.

Many, including international health experts, saw the Cairo pig cull for what it was: an attack on a way of life that had long triggered discomfort among the majority. Pigs are *haram*, forbidden by Muslim dietary laws.[3] The very thought of pigs can elicit disgust and disdain among Muslim Egyptians. For that reason, despite their role in waste management, there was considerable pressure to keep swine out of Cairo. The H1N1 outbreak provided a convenient opportunity to rid the city of an animal to which was attached one of the most powerful taboos in the world.

The slaughter met swift resistance in the form of a citywide strike; the Zabaleen refused to pick up the trash. Reporting for the *New York Times*, Michael Slackman interviewed local Zabaleen:

> "They killed the pigs, let them clean the city," said Moussa Rateb, a former garbage collector and pig owner who lives in the community of the Zabaleen. "Everything used to go to the pigs, now there are no pigs, so it goes to the administration."[4]

Within days, filth piled up in the streets, bringing parts of Cairo to a standstill and exposing the Mubarak government's incompetence. But Moussa Rateb's implication—that the government was the real swine—reflected broader disenchantment with the way that Egypt and other countries in the Near East were being run. People's distrust of those in power and their frustration with their leaders' inability to provide them opportunities for better lives would eventually erupt into a regionwide social movement. The Zabaleen strike itself offered a preview of the Arab Spring and Egypt's revolution two years later, an example of how grassroots mobilization of discontented people could defy seemingly powerful governments.[5]

Although often forgotten in the tangled web of political events and violence that defined the Near East over the next decade—from the ousting of Mubarak, to the eruption of the Syrian civil war, to the rise and fall of the Islamic State, to the increasing tensions between Turkey, Iran, Saudi Arabia, the US, and Russia for influence over the region—Cairo's pig problem reflected in microcosm the greater political, economic, and environmental challenges facing the Near East. In Cairo, pigs acted as a figurehead for class and ethnic conflict. Their presence forced discussions about religious tolerance, respect for Islam's tenets in a Muslim-majority country, and the strength of political liberty. Swine had crept into discussions of public health and the management of waste in the Near East's largest city. And they raised questions in the West about what this taboo on pigs was all about and why it was so important.

The clash between the Egyptian government and the Zabaleen highlights the multifaceted and socially entangled roles that pigs play in the Near East. These 21st-century predicaments reflect a deep and recurring historical theme: although often left out of the popular imagination of daily life in the region, Near Eastern pigs have long been uniquely situated within greater social processes, trapped within the politics of different ethnic and religious groups. In that way, swine offer an underexplored perspective on human cultures in the region. By understanding the pig, we can begin to appreciate the complexities of culture and politics in the Near East.

This is no easy task. Historians, archaeologists, anthropologists, and theologians have long studied and speculated about swine's role in Near Eastern cultures, largely with regard to the origins and significance of the taboo on pigs in Judaism and Islam. For the most part, these scholars have been badly mistaken—not because of a lack of intellectual rigor, but because they have not had solid data on which to ground their arguments. This situation has changed over the past three decades largely in thanks to the work of zooarchaeologists, or archaeologists who specialize in studying animal bones and understanding human-animal interactions in the past.

The Power of Zooarchaeology

Why is it that the pig, an otherwise uncomplicated animal in other cultures, is the focus of so much consternation for Near Eastern peoples? After all, archaeological research documents that pigs were domesticated within the Near East around 10,000 years ago and remained a part of agricultural life throughout the region for millennia. Pork was eaten in abundance by kings, soldiers, merchants, and the poor in the first cities of Sumer, Syria, Anatolia, and Egypt. Pigs and their progenitors, wild boar, were depicted in artwork and writing composed for imperial courts. They had roles in rituals of magic and religion and were part of the tableau of daily life in the Near East for most of recorded history. What changed? How did swine traverse the road from wild boar to domestic pig to an animal so taboo in Judaism and Islam that even mentioning its name can elicit disgust?

This book will tackle these long-standing questions by tracing the history of swine in the Near East from the earliest moments of human prehistory up to the present day. In doing so, it joins other works devoted to understanding the unique position this animal has held. Authors of all stripes have wondered about the pig and its taboo since at least Greco-Roman antiquity. They have put forth many theories—for example, some have claimed that the taboo was a response to pig-borne pathogens like trichinosis; others that the pig is ecologically unsuitable for the Near East; still others that the pig taboo is essentially a confused attempt to understand a symbolically powerful animal, an inversion of sentiments surrounding what was once a holy animal. As we will see, all of these theories are wrong, or at least partially so.

I should warn my readers that this book does not present a single explanation for why the pig got to be the way it is. There is no Sherlock Holmes–style discovery in its pages. Rather, it is a story of converging factors, competing interests and ideologies, and contradictions. It is a tale with many loose ends and much need for future research.

Theories of the pig and its taboo have filled countless pages for almost 2,000 years. But it is only recently, with the aid of zooarchaeology, that scholars have left the realm of speculation and tested their theories empirically. In the process, they have developed a more accurate picture of pigs in the Near East, from domestication to the formation of the taboo to the Zabaleen's and other humans' interactions with pigs in the region today.

Although we zooarchaeologists focus our attention on animals and their remains (i.e., bones), we do so in order to understand human behavior. We do this not because we love animals per se or because we couldn't cut it as veterinarians (although both of those statements may be true), but because animals provide a unique insight into the human experience. They stand at the center of so much of human activity and thought; they provide meat and milk for our tables, labor for our farms, and material for our clothing. On a deeper level, they supply us with metaphors for ourselves. They provide the archetypal characters in the drama of life and death.[6]

Zooarchaeology provides a powerful scientific tool for understanding how pigs and other animals have shaped human history. Excavators recover animal remains from sites—often from ancient garbage dumps where meal refuse was discarded, but also from other deposits like the remains of ritual sacrifices in temples and human graves. Field archaeologists then pass these remains along to faunal specialists (often themselves site directors or excavators). Once on the lab bench, zooarchaeologists identify the animal species to which the recovered fragments of bones and teeth belong; measure these fragments to determine the animals' age, sex, and domestication status; and examine them for pathological lesions to determine what stresses and diseases the animals' may have been exposed to. Some of us examine bones on the microscopic level, sequencing ancient DNA to document population turnovers and unique phenotypic traits, or analyzing ratios of light stable isotopes to reconstruct the ancient environments and diets.

The cumulative work of zooarchaeologists has fostered a breakthrough in understanding past societies. Rigorous scientific approaches to archaeology are only a couple of generations old, and zooarchaeology itself was a relatively marginal subdiscipline until the 1970s and 80s. But since that time, it has become one of the most popular methodological approaches in archaeology. Researchers from around the world have revolutionized the study of animals' centrality to human cultures, from the diversity in hunting strategies and trajectories toward domestication, to the ritual use of animals in human spiritual life, to the fundamental roles that domestic livestock played in early state societies.[7] From zooarchaeologists' tireless efforts, there is now a considerable body of evidence pertaining to all periods of history and prehistory. This is especially the case in the Near East. It is those data I will bring to bear on questions regarding one of the region's most unique animals: pigs.

Pigs as a Lens into the Past

Zooarchaeology is not about telling the stories of dead animals, but uncovering what they meant to the human societies that hunted, herded, and tabooed them. The pig's story, which the reader will trace over the next nine chapters, is a representation of—and a lens through which to perceive—a broader socioeconomic, environmental, and political history. If nothing else, in the pages that follow, I hope to convince the reader that pigs offer a unique perspective on the Near East's long-term social processes. The goal is simple, if indirect: as biomedical researchers study the organs of pigs to better understand human biology, zooarchaeologists investigate the pig to learn about the past, and thereby to understand the human condition.

Swine provide a particularly useful lens for delving into the Near Eastern past because, as the Cairo episode demonstrates, they often find themselves on the battlegrounds (or as *the* battleground) of the politics of identity and piety. Today, Muslims' and Jews' negative attitudes toward these animals often clash with the deep love of pork, ham, and bacon harbored by hundreds of millions of Christians and other people around the world. To the members of each faith, the position opposing their own is ridiculous, insane, and unfathomable. How can you eat something as abominable as *that*? How can you detest something as mundane and delicious as *this*? Addressing these sentiments strikes at the heart of questions of tolerance. Understanding the other side, or refusing to do so, has a long legacy in interethnic and interreligious relations in the Near East and beyond.

Sometimes the conflicting attitudes toward pigs are found within the same community, family, or even individual persons. Recognizing that these contradictions exist today and have existed for millennia is a critical part of the story of swine. Today, as in the past, not every Jew or Muslim abstains from pork, and not every Christian thinks it's okay to eat swine. In Israel, for example, the market for pork has blossomed in recent decades as curious or secular Jews and émigrés from the former Soviet Union have sought out the forbidden flesh—even despite the legal controversy surrounding it and a windstorm of finger wagging from Orthodox rabbis.[8] Love and hate, curiosity and taboo have made the first pork cookbook in Israel (titled *The White Book*)[9] not so much a market success as a salacious offering in a nation bitterly divided between Jewish nationalists and secular cosmopolitans.[10]

Of course, pigs are not unique to the Near East, but instead have a global impact.[11] Today, over a billion pigs are slaughtered every year to satisfy the globe's unrelenting demand for ham, bacon, pork chops, ribs, salami, lard, prosciutto, and pork belly.[12] This would not be possible if swine had not been domesticated, something that occurred at least twice—once in the northern Mesopotamian region of the Near East and once in China. Were it not for early Holocene sedentary hunter-gatherers in the Near East and China and the unintentional consequences of their efforts to manage wild boar populations, there would be no industry in pork. Similarly, had a taboo on pigs not developed in certain corners of the Near East in 1st millennia BC–AD, we would not have the widespread refusal to eat pork by most of the world's 2 billion Muslims, 15 million Jews, 40 million Ethiopian Orthodox Christians, and many other groups. Nor we would have pig-related hate crimes like the one that occurred on December 7, 2015, when worshippers in Philadelphia found a pig's head on their mosque's doorstep, a blatant attempt to belittle and terrorize the city's Muslim community.[13]

These global encounters cannot be understood without a firm grasp on the history and archaeology of the Near Eastern pig. We study the past to learn about ourselves, not only where we come from but who we are. While it is all too easy to view history, especially ancient history, as something remote and unrelated to the present, even a cursory inspection reveals just how much our daily lives have been shaped by millions of decisions made thousands of years ago in faraway places. The millennia-long history of the Near Eastern pig still resonates today. Understanding that story through zooarchaeology brings us closer to appreciating our place within a rapidly globalizing world.

The story of pigs in the Near East that will unfold in the chapters to come is complex and historically contingent. One cannot attribute the changing role of pigs or the rise of the pork taboo to a single underlying factor, although many have attempted to do just that (discussed in Chapter 6). Simply put, the cultural significance of pigs *evolved*. Only by understanding the animal and its biology (Chapter 2) and then tracing its trajectory through archaeology, zooarchaeology, anthropology, and historical texts can we hope to understand how pigs came to be what they are today. We cannot understand this process in a vacuum; the place of pigs within Near Eastern cultures evolved in relation to other social processes. For that reason, the Near Eastern pig is remarkably complex.

The transition from wild boar, an animal hunted infrequently by the Paleolithic peoples of the Near East, to domestic pig in the early Holocene sets the story in motion (Chapter 3). Pigs, along with sheep, goats, cattle, and domestic plants, formed a "Neolithic package" that served as the foundation of the human diet in the region for millennia to come. But from the beginning, swine were unique. They were excluded from most forms of mobile pastoralism, and they produced no "secondary products" such as milk or wool (Chapter 4). On the other hand, pigs' abundant dietary flexibility and their capacity to adapt well to urban environments made them ideal forms of livestock in the Near East's first cities in the 4th and 3rd millennia BC. Pork was one of the main sources of meat for the world's first urbanites, especially those on the lower rungs of the socioeconomic ladder. They also played unique religious roles. People sacrificed piglets to honor fertility deities and their dialectical opposites, the gods of the underworld. Pigs served as substitutes for humans; the gods accepted pork in the place of human flesh.

It was, I argue in Chapter 5, elites' quest for storable and valuable agricultural commodities such as cereal, dairy, and wool that signaled a major shift for pigs. Sheep, goats, and cattle proved useful in these regards; pigs less so. As a result, pigs did not become sources of wealth in the way that these ruminating animals did (and horses, donkeys, and camels later on). While they continued to eat pork, economic and political elites largely excluded pigs from the institutional economies that they founded—the palaces, temples, and manorial estates that formed the nuclei of ancient economic life. Additionally, in some limited cases, some temples began to ban pigs from their premises by the middle of the 2nd millennium BC. The reasons for doing so remain opaque, but the exclusion of pigs effectively transformed the ritual connotations that pigs had carried up to that point. Nevertheless, most people in the Near East continued to eat pork. Among the exceptions were the inhabitants of the Levant, where pig husbandry began to erode beginning in the 3rd millennium BC in favor of wealth-producing ruminant husbandry. Not eschewing pigs per se, the people in this region unintentionally founded food traditions that emphasized beef and mutton, and transmitted those traditions from generation to generation.

By the Iron Age in the late 2nd and early 1st millennium BC, this passive avoidance of pork found fertile ground in the ethnogenesis of the Israelite people to grow into a taboo. It emerged first as a point of conflict with the Israelites' pork-eating neighbors, the Philistines. Later, during a period of political upheaval and existential anxiety, the biblical authors revitalized this taboo as part of their romanticization of an imagined ancestral way of life based on mobile pastoralism and a tribal ethos. It was this image that the biblical authors attempted to promote in what became a religious revolution that laid the groundwork for Judaism. Pigs played no part in this tableau; eating them detracted from the fantasy of living like the ancestors, from resurrecting a glorious past, and from living a pure life devoid of the taint of an ancient enemy's otherness and ritual pollution (Chapter 7).

Political and religious developments during the ensuing twenty-five hundred years accelerated the process by which swine developed into an animal of intractable cultural significance. By the end of the Iron Age, written Jewish Law forbade the consumption of pork. Yet pork was but one of many prohibitions found in the Torah, the first five books of the Hebrew Bible. It was the violent confrontation between pork-avoiding Jews and pork-loving Greeks and Romans that elevated the pig to a new position. The pork taboo emerged as a symbol of Jewish resistance to Hellenic and Roman culture (Chapter 8). Meanwhile, Christians, originally adherents of Jewish Law, abolished the taboos on pork and other meats in order to facilitate a new focus on the spiritual purity of their adherents' communion with Christ. Several centuries later, Islam adopted an evenhanded approach to Christian and Jewish theologies, taking what its founders perceived to be the middle way between them. While most of the taboos outlined in the Torah were abandoned, the Quran kept what its writers saw as most significant—including the taboo on pork (Chapter 9).

Over the centuries that followed, the differences between these three world religions and their approach to pigs grew more pronounced. As pigs became more reviled in Islam and Judaism, pork became more socially significant for Christians. At the extremes, people in all three religions used pigs, in the flesh or as metaphors, in acts of intolerance and swinish bigotry. These episodes have only served to more deeply entrench sentiments surrounding pigs, trapping Jews, Muslims, and Christians within the politics of swine and identity.

The lesson that the story of swine imparts to readers will vary. But one essential point is that taboos, food preferences, and other elements of culture—"social facts," to use the terminology of Durkheim—evolve along complex trajectories and in relation to many factors. One cannot pinpoint a single cause located at a discrete moment in time for such social facts, nor can one identify a specific historical figure responsible for their genesis. For culture and its evolution exist beyond the individual person or his or her ability to truly comprehend it. Culture is a cradle and a medium; it creates each human being. Lest this philosophical approach to history appear too much in line with Tolstoy's fatalism, let me be clear that human agency plays a vital role. To paraphrase Marx, individual humans do themselves create culture and history, but they do not do so in the ways that they think or hope. Actions, social facts, and events that may appear to their direct observers as trivial often radically reshape the conditions of future generations.

The pig, though perceived by many as a humble animal caught in people's social entanglements, a walking larder of pork that can be dispensed with by a butcher or state-appointed health official, holds more power than meets the eye, certainly more than a mass cull disguised as a public safety measure can hope to destroy. As much as present or future generations may want to free themselves from their power, pigs and their history exist as a sort of monolith among the cultures and peoples of the Near East. Shunned or eaten, reviled or idolized, pigs have irreversibly shaped the past and thereby give structure to the future.

2

Animals in a Landscape

The Setting: Geography of the Near East

The Near East, shown in Figure 2.1, encompasses Southwest Asia and the northeast corner of Africa. The region's position between Asia, Europe, and Africa has made it one of the globe's most dynamic mixing pots of peoples for thousands of years. It served as the setting for the first migrations of the genus *Homo* out of Africa, the first contact between *Homo sapiens* and Neanderthals, the mingling of the earliest farming communities, the development and expansion of the first cities and empires, and the meeting point for traders and warriors coming from China, India, Europe, and Africa. In part because of this cultural diversity and its unique location in the Old World, the Near East is the birthplace of three of the world's major religions—Judaism, Christianity, and Islam—as well as numerous other faiths, such as Druzism, Yezidism, Bahá'í, and Zoroastrianism. Dozens of languages are spoken representing three major world language families—Semitic (e.g., Arabic, Hebrew), Turkic (Turkish), and Indo-European (Kurdish, Persian)—as well as those unrelated to the world's major language families, including tongues extant (e.g., Georgian, in the small Kartvelian language family) and extinct (Sumerian).

Supporting this diversity of peoples and cultures are a range of environmental zones—deserts, marine coastlines, lacustrine shorelines, oak-pistachio forests, grassland steppes, marshes, and alpine tundra. Rainfall, which is one of the key determinants of these environments, varies sharply. To the south, the Arabian Peninsula is characterized by extreme aridity, with less than 100 mm of rain falling in an average year, giving rise to deserts that end at the mountainous coasts of the Red Sea and Indian Ocean.[1] (For comparison, Phoenix, Arizona, receives about 200 mm per year.)[2] Deserts extend across the Sinai Peninsula and into Egypt, where the Nile traces a thin green line through the sand. The desert continues northward into Syria, where it gradually gives way to semi-arid grasslands and eventually forested foothills, and eastward until the Euphrates River and its lush riparian ecosystems.[3]

Evolution of a Taboo. Max D. Price, Oxford University Press (2020). © Oxford University Press.
DOI: 10.1093/oso/9780197543276.001.0001.

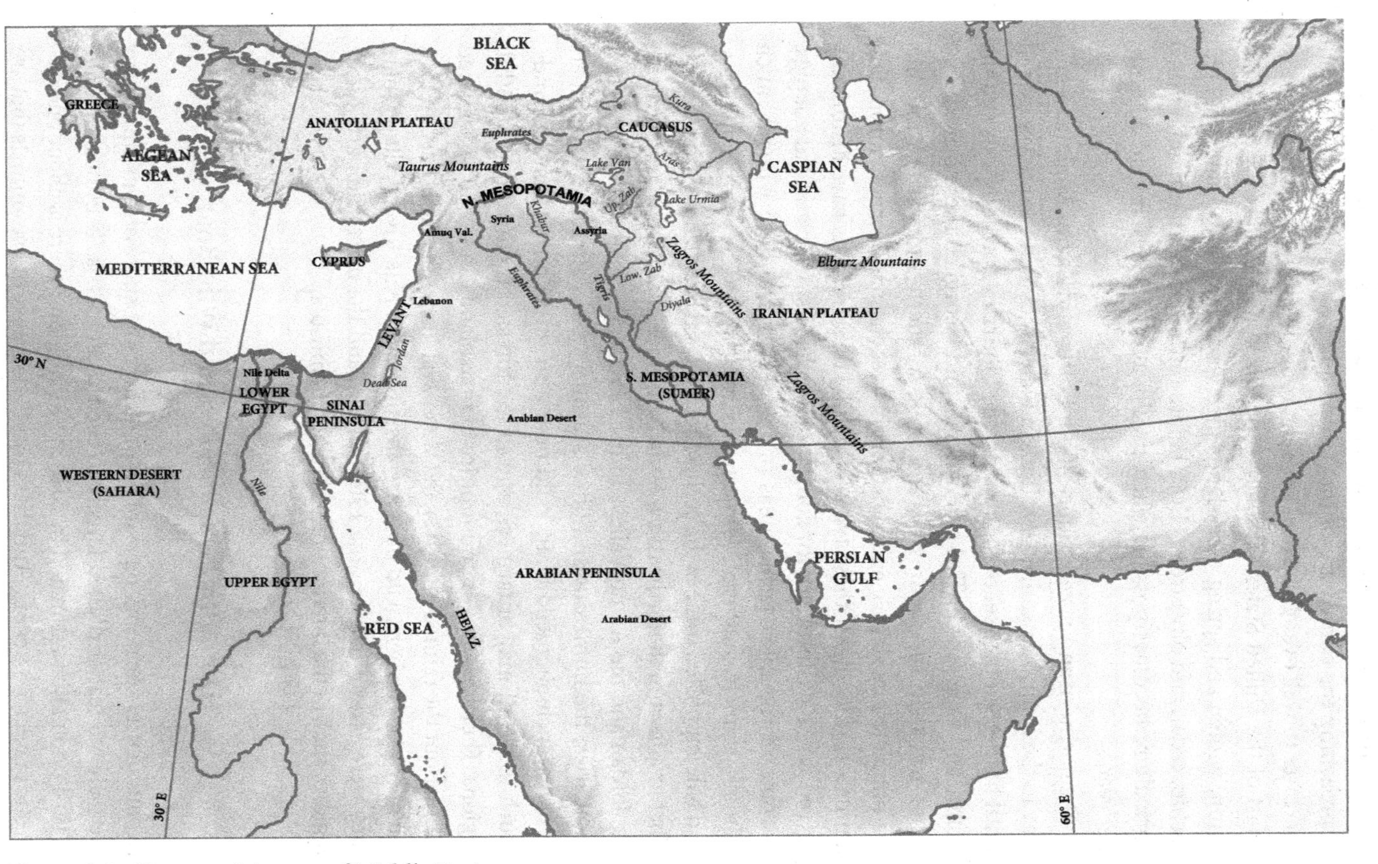

Figure 2.1. Topographic map of Middle East.

Near the coasts of the Mediterranean, Black, and Caspian Seas and approaching the curved arc of the Taurus and Zagros Mountains, rainfall increases to well over 1,000 mm per year, roughly similar to that on the east coast of the United States, although it occurs almost exclusively in the autumn, winter, and spring. Grasslands give way to forests of pine, oak, pistachio, cedar, and olive trees.[4] Even before hitting this wetter band, rainfed agriculture becomes possible when precipitation reaches around 300 mm per year, .

The northern and eastern slopes of the Taurus and Zagros Mountains give way to the central Anatolian and Iranian Plateaus, which themselves extend until hitting another set of mountains: the Elburz in northern Iran and the Pontic and Caucasus in Turkey, Armenia, and Georgia.[5] These plateaus are more arid than the mountains, and their southern and western foothills are populated by steppic and xeromorphic vegetation. Because ruminating animals readily digest these types of plants, these regions have historically served as rich pasturelands for domestic sheep, goats, and cattle, but not pigs.

Some of the most fertile areas of the region are those without much rainfall—namely southern Mesopotamia and Egypt. The Tigris, Euphrates, and Nile supported some of the earliest agricultural heartlands. The northern Khabur alluvium in northeastern Syria is another breadbasket region, as are the Jordan and the Amuq Valleys in the Levant. The major rivers and their tributaries, the numerous and often seasonal wadis, attract large numbers of fauna, including wild boar. Wild boar also thrive on agricultural lands, much to the dismay of farmers, and in the forested mountain and foothill regions. They form large communities in marshlands and swamps scattered throughout the rainfed and river-fed water systems of the Near East. In fact, the only places where wild boar are absent are the Nile River Valley, where they were exterminated in the 19th century, and the most arid parts of the regions, such as the interior of the Arabian Peninsula.

Dramatis Personae: Pigs and Wild Boar

Evolutionarily speaking, pigs are among the most successful mammals on the planet, a feat attributed largely to their usefulness as livestock animals. With a population estimated in 2016 at 981 million,[6] there is about one domestic pig per 7.5 people. And that's just the number of live animals that are counted in official statistics. Pork producers have high turnover rates; a single pig can be born, weaned, fattened, and slaughtered in less than 12 months. For that reason, the number of pigs slaughtered for meat each year is often greater than the number counted as livestock at the end of the year. In 2016, for example, about 1.5 billion swine were slaughtered, more than any other animal besides chickens and 50 percent more than the number counted as livestock.[7] The geographic spread of swine is just as impressive. Domestic pigs are raised on every continent except Antarctica. Wild boar and feral swine (i.e., wild-living animals descended from domestic pigs) are found throughout the earth's temperate regions from Australia to North America.

Domestic pigs descended from Eurasian wild boar, or *Sus scrofa*. The genus *Sus* is classified in the Suidae, a family that includes such distant relatives as peccaries, which are native to the Americas and which split off from the main branch of Suidae about 34–40 million years ago. Within the past 18 million years, the Old World suids have evolved into a number of different genera with diverse characteristics. In Indonesia, one finds babirusa, or "deer-pigs" (*Babyrousa babyrussa*), whose upper tusks emerge through the tops of their snouts. Meanwhile, the beagle-sized pygmy hog (*Porcula salvania*), an endangered species, inhabits eastern India and Bhutan. Other suids, such as warthogs (*Phacochoerus* sp.), giant forest hogs (*Hylochoerus meinertzhageni*), and bushpigs (*Potamochoerus* sp.), are distributed throughout sub-Saharan Africa.[8]

The genus *Sus* evolved in Southeast Asia or Island Southeast Asia sometime in the late Miocene, about 10–5.3 million years ago. Several species still inhabit the region extending across Southeast Asia, Indonesia, and the Philippines: *Sus barbatus*, *S. cebifons*, *S. celebensis*, *S. verrucosus*, and *S. scrofa*, the last of which is the only species found wild across Eurasia.[9] Today, wild boar are an endemic feature of riverine, forest, lacustrine, and marsh environments from Southeast Asia to Siberia and from Scandinavia to northern Africa. Favored sport animals, wild boar have been exported for hunting to other parts of the world, including the Americas.

Sus scrofa's success came at the expense of other members of the *Sus* genus. Prior to the spread of *Sus scrofa* out of Southeast Asia, another suid, *Sus strozzi*, had successfully colonized Europe and the Near East. *Sus strozzi* went extinct one to two million years ago, around the same time that *Sus scrofa* appeared in the paleontological record. The paleontological record reveals a similar story for *Sus minor* and *Sus peii/xiaozhu*, which inhabited China and parts of Europe.[10] The reasons that *Sus scrofa* won out are not known. A likely explanation is that *Sus scrofa* invaded the habitats of other suids and outcompeted them in ways that are not yet understood. But there is also genetic evidence in modern wild boar suggesting that these ancient suid populations interbred.[11] Thus, in addition to outcompeting the other suid populations, Eurasian wild boar may have simply been more prodigious breeders capable of "swallowing" the gene pools of related species. Indeed, such a process appears to be under way today. All species of *Sus* except *Sus scrofa* are currently endangered, in part because *Sus scrofa* have successfully invaded their habitats and hybridized with them.[12] In this way, pig history mirrors human history. Just like *Sus scrofa*—but in a different time and place—anatomically modern humans (*Homo sapiens*) colonized parts of Eurasia inhabited by Neanderthals, a species with which ours interbred and ultimately outcompeted.

If their evolutionary history contains parallels with that of humans, the anatomy of swine also bears some resemblance to our own. Both species are omnivores and therefore possess similar organs associated with digestion. We also share a vulnerability to similar gastrointestinal parasites, such as tapeworm (*Taenia* sp.) and roundworm (*Ascaris* sp.). Pig teeth, like human teeth, are low-crowned, or bunodont. These anatomical similarities make pigs an ideal animal for biomedical testing.

Swine also possess several obvious anatomical features unlike those of humans. They are hoofed mammals (ungulates) with an even number of digits (artiodactyls). They walk on their two middle toes and are therefore "cloven-hoofed," a feature that would later become important to the writers of the Torah. Pigs and wild boar have robust skeletons and thick skulls. Males (boars) possess large, continuously growing canines ("tusks") that present a formidable weapon against predators and other males.

Pigs are intelligent, curious, and social animals, features that scholars have long recognized.[13] Swine are exploratory, seeking out environmental novelties and stimulation. They possess brain structures and biochemistry broadly similar to those of humans, and for that reason pigs are occasionally subjects of cognitive studies.[14] In the wild, females (sows) form small herds, called "sounders," of around two to five mature sows and their offspring. In the wild, young males leave the sounder around sexual maturity—usually around one year of age—and will seasonally compete for opportunities to join a sounder and breed with its females.[15]

Pigs are prodigious breeders, a fact that has made them valuable as livestock. Males and females reach sexual maturity around 1 year of age and can live past 10 years.[16] Although wild boar and feral swine typically have only one litter per year, livestock keepers can reliably achieve two litters per year by taking advantage of two key features. First, gestation time is short—about 114 days. Second, sows can become pregnant within 4–6 weeks of giving birth.[17] Amplifying the productivity of swine-keeping operation are pigs' sizable litters. In the wild, sows typically give birth to 4–6 piglets, but this number is considerably higher in domesticated breeds, especially those that have undergone positive selection for larger litter size and higher teat counts.[18] Depending on the breed and management style, a sow-boar pair can easily produce 25, 100, or more piglets in their lifetimes. This allows for potentially explosive population growth unmatched by single-birth animals like sheep, goats, cattle, horses, and humans.

Pigs are also differentiated from most of the other domestic animals by their diet. With the exception of pigs, the earliest livestock species in the ancient Near East—sheep, goats, and cattle—were ruminants, which extract calories from high-fiber foods through foregut fermentation. Ruminants' multichambered stomachs evolved to hold masticated plant matter, which is often regurgitated and chewed a second time, so that microbes can break down cellulose and convert it into nutrients the animal can absorb. Pigs do not have this ability. They cannot subsist on grasses alone but must seek out high-calorie foods like nuts, tubers, fruits, and seeds. Like humans, chickens, and dogs, pigs are omnivores. They eat insects, worms, small mammals, and birds when the opportunity arises.[19] Not particularly picky, pigs and wild boar also eat carrion and the feces of other animals, including humans.[20] Pigs are also fond of cultivated crops, and wild or feral swine can cause significant damage to crops.[21] For example, in 2007, feral pigs cost farmers in the US about $1.5 billion in damages.[22]

Pigs drink a variable amount of water depending on their age, sex, and health status. Most farmers provide water *ad libitum* and do not keep tabs on how much is going to their livestock, but agricultural scientists have made these calculations (see Table A.1 in the appendix). In general, 5–20 liters of drinking water per day per pig is typical, with more required at higher ambient temperatures.[23] Pigs' daily drinking water needs are greater than those of sheep or goats but considerably less than those of cattle or horses. However, cattle and horses are more efficient consumers of water per kilogram of body mass. Additionally, to enable proper thermoregulation, pigs must have access to water or mud to wallow in when temperatures rise above 30°C.[24]

Pig Domestication

Researchers differ in their definition of the term "domestication" as it applies to animals, each stressing different aspects of this singular form of human-animal relationship.[25] At its core, domestication is an evolutionary process by which a population of animals adapts on a genetic level to the unique ecological niche created when a human society attempts to manage that population for the provision of food, companionship, or some other benefit. Managing an animal population means exerting control over the population's mobility, reproduction, social structure, and/or diet. Thus, management has a direct impact on evolutionary selection pressures.

Most archaeologists agree that management, and the selection pressures it entails, first emerged from one of two situations. Humans may have pursued hunting strategies designed to secure more reliable and predictable sources of meat. Alternatively, humans may have managed animals that had already invaded habitats populated by humans as commensals, perhaps attracted to garbage or safety from predators. Or perhaps a combination of both was in play, as I suggest for the pig. In either case, the domestication of pigs and the other earliest domesticates was not intentional, as early thinkers such as Darwin suspected.[26] Rather, by interacting with animals and altering the set of selection pressures on certain populations to maximize their access to meat, people inadvertently selected for new behavioral and physiological traits.[27]

Selection for the specific traits of domestic animals was a complex process involving feedback between human and animal partners. The processes of domestication were likely unique from context to context. In general, new biological traits probably arose from a combination of the relaxation of selection pressures on wild populations and specific adaptations by animal populations to life among—and exploitation by—humans. For example, one hallmark of domestic animals, variable coat colors, might have persisted in domestic populations simply because camouflage was no longer necessary, while the behavioral traits for which domestic animals are best known (being less afraid of humans or other animals) have helped animals adapt to their roles as livestock. It is important to recognize that traits like these not only developed in response to exploitation by humans, but also facilitated it. Friendlier and more docile animals were easier for humans to manage in larger numbers. These adaptations pushed humans to reconsider their relationships with animals, initiating a revolutionary cultural transformation that flipped the logic of hunting on its head: people shifted from a focus on obtaining meat from dead animals to acquiring and maintaining live ones. The process of domestication thus involved feedback between human and animal partners, as well as between cultural and biological traits, that propelled a continual ratcheting up of human exploitation, animal adaptation, and human adaptation to new opportunities presented by behaviorally and physiologically modified animal populations.

From an evolutionary standpoint, the domestication process proved incredibly successful for animal populations. By living in the "human niche," animals essentially piggybacked onto the success of human populations.[28] But domestic animals are not free-riders. They contribute to the success of human populations by providing reliable and movable sources of food, traction power, and clothing—not to mention companionship. Animals were central to the complex economic systems that emerged in the ancient Near East. Domestication is therefore best thought of as a unique form of symbiosis with vague parallels to other forms of mutualism known to biologists.[29]

Across species, researchers have noted that many domestic animals share similar traits. This is referred to as the "domestication syndrome," and the traits associated with it are shown in Figure 2.2. They include floppy ears and tails, variable coat colors, increased fertility, and shorter snouts. They also include smaller teeth and, at least initially, smaller body size. Domestic animals tend to have smaller brains than wild animals. The brains of domestic pigs, for example, are about 35 percent smaller than those of wild boar, with major reductions in those parts of the brain associated with memory and emotional response.[30] These changes go hand in hand with behavioral ones: domestic animals tend to be less frightened of humans and other animals, are generally more affable, and exhibit a "decline in environmental awareness"[31]—a loss of ability to detect and respond to potential environmental hazards and opportunities. These behavioral changes, which are probably the most significant adaptation by domestic animals to the "human niche," stem from neurochemical alterations that biologists have begun to map out in the genomes of pigs and other animals.[32]

Recently, it has been suggested that all, or at least most, of the unique traits of domestic animals reflect initial and continuing selection for tameness. In one famous experiment, the Soviet scientist Dmitri Belyaev domesticated wild foxes—replicating the suite of physiological traits of the domestication syndrome—by selectively breeding only those individuals that displayed less aggressive/fearful responses to human handlers.[33] In

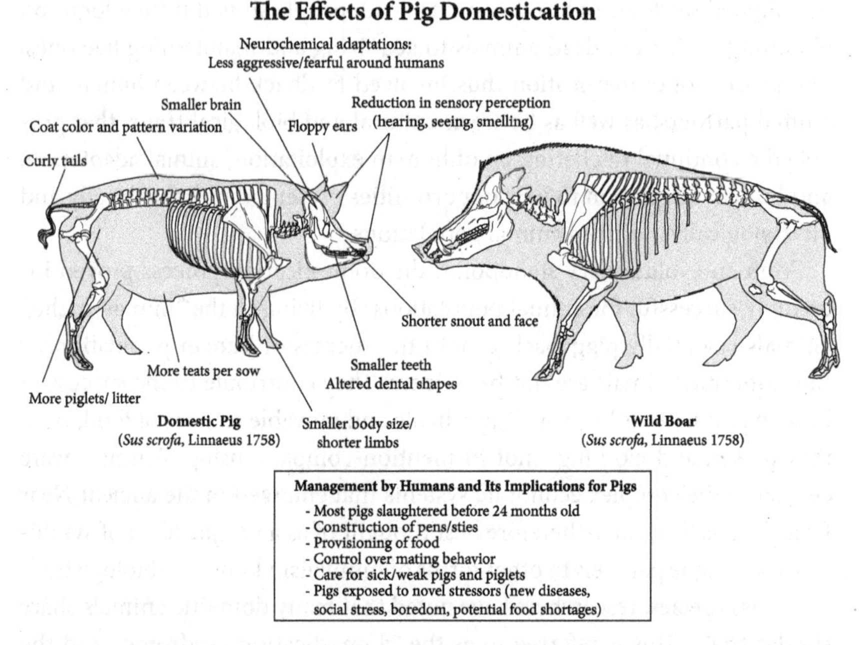

Figure 2.2. Physiological and behavioral changes in domestic pigs.

light of Belyaev's experiment, recent work has hypothesized that selection for tameness affected neural crest cells, which are found in fetal animals.[34] These cells migrate during maturation of the fetus to a number of locations, including those where hair, tooth, skeletal, and connective tissues develop—locations where we see many of the traits associated with the domestication syndrome. Thus, by selecting for tameness, humans may have unintentionally impacted neural crest cell development and thereby created many of the hallmarks of domestic animals. This theory remains to be tested, but if true, it would provide a grand unifying biological theory of animal domestication. On the other hand, it would not explain the cultural changes that also drove domestication.

The specifics of the process of pig domestication are still unclear. Around 10,000 years ago, certain populations of *Homo sapiens* and *Sus scrofa* formed close relationships that probably developed along two lines. First, some wild boar were able to thrive within the environmental niches carved out by increasingly sedentary human communities. In these niches were garbage dumps and, increasingly, fields of cereals and other cultivated plants, some of which were already domesticated or in the process of becoming so. Wild boar would have been attracted to these new sources of food. But as agricultural pests and/or trash heap commensals, wild boar would have needed to tolerate the proximity of humans. This may have created a unique selection pressure for greater tameness among some of them.

Second, and at the same time, some people began to hunt wild boar more intensively. That is, they were relying more heavily on hunted pork than their ancestors had. To enhance their chances of success, hunters targeted younger animals, which were smaller and less aggressive. At some point, people began directly interfering with wild boar populations in order to make pork an even more reliable resource. By allowing mature females to live and instead focusing on juveniles and males, hunters facilitated the growth of the local wild boar populations. Meanwhile, in an effort to expand the range of wild boar populations, people transported wild boar to islands and other new habitats. In doing so, they increased the availability of pork regionally..

Close contact with humans as commensals and managed game probably began a process of selection for tameness and other genetically controlled traits advantageous to regular interaction with people. While this biological transition was happening, a cultural one was occurring as well. For the humans who increasingly relied on wild boar for meat, a perceptual shift from dead to live animals owned by humans began to take shape; hunters grew less concerned with tracking and killing wild boar than with keeping *their* swine alive until the appropriate time for slaughter.[35] In this way, game management—controlling populations of animals in order to facilitate hunting—became herd management, or animal husbandry, with the intent of keeping and breeding animals as the property of people.[36]

Yet contact between humans and wild boar did not always lead to domestication. Evidence for close relationships between ancient humans and wild boar in the early Holocene is widespread—for example, in the Near East/eastern Mediterranean,[37] Japan,[38] northern Europe,[39] and Italy.[40] But for reasons that are not entirely clear, only in some of these contexts did domestic pigs evolve. Zooarchaeologists and geneticists have established two geographic regions where pigs were, without question, domesticated independently: northern Mesopotamia around 8000 BC and China around 6500 BC.[41] Pigs are one of only a few animal species for which we have concrete evidence of independent domestication in two different cultural settings.

From their original centers of domestication, pigs spread to other areas of the world. From the Near East, people brought pigs into Europe and North Africa.[42] From China, people brought pigs into Southeast Asia, Indonesia, and the Pacific as far as Hawaii.[43] In each of these cases, domestic pigs bred—or were bred—with local populations of wild boar. The resulting *hybrids* (a Latin derivative used to describe the offspring of, in this case, a domestic pig and a wild boar) were often better livestock animals than their parents. This has resulted in a genetic palimpsest reflecting both persistent gene flow between wild and domestic populations as well as a continuous selection for domestic phenotypes.[44]

In the past 300 years, it is the descendants of the Chinese domestic pigs that have enjoyed the most reproductive success. Beginning in the 18th century, agricultural scientists in Europe realized that Chinese breeds gained weight more rapidly and produced more piglets than European breeds.[45] They began importing these pigs to Europe, where they interbred them with local stocks and eventually created the major meat-producing breeds we know today, such as the Large White, Berkshire, and Duroc.

Zooarchaeologists have devoted much research over the past two decades to detecting these instances of pig domestication and spread. Of the traits associated with the domestication syndrome, only a few are imprinted onto bones and other hard tissues and can therefore leave traces in the archaeological record. The reduction in tooth and body size is the major zooarchaeological indicator of pig domestication.[46] In recent years, however, researchers have begun to track genetic features, like novel alleles for coat color, through the analysis of ancient DNA.[47] Other indicators show changes in the exploitation of suids by humans rather than the biological changes indicative of adaptation. For example, some scientists have tracked the incidence of pathologies on bones and teeth (i.e., osteological signatures of disease, malnutrition, or injury), which reflect the application of novel stresses to domestic animals.[48] Others have examined chemical (carbon, oxygen, and nitrogen isotope) signatures of dietary change.[49] Still others have reconstructed the demographic profiles of ancient swine populations to document shifts related to human management.[50] The cumulative result of all of these examinations is that zooarchaeologists have been able to determine where and when pig domestication occurred. It has also allowed some to explore how pig domestication happened, as we will see in Chapter 3.

Raising Pigs

I find the diversity of ways that people have raised swine in cultures around the world to be the most fascinating aspect of this animal. The number of unique forms of pig management reflects both human ingenuity and pigs' incredible ecological and dietary flexibility. Although there are dozens of ways to manage pigs, it is possible to describe two major forms of husbandry: intensive and extensive.

Intensive husbandry provides perhaps the most iconic image of swine management for American and British readers—the backyard sty inhabited by a couple of pigs drowsing, rolling in mud, and awaiting their next meal of kitchen slop. Under these conditions, the animals have less flexibility and independence. Humans dictate their social structure, living spaces, diet, and reproductive partners, if they have any—most males are castrated before puberty. For this level of control to operate effectively, human caretakers must understand their livestock's needs. Without proper care, intensively managed pigs can suffer from poor diets, epidemic diseases, and even boredom and physiological stress.[51]

Confining pigs to smaller spaces and controlling their diets has one major advantage: it is an incredibly efficient method of pork production. Intensive husbandry minimizes the loss of animals to predators and it allows for greater control over pigs' well-being, meaning that fewer piglets die from hypothermia or other preventable causes. Intensive husbandry also decreases the time necessary for pigs to reach slaughter weight because it eliminates the need for swine to search for food, avoid predators, and seek shelter from the elements. Instead, they can direct their consumed calories toward growth and fat accumulation.[52] Finally, intensive husbandry enhances herd growth by limiting the time it takes for sows and boars to reach puberty and by decreasing the age at which piglets can be weaned. Sows can become pregnant earlier in life and more frequently give birth to two litters per year.

Modern industrial farms have taken intensive husbandry to an extreme, supplying swine with scientifically tested food regimens, pumping them full of antibiotics to increase weight gain and eliminate the spread of disease, and housing them in facilities that can accommodate hundreds of pigs. This type of production is, as Upton Sinclair described it in *The Jungle*, "porkmaking by machinery, porkmaking by applied mathematics." Although large-scale pig operations existed in the ancient past,[53] modern factory farms have their roots in the Industrial Revolution and were designed to meet the demands of burgeoning urban populations working long hours for capitalist enterprises.[54]

Efficient, factory-style production rarely takes pigs' well-being into consideration. However, recent legislation in the US has focused on gestation crates (or farrowing crates), which are metal cages designed to prevent a sow from rolling over on her piglets.[55] Critics contend that long-term confinement in gestation crates is torture; hog producers argue that it keeps pork prices low on supermarket shelves. This ongoing debate between animal rights activists and major pork producers grabbed headlines in 2014 when New Jersey governor Chris Christie vetoed a bill designed to eliminate gestation crates in the state.[56] Readers and voters alike must weigh in on the ethics of modern meat production, something from which most consumers are alienated. As for the pigs, although they have evolved from their wild boar ancestors to sustain higher levels of stress, industrial-scale production pushes them beyond their "limits of endurance."[57] There can be no denying that cheap pork comes at the price of pigs' welfare.

Extensive husbandry allows pigs greater freedom to search for food and develop social structures. One finds some of the most extensive forms of pig husbandry on New Guinea, an island remarkable for its incredible diversity of pig husbandry practices,[58] as well as in parts of the Mediterranean, such as Greece, Sardinia, and Corsica.[59] In some New Guinea communities, villagers (usually the men) own pigs that are allowed complete freedom to wander around villages or into the bush. These pigs have almost total control over what they eat, where they go, where they sleep, and with whom they mate. If their caretakers (usually the women) impose any restrictions on pigs' behavior, it is often limited to fencing them out of certain areas (e.g., vegetable gardens) or applying hobbles to encumber their movements.[60] Similarly, in parts of the Mediterranean, free-range pig owners limit their animals' potentially destructive eating habits by inserting a ring into their snouts; the ring administers a small shock of pain when the pig roots in the ground.[61] In both cases, these free-ranging pigs' contact with humans occurs primarily during infancy and development, on their trips to their owners for supplemental food, and at slaughter.

Extensive husbandry takes advantage of otherwise unused resources, especially in forested environments, and requires little labor investment. There are, however, some substantial downsides. Litter sizes tend to be smaller and females reach puberty later than under intensive management. Additionally, the number of piglets lost to predators and hypothermia is higher.[62] There are also social risks. Free-ranging pigs may wander onto another person's property and cause damage. Pigs may be poached or stolen, which can lead to quarrels between neighbors or even violence. The infamous 19th century feud between the Hatfields and the McCoys is one example of swine-inspired strife. In New Guinea, some of the most common causes of inter-village hostilities are thefts of free-living pigs and crop damage by roving swine.[63]

One way to avoid some of the social conflicts is to enlist a swineherd, a professional trained to care for pigs, protect them from predators and poachers, and direct them toward feeding places that won't cause damage to farmland. But a drove of pigs is not as easy to control as a herd of sheep. A single swineherd can manage only a few dozen pigs at a time and usually over short distances, although there are ethnographic examples of pigs being driven over longer distances—sometimes as much as 100 km.[64] With some coordination, extensive husbandry can be made into a large-scale endeavor. Every autumn in medieval England, swineherds turned out thousands of swine to fatten on the nuts produced by hardwood forests, a system known as *pannage*.[65] Even today, swineherding remains a major form of pig production in Spain and Portugal,[66] Sardinia and Corsica,[67] and Greece.[68] Readers who have indulged in *jamón ibérico*, the meat of Black Iberian pigs fattened on acorns, know just how fruitful extensive husbandry can be.

A form of extensive husbandry is *settlement scavenging*, or free-ranging within human settlements, especially larger towns and cities. Similar to the sty-raised pigs owned by the Zabaleen of Cairo, pigs raised under this form of husbandry remove waste and recycle it into pork. For example, in modern-day Calcutta and Agra in India,

> [h]alf wild and half domesticated [pigs] move from open-air dumps to piles of refuse, snuffling around among rotten vegetable peelings, sheep bones, decomposing fruit and similar garbage, wallowing and snorting in the open drainage channels of towns and villages, rummaging unhesitatingly amidst the excrement and mud of the sewers in search of some titbit. They manage to eke out a fairly good living, judging by the size of the vast majority of these beasts.[69]

Pigs served similar trash-disposal roles in other cities, such as 19th century New York. In fact, much like Cairo in 2009, Manhattan was the stage of a major showdown between pig owners and the state in the 1850s—a dramatic series of events described in contemporary newspapers as the "Piggery Wars."[70] As many residents of New York complained at the time, free-ranging pigs can spread disease, damage property, and injure pets, small children, and the elderly. Nevertheless, the practice of allowing pigs to fend for themselves in human settlements persists to this day in sub-Saharan Africa, Southeast Asia, and South America, among other places.[71]

"Pig Principles" and Types of Data

Building on the general information about pigs provided in this chapter, the rest of this book will examine how pigs interacted with humans in the Near East. Much has been written about this topic. Categorizing and comparing this research is a complicated endeavor because of the diversity of scholarly approaches and the tendency of these approaches to talk past one another. As a guiding framework, I turn to the "pig principles" laid out two decades ago by zooarchaeologists Brian Hesse and Paula Wapnish.[72] While some have exaggerated the importance of these "principles," they nevertheless have had a major impact on the Near Eastern pig literature and thus provide a useful starting point.

1. Because of their need for water and shade, pigs are less adapted to arid environments than ruminants, especially sheep and goats.[73]
2. Deforestation can make certain forms of extensive pig husbandry less viable.[74]
3. Because it is cheap and does not require access to pasture, pig husbandry is often more prevalent among lower socioeconomic classes.[75]
4. Pigs are not as mobile as sheep, goats, cattle, equids, and camels. For that reason, nomadic pastoralists do not (often) raise pigs.
5. Pigs reproduce quickly and a large herd can be produced from a few "starter" animals. Pigs are thus ideal animals to accompany humans in the initial settlement of a territory.
6. Pigs carry diseases that can be transmitted to humans, especially tapeworm and trichinosis.
7. In contrast to ruminants, pigs have dietary needs similar to those of humans and therefore can be said, in an ecological sense, to compete with humans for resources.[76]
8. Because the pig—a nonruminating and omnivorous hoofed mammal—is unique among the animals of the Near East, some people may have found it symbolically ambiguous and therefore dangerous.[77]
9. Because pigs reproduce quickly and litter sizes vary, it is difficult for centralized institutions to tax or regulate them.[78]

10. Pigs do not provide "secondary products" like wool or dairy, which can be stored and traded over long distances. This makes pig production an unattractive undertaking for elites and their institutions.
11. Because of principles 9 and 10, pigs may be less common in urban centers than rural hinterlands.[79] On the other hand, because pigs adapt so well to urban environments, they are often *more* common in cities than in rural areas.

I argue that no single pig principle can adequately explain the history of this animal or the development of the pig taboo, despite scholars' frequent attempts to do so. Instead, in the chapters to come, I will show how some of these principles, as well as a few other factors, helped shape swine's historical trajectory at specific moments in time. In addition, I will trace Near Eastern cultural attitudes toward pigs beginning almost 2 million years ago and ending with the present day.

I rely on several types of data. None of them is perfect. Iconographic data show the contexts and ways in which people depicted pigs. However, they are notoriously difficult to interpret, especially in the absence of historical texts attesting to their significance. Ancient texts provide another body of evidence. Writing first developed around 3000 BC. The texts recovered from archaeological sites and, in some cases, passed on to modern readers through copying over the millennia shed light on the ways in which pigs figured into economic and ritual activities. However, these documents were primarily written by and for the elite. They offer an incomplete and biased perspective on ancient cultures.

The vast majority of the information in this book derives from published zooarchaeological data. While animal bone data do not come with the interpretive baggage that accompanies texts and iconographic images, other problems affect them. Issues include differential deposition, preservation, and recovery—that is, biases concerning how and where bones get into the archaeological record, which ones survive intact to the present-day, and which ones archaeologists ultimately recover and study. This is particularly problematic when one compares material from different contexts. This book includes data spread across different time periods from many sites. The sites themselves were excavated by archaeologists who possessed varying levels of scientific interest in collecting and studying animal remains. Evaluating patterns in such diverse faunal assemblages often feels like comparing apples to oranges to dates to bananas.

I have tried to overcome these issues by avoiding problematic comparisons. One way of minimizing biases is to situate zooarchaeological data within their specific temporal contexts. I have also tried to be careful about the tabulation of animal taxa, as the size and composition of bones can influence preservation and recovery. I rely heavily on the ratios of pig remains to those of other animal species. For the most part, I calculate these proportions as percentages of pigs relative to medium and large mammals—that is, animals that reach about 10 kg as adults (about the size of a fit beagle). Over time, four species came to dominate this category: domestic sheep, goats, cattle, and pigs, although domestic dogs and equids also made variable, but typically small contributions. For that reason, beginning in Chapter 5, I switch from discussing the relative abundance of pig remains compared with other medium and large mammal bones to the proportion of pigs as a percentage of the combined total of the four main domestic livestock species.

Finally, I have attempted to include debates about the data and the claims made. While no dataset is perfect and no analysis free from error or bias, I hope that scholars and general readers find in the following pages a balanced perspective that draws on multiple viewpoints and sufficiently contextualizes the information and its problems.

3

From Paleolithic Wild Boar to Neolithic Pigs

In the beginning, the Near East was devoid of both pigs and humans. The genus *Sus* evolved in Southeast Asia and arrived in the Near East around 1 million years ago.[1] Humans came from the other direction. The genus *Homo* evolved in Africa around 2.5 million years ago, and by 1.8 million years ago, *Homo erectus* had successfully colonized the Near East.[2] Because these and other species of *Homo* made stone tools, the period from 2.5 million to 200,000 years ago is known as the Lower Paleolithic (literally "old stone" age). It corresponds to a geological epoch, the Pleistocene, characterized by cooler temperatures across the globe.

By the Middle Paleolithic (200,000–40,000 years ago), two hominin species had replaced *Homo erectus* in the Near East: Neanderthals (*Homo. neanderthalensis*) and our ancestors, anatomically modern humans (*Homo sapiens*). The early career of our species in the Near East was not promising. Throughout the Middle Paleolithic, Neanderthals were the dominant hominin and the small Near Eastern populations of *Homo sapiens* repeatedly went extinct. Something changed around 60-50,000 years ago. Modern humans began to make more complex stone tools, developed elaborate and diverse ritual behaviors, and displayed a near-universal tendency to manipulate their environments—they became "the ultimate ecosystem engineers."[3] These traits led to larger, more stable, and more successful populations that drove Neanderthals to extinction.

Modern humans lived as hunter-gatherers from their earliest arrival in the Near East until the end of the Epipaleolithic period around 9700 BC. As technological advances and environmental impacts accumulated, human populations developed new ways of exploiting the world around them. Populations increased and, in some places, hunter-gatherers adopted a more sedentary way of life, especially during the Natufian period (12,500–9700 BC). Domestic dogs—perhaps first domesticated in Europe—had appeared in the Near East by 11,000 BC.[4] The climate was also changing: the cold and

Evolution of a Taboo. Max D. Price, Oxford University Press (2020). © Oxford University Press.
DOI: 10.1093/oso/9780197543276.001.0001.

dry climate gave way to a wet and warm period from 12,500 BC to 10,900 BC, during which human populations thrived. It briefly returned to cold and dry conditions in the Younger Dryas (10,900–9700 BC), forcing the larger and more environmentally manipulative societies to change their subsistence practices.[5]

The amelioration of the climate, the easing of the Younger Dryas into the warmer and more stable Holocene, set the stage for the Neolithic (9700–5200 BC). People settled into permanent villages and began to make houses out of mudbrick. While they continued to harvest local plants and hunt animals as their ancestors had done, they did so more intensively. Over time, the cumulative effects of these types of activities around sedentary villages led to the selection of unique mutations in some populations of plants and animals—that is, domestication. Beginning around 8500 BC, populations of cereals, like barley and wheat, lost their seed-shattering dispersal mechanisms and became domestic. Shortly thereafter, animals from a select group of species, including *Sus scrofa*, developed the characteristics of the "domestication syndrome." By 7500 BC, the list of domesticates included barley, wheat, chickpeas, peas, lentils, fava beans, flax, vetch, sheep, goats, cattle, and pigs. A farming economy had been established. It would become the way for most people living in Anatolia, Mesopotamia, and Levant.

Reluctant Hunters of the Lower and Middle Paleolithic

The earliest interactions between *Homo* and *Sus* were of reluctant hunters encountering ferocious prey. *Homo erectus* pursued *Sus strozzi* and, later, *Sus scrofa*. However, despite the fact that *Homo* and *Sus* populations inhabited similar ecosystems, evidence for Lower Paleolithic wild boar hunting is scarce. Lower Paleolithic faunal data consistently show that *Sus* remains make up a small proportion (usually less than 1 percent) of the recovered medium and large mammal bones.[6]

The big-brained humans of the Middle Paleolithic Near East, Neanderthals and *Homo sapiens*, were better equipped to hunt and process the carcasses of various species of animals. Nevertheless, they continued to avoid wild boar, generally preferring to go after fallow deer, roe deer, wild goats, and gazelle, as well as other big mammals like horses, onagers, and camels.[7] Across the Near East, wild boar remains consistently make up only about 1–5 percent of medium and large mammal bones from Middle Paleolithic sites.[8]

However, people did occasionally work wild boar into their burial ceremonies and other rituals. Excavators at Skhul V in Israel, for example, uncovered a human skeleton holding a wild boar jaw.[9]

At only one site is there evidence that Neanderthals or anatomically modern humans enjoyed more frequent success at boar hunting in the Middle Paleolithic: Üçağızlı II Cave in southern Turkey. Wild boar make up around 9 percent of the medium and large mammal remains found there. This atypical pattern likely reflects the unique geography of the site, which lies near several steep-walled canyons. By positioning themselves at key locations, even a small number of Middle Paleolithic hunters armed with thrusting spears and rocks would have been able to target, trap, and dispatch wild boar and other prey with relative ease.[10]

With the exception of their boar hunting at Üçağızlı II Cave, ancient humans seem to have avoided wild boar. The reason for this is probably obvious to anyone who has encountered these animals in the wild: they are fast and, when frightened, can become aggressive. Once cornered, wild boar are apt to turn and charge their predators, using their razor-sharp tusks and massive head like a spiked club. Killing these animals is also no easy task. Wild boar are armored by tough hides and dense, heavy skulls that can absorb direct blows. Older boars develop a layer of hard cartilage over their rib cages. This "shield," as modern hunters call it, can deflect or outright stop arrows and even bullets.

The ferocity of wild boar makes for dramatic hunts. Without the aid of projectile weapons or traps, the final confrontation is a bloody struggle akin to hand-to-hand combat between two warriors. For that reason, wild boar have long symbolized masculinity throughout Eurasia.[11] *The Odyssey* provides perhaps the most vivid description of such an encounter, when a young Odysseus, impetuous and pugnacious, rushes to attack a boar:

> Here, as the hunters closed in for the kill,
> crowding the hounds, the tramp of men and dogs
> came drumming round the boar—he crashed from his lair,
> his razor back bristling, his eyes flashing fire
> and charging up to the hunt he stopped, at bay—
> and Odysseus rushed him first,
> shaking his long spear in a sturdy hand,
> wild to strike but the boar struck faster,
> lunging in on the slant, a tusk thrusting up

over the boy's knee, gouging a deep strip of flesh
but it never hit the bone—
  Odysseus thrust and struck
stabbing the beast's right shoulder—
  A glint of bronze—
the point ripped clean through, and down in the dust he dropped,
grunting out his breath as his life winged away.[12]

The danger of wild boar explains their relative infrequency in Lower and Middle Paleolithic diets. Odysseus's thrilling battle with the boar probably mirrored the nightmarish experiences of more than a few ancient hunter-gatherers. Although direct evidence for hunts gone south is hard to come by, one can imagine that more than a few of the hunters who decided to pursue wild boar ended up mauled or killed. The risk posed to Lower and Middle Paleolithic hunters, who possessed only thrusting spears, effectively prevented suid hunting on a large scale when there were other, less dangerous options available.[13]

The Upper Paleolithic and Epipaleolithic

Homo sapiens in the Upper Paleolithic, beginning a little over 40,000 years ago, made an impressive number of technological advancements. They developed blade and bladelet lithic production techniques used bone and antler as raw materials, fashioned grinding stones for processing other plants, and deployed projectile weapons for the first time.[14] These technological innovations allowed modern humans to exploit their environments with greater efficacy. For example, they could hunt difficult-to-catch animals, such as hares, and more frequently target prime-aged big game mammals, which are quicker and more difficult to kill but provide larger packages of fat-rich meat.

In theory, these technological changes also made wild boar hunting more practical. Projectile weapons allowed hunters to strike their prey from a safe distance. The use of nets and snares, meanwhile, made it possible for them to deliver a killing blow without risking death or injury.[15] Nevertheless, equids, gazelle, deer, and wild goats and sheep remained the major sources of meat.[16] *Sus* bones rarely exceed 5 percent of the recovered medium and large mammal remains at Upper Paleolithic sites.[17]

Subtle changes in boar-hunting strategies may have been taking place. Although the numbers of wild boar bones recovered from archaeological sites do not change much, the ages of the animals killed do. During the Upper Paleolithic, hunters appear to have targeted younger (and therefore smaller) animals more frequently. They may have even sought out on occasion farrowing sows and their litters.[18] This was a safer method of procuring meat that marked an important milestone in human-suid relations.

Wild boar hunting continued to evolve through the Epipaleolithic. The early phases of the Epipaleolithic saw little change in its popularity, despite transformations in other food procurement strategies. At the site of Ohalo II, located on the shores of the Sea of Galilee, hunter-gatherers intensively collected and possibly even stored and cultivated wild cereals by 21,000 BC.[19] However, Ohalo II's inhabitants focused primarily on gazelle and deer for their meat. They ate almost no pork, despite attraction of wild boar to lacustrine environments.[20] Similarly, at other sites across the Near East, *Sus scrofa* continued to comprise less than 5 percent of the medium and large mammals.[21]

The first sign of change occurred in the Natufian period, ca. 12,500–9700 BC. The Natufian is often portrayed—correctly or not—as the harbinger of agriculture, a sort of proto-Neolithic that was interrupted by the Younger Dryas climate downturn at 10,900 BC.[22] Natufian hunter-gatherers developed improved methods for hunting and trapping small game and ungulates. They also left behind stone sickle blades, evidence of wild cereal harvesting.[23] Driving, or driven by, these alterations to the subsistence economy was an increase in human populations. In some places, people settled down into more or less permanently occupied base camps. Rituals grew more elaborate, especially those surrounding burials.[24]

In terms of meat procurement, people living at Natufian settlements focused heavily, and in some places almost exclusively, on gazelle hunting.[25] But one Natufian site contrasts with this pattern: 'Ain Mallaha, located in northern Israel. There, wild boar account for around 20 percent of the medium and large mammal bones in the later phases of the site's occupation, dating to the 11th millennium BC. The wild boar bones include the remains of fetal individuals and a relatively high proportion of animals less than three years old. Taken as a whole, this evidence is suggestive of an intensive hunting strategy in which wild boar were increasingly relied on for meat and were targeted in such a way as to minimize the risks to the hunter.[26]

A number of factors, in combination or by themselves, might have made boar hunting attractive to the Natufian inhabitants of 'Ain Mallaha. First, 'Ain Mallaha was located in a marshy area near the now-drained Lake Hula, a place likely to attract and sustain sizable populations of wild boar. Second, the site of 'Ain Mallaha was a more or less permanent base camp for several Natufian hunter-gatherer families.[27] The presence of a sedentary human occupation, and especially the organic waste that such groups deposit, may have attracted wild boar to the settlement, pulling them into closer orbit with humans.

A third factor may also explain the unexpectedly high proportion of wild boar bones in the assemblage. 'Ain Mallaha is well known for containing some of the earliest evidence for domestic dogs in the Near East, including the burial of a puppy alongside an adult human.[28] Dogs, historically and today, are vital features of boar hunts. Packs of hounds can chase and track wild boar through forest undergrowth much better than any human hunter. Once cornered, dogs keep the boar at bay until the hunter can dispatch it. And if the boar charges, it is more likely to gore a canine than a human. There is no doubt that wild boar hunting would have become a more feasible and less dangerous option once dogs were introduced. Although it remains difficult to find direct evidence for the use of dogs as hunting animals in the archaeological record, the fact that the presence of domestic dogs corresponds to an uptick in wild boar hunting at 'Ain Mallaha might not be a coincidence.

'Ain Mallaha was abandoned around 10,000 BC. But both the increasing intensity of wild boar hunting—probably aided by dogs—and wild boar's attraction to human settlements would play central roles in the domestication of pigs in the following millennia, when human populations across the Near East grew sharply and the number of sedentary settlements increased.

The Road to Domestication: The Pre-Pottery Neolithic A

By 9700 BC, the Younger Dryas had ended and the climate in the Near East had stabilized. Sedentary lifestyles, which were largely abandoned in the 11th millennium BC, became favorable once again and human populations increased. At the same time, people across the Levant, central/southern Anatolia, and northern Mesopotamia began to change how they acquired food.[29] These changes set in motion domestication, that coevolutionary process by which organisms adapt to human exploitation, and humans modify their behaviors to protect and promote those animal/plant populations.

Archaeologists label the societies that inhabited the Near East beginning around 9700 BC as "Pre-Pottery Neolithic" (PPN) since they did not make ceramic vessels, but practiced sedentism and eventually domesticated plants and animals—cultural traits traditionally categorized as Neolithic. Archaeologists further distinguish the PPNA (9700–8500 BC) from the PPNB (8500–7000 BC). The PPNA saw the initial foundation of villages and the tending of wild cereals and legumes on a large scale, as well as the first steps toward developing relationships with populations of animals that would eventually become domesticated.[30] By the beginning of the PPNB, plants had acquired the mutations transforming them into domesticates, and animals soon followed.

There is considerable debate about why people domesticated pigs as well as other plants and animals in the PPN. Some scholars, like zooarchaeologist Melinda Zeder,[31] argue that humans are naturally experimental and will try to improve the reliability of their resource base whenever they can.[32] We are niche constructors, "ultimate ecosystem engineers."[33] For Zeder and others, domestication happens when humans have occupied a place long enough for niche-constructing behaviors to have a cumulative effect on ecosystems and, significantly, when climatic conditions make it possible for people to settle down and intensify pressure on local environments. Other scholars have argued that humans are more conservative and exhibit ingenuity only when pushed to do so. They argue that groups of people will not change their subsistence behavior unless compelled by external factors, such as environmental/climate change, resource depression, or population pressure.[34]

In many respects, this debate hinges on philosophical differences. How self-directed are humans? How constrained are societies by their surroundings? Are humans open to change or are they inherently conservative? Such debates often come down to different outlooks. But what is valuable to the study of domestication—and, indeed, large-scale processes of social change in general—is that the different perspectives provide unique frameworks for collecting, interrogating, and understanding complex datasets. The dialogue between them, moreover, helps scholars home in on the details and think critically about their assumptions.

Recognizing the value of both perspectives, several recent papers have sought to reframe the discussion in terms of how both human ingenuity and external pressures led to changing subsistence practices at the beginning of the Pre-Pottery Neolithic.[35] Such a model highlights the importance of feedback between the environment and human behavior. For example, an improved climate would have allowed hunter-gatherers to intensify food procurement, which would have enabled greater population growth, increased population pressure, and incentivized people to settle into sedentary villages. This, in turn, would have forced or inspired neighboring communities to intensify their subsistence practices as well, further increasing population density and pressure on resources. The resulting feedback cycle would eventually lead people to invest more time and energy into managing plants and animals, unintentionally selecting for domestic varieties.[36]

But to focus only on human behavior in the process of domestication is to ignore the important fact that certain plants and animals are better suited for adaptation to human management. That is, the natural behaviors of some organisms enable them to fit within the human niche and benefit from it, in large part by inspiring humans to invest in raising and nurturing them.[37] These "pro-domestic" species, for example, tend to exhibit considerable phenotypic plasticity—that is, they can alter the expression of behavioral and physiological traits depending on environmental conditions.[38] Recognizing the importance of species' natural behaviors and physiologies, some scholars have reframed the discussion of domestication, suggesting that these species, in fact, "chose" us.[39] However, while it is certainly important to recognize the animal or plant partner in domestication, the process is ultimately driven by human cultural factors. No natural plant or animal behavior would lead to domestication were it not for two fundamental and innate features of *Homo sapiens*: our ability to initiate goal-oriented behavior[40] and our compulsion to seek out relationships with other species.[41]

A subset of wild boar began its transition to domestic pig in the PPNA. While the period ended centuries before pigs developed the characteristics of the domestication syndrome (at least as far as zooarchaeologists have been able to detect), there is considerable evidence that interactions between humans and wild boar were changing beginning around 9700 BC. First, wild boar bones are found in proportions over 10 percent at a number of PPNA settlements.[42] As we speculated about Natufian 'Ain Mallaha, it is likely that (1) wild boar were drawn to sedentary villages and their garbage, and (2) that people were able to hunt wild boar more successfully with dogs. Second, PPNA hunters increasingly focused on killing animals less than three years old. This intensive hunting strategy, with roots in the Epipaleolithic or even Upper Paleolithic, increased the likelihood of obtaining pork and decreased the likelihood of being gored. It also may have been an early form of game management designed to kill off excess males and thereby increase the number of farrowing females within local ecosystems. Early forms of game management also involved the transportation of live wild boar to new locations—stocking them with favored game animals for future hunting. Finally, these new forms of human-swine relations probably inspired novel types of ritual behavior in PPNA communities. PPNA sites provide some of the earliest evidence for large-scale feasts involving pork.

The Intensification of Wild Boar Hunting at Hallan Çemi

Sometime between 9700 and 9300 BC, a group of hunter-gatherers decided to make their permanent home at Hallan Çemi in the foothills of the Taurus Mountains in southeastern Turkey.[43] They constructed several round huts, and within a few generations had deposited a large amount of animal bones and other garbage at the settlement. Many of these bones were found in a large pit at the center of the site, probably an area in which people congregated for large feasts.[44] Around 25 percent of the medium and large mammal bones were from wild boar.[45]

At the time of discovery, the high percentage of wild boar bones at Hallan Çemi caused quite a stir. The number of *Sus scrofa* remains and the high proportion of juvenile animals represented in the assemblage suggested that the inhabitants of Hallan Çemi may have been raising and breeding wild boar. Specifically, the excavators argued[46] that the inhabitants of Hallan Çemi kept small herds of females, which were allowed to mate with wild males—a situation similar to that observed ethnographically in parts of New Guinea, among other places.[47] The announcement sent ripples through zooarchaeological channels. The *New York Times* picked up on the story, proclaiming that pigs were domesticated at Hallan Çemi almost 12,000 years ago.[48]

The initial excitement elicited by reports of the earliest domestic pigs at Hallan Çemi turned out to be misguided. There is no evidence for morphological change in the Hallan Çemi suids.[49] There is no significant difference between the dental size of modern Turkish wild boar and that of wild boar from Hallan Çemi, which is what one would expect for a population of wild boar undergoing the process of domestication. The data suggest, in other words, that the wild boar from Hallan Çemi did not possess the characteristics of the domestication syndrome.

The swine at Hallan Çemi may not have been domesticated, but the zooarchaeological data nonetheless speak to the changing nature of human-suid relations in the PPNA. These changes were ultimately tied to the process of domestication. The abundance of wild boar remains, for example, suggests greater confidence in boar hunting. Meanwhile, the demographic profiles indicate that people were targeting animals less than three years old, which, as mentioned above, might reflect a game management strategy.[50] There were also a large number fetal or newborn piglet remains. The excavators' interpretation that the inhabitants of Hallan Çemi kept sows[51] would indicate that people were already experimenting with breeding wild boar in captivity long before the first appearance of domestic pigs.

One could interpret the demographic data differently, however, and argue that they represent a unique form of hunting, not a management strategy. For example, the high proportion of juveniles could indicate a hunting strategy designed to obtain meat from younger—and thus smaller and less dangerous—animals.[52] Or perhaps hunters targeted farrowing sows, which would be particularly vulnerable and easy prey since female swine leave their sounders to give birth.[53] Another alternative is that sows, their young, and juvenile animals, which are often less wary of the dangers posed by humans, frequently wandered into the settlement. If so, then the inhabitants of Hallan Çemi may have killed these commensals in order both to eliminate a pest and to obtain meat from a readily available source.

Dogs may have aided these hunts. For example, they could sniff out farrowing females and keep a sow at bay while the hunters dispatched her and her litter, perhaps leaving the hunters the opportunity to capture a piglet and raise it back at the settlement.[54] Several canid bones in the Hallan Çemi assemblage were tentatively identified as those of domestic dogs.[55] Moreover, an image of what appears to be a dog wagging its tail was carved on a stone bowl that excavators recovered from the site.[56] It is therefore likely that, if Hallan Çemi's inhabitants were experimenting with new forms of boar hunting, or perhaps with capturing live animals, dogs were key elements.

The data from Hallan Çemi currently offer no clear indication about which of these alternatives is correct. Future research will be needed to determine exactly how humans and wild boar interacted in the PPNA. But the combination of intensive resource procurement and innovative approaches to animal populations represent important developments in human-suid relations.

Even if they remained hunters, at some point the inhabitants of Hallan Çemi may have decided that a more efficient strategy would involve controlling the movements of sounders. This incipient game management strategy would represent an important step in the domestication process, signaling a shift in how humans thought about and exploited *Sus scrofa*. In fact, exciting finds from the island of Cyprus show such a process in action.

Taking Boars on Boats to Cyprus

Another piece of evidence pertaining to the developing relationships between populations of wild boar and people in the PPNA comes from the island of Cyprus, where seafaring hunter-gatherers first landed in the late Epipaleolithic.[57] Soon after people arrived, animals appeared on Cyprus that were not native to the island. One of these invasive species was wild boar. As Cyprus is separated from the mainland of the Near East by about 70 km, the only possible vector for transmission of wild boar and other nonendemic animals is humans. This suggests that people were stocking the island with the animals they wanted to hunt or live in their settlements, such as wild boar, wild cats, foxes, and domestic dogs. Later, they brought deer and (initially wild) sheep, goats, and cattle.[58]

The faunal evidence indicates that wild boar were among the first animals introduced to Cyprus. The earliest published evidence comes from the cave site of Akrotiri Aetokremnos, where excavators recovered 18 wild boar bones and teeth.[59] The team was able to directly radiocarbon-date the bones to between 9700 and 9400 BC, a date contemporaneous with the occupation of Hallan Çemi.[60] Akrotiri Aetokremnos thus provides some of the clearest evidence that humans were pursuing a game management system involving the capture and transportation of wild boar in the millennia prior to the initiation of fully fledged controlled breeding and domestication.[61] Such quintessentially human niche-constructing behaviors secured the earliest Cypriots a reliable source of meat for centuries to come.[62]

Wild Boars in Rituals

A final piece of evidence for evolving human-suid interactions in the PPNA is the increasingly important role of wild boar in rituals. Throughout the

Pre-Pottery Neolithic, human communities introduced significant changes in their ritual practices; on a regional level, they invested more time and resources into ceremonial activities. The reasons for this shift in ritual behavior are complex, but likely stem from the unique challenges facing Neolithic peoples. For example, sedentism and population growth forced larger numbers of people to live closer to one another, thus fostering social tension. The subsistence economy, increasingly dependent on management/cultivation strategies and a delayed-returns approach to acquiring food, was another source of anxiety, one that manifested itself both at the individual and social level. Rituals helped relieve these stresses, diverting this anxious energy into community-building activities.[63] Wild boar played a role in these rituals. While zooarchaeologists cannot completely understand the social and religious significance of wild boar, we can see evidence of a more prominent role for these animals in two types of activities: feasting and symbolic representation.

Feasting, or the communal consumption of food, is an important component of public ritual cross-culturally.[64] Feasts, especially those involving large groups of people, can create a sense of togetherness and a shared sense of participation in socially meaningful activity. This promotes social cohesion. On the other hand, in some contexts, feasting can exclude and alienate people, such as members of other ethnic groups, genders, and social classes. Feasts can also be a means by which more ambitious individuals, playing the role of feast-giver, can display and promote their power and wealth to the community.[65]

Meat is a common component of feasts in many cultures, past and present. Generally speaking, humans crave the taste of meat. They devote much time and many resources to acquiring it. But meat also tends to be quite symbolically loaded. The nature of its procurement (hunting or slaughtering), preparation (butchery and cooking), and consumption are often highly social activities.[66] Hunting and slaughtering an animal, in particular, are emotionally charged undertakings, acts of violence that, in many cultures, warrant praying to the god(s) or requests for forgiveness from the animal's spirit. In other cultures, people have devised different ways to overcome this guilt: rituals that regulate how to kill animals (e.g., kosher or halal slaughter), taboos that restrict consumption or harm of certain animals, a conceptualization of a hierarchy of beings, and metaphors that recast hunting as an act mimicking sexuality or warfare.[67] Despite these cultural constructs, the trauma of killing another animate being, not to mention the inner turmoil caused by contradictory impulses to feel compassion toward animals and at the same time to lust after their flesh, lends meat its power.

Zooarchaeological indications of feasting can be difficult to differentiate from those derived from more quotidian meals. However, the relatively rapid deposition of a high volume of animal bones bearing cut marks and other evidence for human consumption can be taken as a sign of feasting. One example of this type of deposit was found at Hallan Çemi, where the sheer number of wild boar and other animal bones within a pit at the center of the settlement argues for large-scale feasting.[68]

Another example of feasting derives from a spectacular find at the site of Tappeh Asiab in Iran, dating to around 9400 BC.[69] There, in a single discrete deposit, excavators found the skulls of 19 wild boar as well as numerous long bones bearing cut marks from ancient butchers. The 19 animals, the majority of whose ages ranged from three months old to three years old, would have produced at least several hundred kilograms of meat, enough for hundreds of people. Such an event not only would have included all the inhabitants of Asiab, but also could have attracted people from outside the community, whose presence would have reinforced friendly relations and perhaps political allegiance. The symbolic importance of this particular feast at Asiab is demonstrated by the care with which the remains of the consumed animals were deposited. The wild boar skulls were neatly packed together and laid down along an east-west axis along with the boars' long bones, the skull of a bear, and deer antlers .[70]

While pork was good to eat, wild boar were also potent symbols. We can only speculate on what wild boar meant to prehistoric peoples, such as the feast participants at Tappeh Asiab, but the fact that hunting them was a formidable enterprise would have lent itself to connotations of ferocity, masculinity, and power. Some of the most tantalizing—if cryptic—evidence comes from the site of Göbekli Tepe. Göbekli, located near the modern city of Urfa in Turkey, was a large cultic site used by hunter-gatherers around 9200–8500 BC. The site consists of a series of standing T-shaped stone pillars contained within sunken circular buildings. Archaeologists have excavated about 20 of these "enclosures," each measuring 10–20 meters in diameter.[71] One of the most striking features of Göbekli Tepe is the array of animal imagery carved into the stone pillars. Dozens of individual beasts, including snakes, foxes, bears, wild sheep, aurochsen, and wild boar, are represented in various poses alongside geometric designs and the occasional human phallus. Interestingly, the enclosures seem to be themed around certain animals or groups of animals. Snakes, for example, dominate Enclosure D. Enclosure C, on the other hand, has a heavy concentration of wild boar. Like the one shown in Figure 3.1, many of these wild boar are visibly male and are depicted with bared teeth and menacing tusks.

Some archaeologists have indulged in a healthy dose of speculation on the imagery from Göbekli. The enclosures may, for example, have served as shrines in a cult of the dead.[72] There is also a detectable masculine theme to the rituals, such as the representations of human phalluses and large-tusked boars. Perhaps the rituals carried out at the site drew upon the symbolic imagery of the hunt, a masculine activity that was beginning to be replaced by animal herding. While these and other hypotheses about Göbekli amount to little more than informed guesswork, the symbolic representation of wild boar suggests that people in the PPNA were beginning to think about these animals in new ways, drawing on them as symbols of human values.

Figure 3.1. Stone pillar from Göbekli Tepe (Enclosure C) decorated with bas-relief image of a male wild boar and other animals.

Domestication by Two Pathways

The examples from Hallan Çemi, Cyprus, Asiab, and Göbekli Tepe show how human-suid relationships were evolving in various corners of the Near East in the PPNA. These new relationships would lay the foundations for domestication in the PPNB. Zooarchaeologist Melinda Zeder[73] has suggested two main pathways for early animal domestication.[74] The first is the "prey pathway," in which people begin to control animal herds, first as a form of game management, later by more actively interfering in animals' daily lives. Once management becomes intense enough, people cross the threshold to animal husbandry and unintentionally select for mutations that underpin the characteristics of the domestication syndrome. The second route to domestication is what Zeder has called the "commensal pathway," in which animals take the initiative by invading human settlements and adapting to them as commensals. Only later, when humans recognize the economic value of these animals, did they begin to raise and breed them.

There is good reason to believe that wild boar followed *both* Zeder's prey and commensal pathways to domestication. A dual pathway model particularly makes sense when we consider the role of cultivated cereals and legumes.[75] In terms of the prey pathway, there is solid evidence at Hallan Çemi and Akrotiri Aetokremnos that people had begun to hunt or manage wild boar with new tactics and, in some cases, capture and transport them. At some point, sedentary PPNA hunter-gatherers probably got the idea that providing animals with food would ensure that the animals stayed healthy and close at hand. Providing food would also habituate young animals to people, such that they would conceive human caretakers as trusted herdmates. This critical step, called imprinting by animal ecologist, solidified bonds between humans and individual animals. It remains an important feature of husbandry today. As the bond tightened, some people were inspired to consider animals not just as sources of meat, but also as property. This fostered even greater investment in animal care. The plants raised by people at PPNA settlements, themselves undergoing domestication, were therefore key elements in the transformation of prey into livestock. Cultivated plants provided a reliable source of fodder that enabled humans to pursue new forms of suid management..

We know less about the commensal pathway that wild boar also took towards domestication. This is because it is more difficult to detect with traditional zooarchaeological techniques. However, we can be reasonably sure that commensalism took place because of what we know about the behavior of wild boar: their willingness to infiltrate human settlements to eat garbage and cultivated crops. Indeed, ravaging PPNA horticultural fields would have made wild boar quite a pest, but one that could be drawn upon as a reliable food source by humans. For the commensal wild boar, there would be selection pressure to be less wary of people—the better to eat humans' cultivated plants and settlement refuse. These commensal populations would have stayed close to villages and probably interbred quite freely with managed wild boar. Ultimately, they helped create a gene pool that, existing under a unique set of selection pressures, diverged from those of wild boar experienced minimal contact with humans.

The reader might be wondering at this point why pig domestication took so long if all the pieces were in place by the PPNA. In experimental studies, animal populations can display domestication syndrome phenotypes after just 10–20 generations of selection.[76] For wild boar, that amounts to less than half a century, but even by the early phases of the PPNB—1,000 years after the occupation of Hallan Çemi—there was no evidence for domestic pigs. The reason is not entirely clear, but two factors are important to consider. First, we don't know how many times domestication failed before it succeeded. How many times, in other words, were people on their way to domesticating pigs, but then switched hunting strategies or experienced a catastrophe that wiped out their herd, such as a disease jumping from people to wild boar? Second, managed animals or scavenging commensals could still mate with other wild boar, those who were not drawn into the human orbit. In fact, people probably relied on strategically restocking their herds with wild-living animals. This created continuous gene flow from the wild. For domestication to occur, the selection for domestic mutations had to be intense enough and over a large enough area to overcome this "out-breeding" problem. In the PPNA, the selection pressures may not have been quite high enough and/or human populations may not have been dense enough to cause pig domestication.

First Domestic Pigs: The Early and Middle PPNB

By the 9th millennium BC, populations of wild boar began to develop and sustain the mutations that would transform them into domestic pigs, selecting for the traits of the domestication syndrome. Along with pigs,

PPNB villagers living in the foothills of the Zagros and Taurus Mountains also domesticated sheep, goats, and cattle.[77] The appearance of domestic animals at this time was no doubt related to regional population increase, sedentism, and the presence of now-domesticated cereal and legume crops.

Current zooarchaeological evidence indicates that domestic pigs were present by the Middle PPNB. Several sites in southern Anatolia and northern Mesopotamia show an increase in the proportion of *Sus scrofa* remains between the Early (8700–8200 BC) and Middle PPNB (8200–7500 BC).[78] More intriguing, measurements from *Sus scrofa* remains at some Middle PPNB sites indicate that the individual animals possessed small bodies and teeth, features consistent with those of the domestication syndrome.[79] The best evidence comes from the site of Çayönü Tepesi in southeastern Turkey, which was occupied for around 3,000 years from the PPNA through the end of the PPNB.[80]

Çayönü's long and continuous occupation from about 9500 to 6500 BC makes it an ideal site for studying the process of pig domestication. It is also significant that almost 30 percent of its sizable faunal assemblage (and over 38 percent of the medium and large mammals) derive from *Sus scrofa*, enabling a more detailed study of pig domestication than at contemporaneous sites where suids are less common and bone assemblages smaller. From the earliest occupation in PPNA, the inhabitants of Çayönü hunted wild boar in large numbers and, as at Hallan Çemi, they focused on young animals. As at Hallan Çemi, around two-thirds of animals were killed before they were two years old. But at Çayönü, almost 60 percent of the wild boar were less than one year old at death, compared with 40 percent at Hallan Çemi. The high proportion of very young animals is strikingly similar to that observed in later contexts in which domestic pigs were being raised.[81]

In the centuries that followed the PPNA, the inhabitants of Çayönü gradually exerted greater effort to control and feed the suid populations, interacting with the animals more frequently and regularly. Demographic patterns indicate these changes. While the proportion of animals killed at less than one year of age remained stable over time, the villagers exploited fewer old animals; by the abandonment of the site in the 7th millennium BC, around 85 percent of the swine had been killed prior to reaching two years of age.[82] This is exactly the kind of demographic profile one would expect for a herd of domestic swine in which humans cull animals not needed for reproduction once they reach an ideal weight.

Other evidence indicates that swine management was becoming more intensive over time at Çayönü. The incidence of hypoplasias on suid teeth was much higher in the Early and Middle PPNB at Çayönü (43 percent of teeth affected) than in the PPNA at Çayönü (22 percent) and at Hallan Çemi (21 percent).[83] Hypoplasias are linear or pitlike depressions in tooth enamel caused by a temporary cessation of growth. Enamel growth halts when an animal experiences certain forms of physiological stress, especially those related to insufficient diet or acute illness. The proportion of pigs' teeth affected by hypoplasias is thought to increase as people ramp up their management of pigs, which may cause the animals to experience dietary shortages or be exposed to new pathogens from their human hosts.[84] Alternatively, the increase in hypoplasias might reflect increased rates of survival of such stressors through human intervention. While this may seem paradoxical, the appearance of hypoplasias requires both the application of stressor (e.g., illness) and the survival of that stressor.[85] Humans caring for sick or underfed piglets, which would otherwise die in the wild, could lead to an increase in the rate of hypoplasias. In either case, however, the data from Çayönü suggest that human interference in suid populations increased after the PPNA.

Finally, the biometrical data provide some of the most convincing evidence for domestication. Measurements taken on suid teeth and bones from Çayönü exhibit a subtle but significant reduction over time. The decrease in size, shown in Figure 3.2, was gradual. The trend first appears in the Early PPNB phase,[86] but continues on through the entire span of the site's occupation, with a brief reversal in the trend in the Late PPNB, which might relate to the temporary abandonment of the settlement. Taken together, the demographic, pathological, and metrical data show a clear trend: beginning in the early phases of the PPNB, the inhabitants of Çayönü and other settlements in the foothills of the Taurus Mountains controlled suid populations over a long enough period of time and in a manner sufficiently intense to enable the slow development of domesticated pigs.

To the untrained eye, the domestic pigs of the PPNB would have more closely resembled wild boar than the sheepherding hero of *Babe*. The changes were subtle, at least at first. And the process was slow enough that people may not have been aware of any major changes in suid physiology or behavior. But by 7500 BC, villagers across northern Mesopotamia were raising pigs that were significantly smaller, shorter-faced, and more docile than their wild boar ancestors.[87] Villagers at Çayönü and other places also probably considered these animals property, of individual families or of the community as

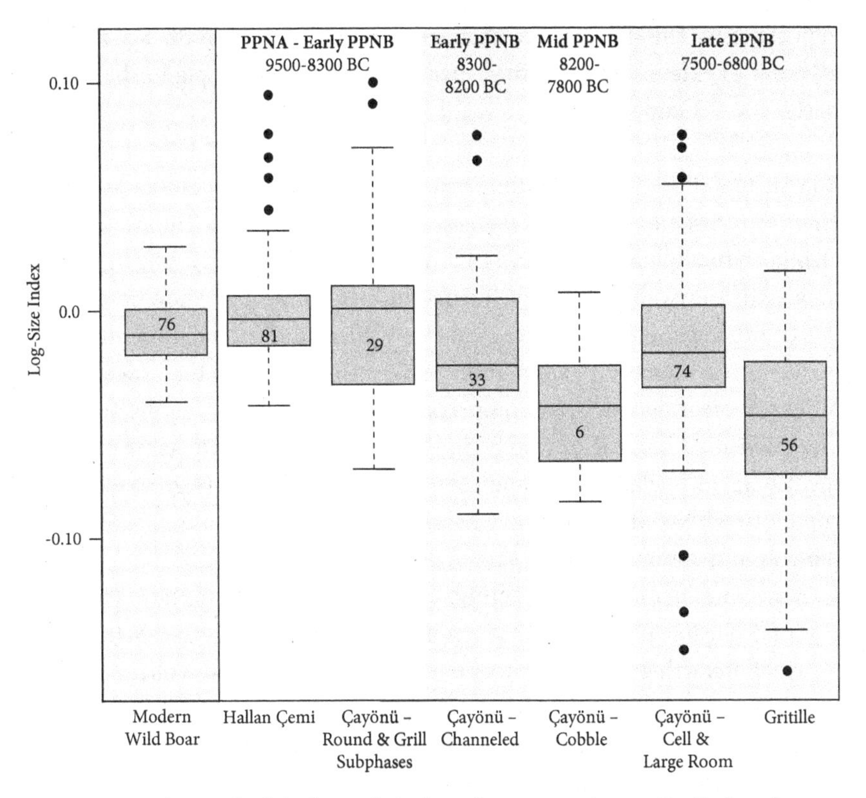

Figure 3.2. The gradual decline of pig dental size over time at Çayönü and nearby sites. Numbers inside boxplot squares indicate sample size. This figure uses the log-size index of molar breadth measurements. This method compares observed values against those of a standard, in this case the mean values of the modern wild boar presented in the figure.

a whole. That is, they thought of pigs as belonging to people rather than as independent beings.

While they probably considered pigs property, the first pig breeders probably did not practice intensive husbandry. If wild boar were initially penned during the process of domestication, we would expect a rapid onset of the domestication syndrome as domestic pigs were reproductively cut off from wild populations and subjected to intense selection pressures. We would also expect the virtual elimination of older animals and especially older males, which would have been particularly difficult to keep in pens. For the same reason, we would expect the average age at death to decline suddenly, not gradually. In fact, the Çayönü data show the exact opposite of these expectations. The slow pace of changes in morphology and the age at death all point to extensive husbandry—perhaps reminiscent of pig husbandry in parts of New Guinea today.[88]

To conclude, for over one million years, suids were an unpopular target for *Homo erectus*, *Homo neanderthalensis*, and *Homo sapiens*. Things began to change in the Epipaleolithic and PPNA, when people began living in permanent settlements and practicing horticulture. Responding to new opportunities (more dependable resources) and new dilemmas (population pressure), people intensified their exploitation of the animals within their local environments, creating new selection pressures for sheep, goats, cattle, and pigs. The intensified hunting of wild boar, which transitioned into game management (e.g., at Hallan Çemi) and then herd management (e.g., at Çayönü), coincided with wild boar being attracted to human settlements; pigs followed the prey and commensal pathways to domestication.

While human niche-constructing activities were vital to initiating the process of domestication, we cannot ignore the unique biology and behavior of pigs that made them amenable to life among people. There's an old joke about how many psychiatrists it takes to change to a lightbulb: just one, but the lightbulb has to *want* to change. The same principle can be applied to domestication. Although humans and the unique ecosystem niches they create are the driving factors behind domestication, the target species themselves must be able to find a way to thrive in these new settings.

The dual prey-commensal pathway created a unique opportunity for wild boar to live in the "human niche"[89] and, beginning in the 9th millennium BC, to evolve into domestic pigs. In the process, *Sus scrofa* became a more reliable source of food for humans, whose populations were growing at a rapid clip. As a result, humans intensified their management practices, which in turn ramped up the selection for genes and phenotypes advantageous to human control. This runaway feedback loop, taking centuries to play out, had radical consequences. *Sus scrofa* transitioned from a relatively ignored prey species to a major livestock animal. Within a few millennia, *Sus scrofa* had become one of the most successful mammals on the planet with a global population in the hundreds of millions.

For humans, the domestication of plants and livestock had no less dramatic consequences. Birthrates increased as more food became available. However, their nutritionally deficient and starch-rich diets, the crowded and unsanitary conditions of their villages, and the backbreaking labor involved in farming led to a health decline among the first farmers.[90] Nevertheless, the explosive population increase initiated by plant and animal domestication helped establish farming as the dominant way of life in the Near East. By the close of the Neolithic, societies were bound, for better or worse, to agriculture. This was truly revolutionary, for it was agriculture that set the stage for the development of cities, writing, mathematics, and the other hallmarks of Near Eastern civilization. It also helped spawn economic inequality, imperialism, slavery, and environmental degradation.

4
Out of the Cradle

Pigs and the other domestic animals and plants formed the nucleus of a "Neolithic package" that had emerged and spread throughout the Fertile Crescent (i.e., Mesopotamia and the Levant) and Anatolia by the Late Pre-Pottery Neolithic B (7500–7000 BC).[1] By around 7500 BC, a number of settlements had grown to such an extent that archaeologists have labeled them "megasites." Encompassing 5–10 hectares or more, villages like 'Ain Ghazal in Jordan and Çatalhöyük in Turkey were home to several hundred or even thousands of people.[2] And while, after 7000 BC, people abandoned the megasites,[3] the agricultural way of life persisted. Over the course of the 7th and 6th millennia BC, farming spread into eastern Europe, the Caucasus, Egypt, the Iranian Plateau, and South Asia.[4] However, the spread of domesticates out of the Fertile Crescent was uneven and piecemeal. Some parts of the package, especially pigs, were adopted much later than others.

Around 7000 BC, people in northern Syria and Iran began to make ceramic vessels and use them for cooking, storage, and the presentation of food.[5] This development marked the beginning of the Late Neolithic. Pottery allowed people to decorate utilitarian objects with stylistic motifs on an incredible scale, imbuing ceramics with creative elements by painting, incising, and molding them into different shapes. Many of these styles persisted over long periods of time, enabling archaeologists to reconstruct a number of ceramic traditions. Some, like the Halaf (5900–5200 BC) in northern Mesopotamia, were spread over a large area. Others, like the Yarmukian (6400–5800 BC) in the southern Levant, were regionally circumscribed.

Neolithic communities were more or less egalitarian societies in which wealth and status were largely independent of birth. There were no substantial differences between the members of each community with respect to the distribution of prestige goods, burial practices, and house size. Communities may have even enforced an egalitarian ethos.[6] But egalitarianism began to crumble in the 5th and 4th millennia BC—the Chalcolithic period. Evidence for social inequality includes the division of settlements into centers and peripheral villages; increasing differentiation in burials; and the appearance

Evolution of a Taboo. Max D. Price, Oxford University Press (2020). © Oxford University Press.
DOI: 10.1093/oso/9780197543276.001.0001.

and concentration of prestige goods like precious stones obtained from places as far away as Central Asia. The emergence of economic specialization, the basis for a division of labor, was connected to the rise of social hierarchy. Warfare, a symptom of competition between territorial societies and their ambitious elites, also increased.[7]

Archaeologists refer to groups that display differences in heritable social status and specialization as "complex societies." Like agriculture, complex societies would spread and in a few millennia define the political landscape of the Near East. By the early 4th millennium BC, cities and states began to appear in Mesopotamia.[8] By around 3600 BC, the world's first colonial system, known as the "Uruk Expansion," integrated parts of the Fertile Crescent into an economic sphere centered in southern Mesopotamia.[9]

The development of complex societies was aided by changes in agricultural practice in the Near East. People domesticated new crops—olives and grapes—and adopted domestic donkeys from Africa as pack animals.[10] Additionally, cereal production intensified as people applied more labor in order to extract more calories, even though it meant lower marginal returns.[11] This included the adoption of techniques designed to enhance productivity, such as adding manure to the soil.[12]

For animals, the "secondary products revolution"[13] was a key turning point. A "secondary product" is a resource that one can extract from an animal without killing it.[14] Examples of secondary products include milk and wool, but also traction power—the use of animals for pulling carts and plowing fields—which greatly amplified the ability to produce and transport large amounts of grain. People had used secondary products since the Neolithic,[15] but beginning in the Chalcolithic, communities began to change how they managed animals in an effort to maximize the output of these valuable and transportable goods.[16] This economic development, as we will see, engendered a shift in the perception of animals no less revolutionary than the Pre-Pottery Neolithic reimagining of prey as livestock.

The (Delayed) Adoption of Pig Husbandry in the Near East

The spread of animal husbandry across the Near East was a piecemeal process. Domestic sheep and goat husbandry spread relatively quickly, reaching western Anatolia, Crete, the Levant, and southwestern Iran by around 7000

BC.[17] Domestic sheep and goats appeared in the Arabian Peninsula in the 7th millennium BC[18] and in Lower Egypt in the 6th millennium BC.[19] However, the spread of cattle and especially pig husbandry was a much slower process, despite the fact that these animals were fundamental features of the agricultural economy in northern Mesopotamia in the PPNB and Late Neolithic.[20]

The spread of domestic pigs was not only slow, but also uneven. For example, domestic pigs appeared in the southern Levant by the early 7th millennium BC,[21] but not in central Anatolia until the 5th millennium BC—3,000 years after domestic sheep and goats and 2,000 years after cattle.[22] Though the delay was less severe in other regions, pigs still did not appear for centuries after sheep, goats, and often cattle. Domestic pigs first appeared in northwest Anatolia, by 5800 BC—about 700 years after sheep, goats, and cattle.[23] They were not introduced to central and southern Mesopotamia until the 6th millennium BC[24] and were not in Iran until the 5th millennium.[25] In Egypt, pigs arrived only once people began practicing cereal agriculture, something they resisted until the 5th millennium BC—again, centuries after sheep, goats, and cattle.[26]

The reasons for the slow spread of domestic pig husbandry are not entirely clear. One possibility is that sheep, goats, and, to a lesser extent, cattle are more mobile than pigs. The ruminants can be herded over long distances and eat grass along the way. Sheep and especially goats fare better in arid environments. The vast tracts of arid grasslands and deserts of the Near East may therefore have presented a barrier to the spread of domestic pigs. Indeed, once people had adopted pig husbandry, the relative importance of pork mapped onto well-watered environments: pigs represent 20–50 percent of medium and large mammal remains from Neolithic sites in the marshlands of the Nile Delta and southern Mesopotamia.[27] The same is true for the Levant, where pig husbandry flourished in the oak-covered hills of the Galilee, the Hula Valley, and the Jordan River Valley.[28] But in the arid grasslands and dry plateaus, the landscapes that define central Anatolia, Iran, eastern Jordan, and the Syrian steppe, people kept very few or even no pigs.[29]

But the environment offers only a partial explanation for the slow spread of pig husbandry. Even in well-watered regions, there was a considerable delay between the first appearance of domestic ruminants and the first appearance of domestic pigs. Cultural factors were also at play.[30] We don't know what these were, exactly, but we can reasonably speculate. For example, it is possible that sheep and goat husbandry allowed the first herders to maintain some semblance of their hunter-gatherer ancestors' mobile lifestyle. Or perhaps keeping pigs as livestock represented too great a contradiction to boar hunting and its connotations of masculinity. Bringing swine into the household may have been perceived as feminizing these animals, a bridge too far for early livestock keepers adjusting to a new way of life with new gender roles.

The First European Pigs

Unlike the situation in the Near East, the spread of agriculture into Europe involved all the animals of the Neolithic package at once. Sheep, goats, cattle, and pigs arrived more or less contemporaneously throughout Europe between about 7000 BC and 5500 BC.[31] There are a number of potential reasons that pig husbandry was adopted more readily. First, the environment of Europe is significantly wetter and more temperate than that of much of the Near East. There were few places in Europe in which pig husbandry could not flourish. Second, the spread of agriculture into Europe was closely tied to the diffusion of people.[32] Rather than local hunter-gatherers adopting farming, picking and choosing from the Neolithic package, in Europe it was pioneering farmers, originally from the Near East, that brought the agricultural way of life.

Research on ancient suid DNA has shed more light on the spread of pigs into Europe. Mitochondrial DNA (reflecting maternal ancestry) and nuclear DNA (reflecting both maternal and paternal ancestry) extracted from pig bones from Neolithic sites indicate that the first domestic pigs in Europe descended from Near Eastern pigs. Their ancestors were the livestock that the pioneering farmers brought with them into the new continent. But beginning in the 4th millennium BC, DNA extracted from domestic pigs matched that of *European* wild boar.[33] In other words, there was a genetic turnover in which Near Eastern-derived genes were replaced by European-derived ones. Today, none of the major domestic pig breeds in Europe contain any detectable Near Eastern ancestry.[34]

The likely explanation for this genetic turnover is that there was frequent hybridization between local wild boar in Europe and domestic pigs. There is good evidence that early European stockbreeders routinely captured wild boar or otherwise allowed them to breed with their pigs.[35] Hybridization was part of a successful husbandry strategy that, whether the herders were aware of it or not, eliminated the deleterious effects of inbreeding and ensured long-term reproductive success.[36] However, it is not clear why European pig genes were more successful over the long term. Because the genetic turnover was so widespread (even extending into the Near East, as we will see in Chapter 7) genes found in European wild boar must have conferred some type of advantage to swine living under pig husbandry systems. We can speculate what these were—slightly higher fecundity, larger body size, tolerance to cold weather, to name a few. But we won't know the cause of the genetic turnover until future research documents which phenotypes underlay the success of European-derived lineages.

New Approaches in the Late Neolithic

Back in the Near East, major changes in agricultural production were afoot in the Late Neolithic (7000–5000 BC).[37] Farmers made innovations to adapt to new challenges, such as those posed by the "8.2 ka event," an episode of global cooling that brought arid conditions to the Near East.[38] People also improved upon existing farming techniques in order to generate surpluses. In terms of animals, Late Neolithic peoples in the Levant, Mesopotamia, and Anatolia began to exploit animal secondary products.[39] They also began to intensify pig husbandry.

Penning Pigs

Zooarchaeological data from Pre-Pottery Neolithic sites like Çayönü indicate that pigs were raised in an extensive manner during the Pre-Pottery Neolithic. Pig owners kept their animals under a loose form of management, allowing them considerable freedom to wander in search of food for much of the year. By the Late Neolithic, this situation began to change. People in northern Mesopotamia began confining pigs more permanently to pens.

Evidence for this shift to intensive pig husbandry is fourfold. First, there was a sudden decrease in pig size, especially the size of teeth. At Çayönü, the initial domestication of pigs caused a subtle and minor decrease in skeletal and dental measurements. In the final phase of the Neolithic occupation, the pigs at Çayönü were only about 5–10 percent smaller than their wild boar ancestors after more than 2,000 years of game management and husbandry. But metrical data from several 6th millennium sites indicate that pig teeth

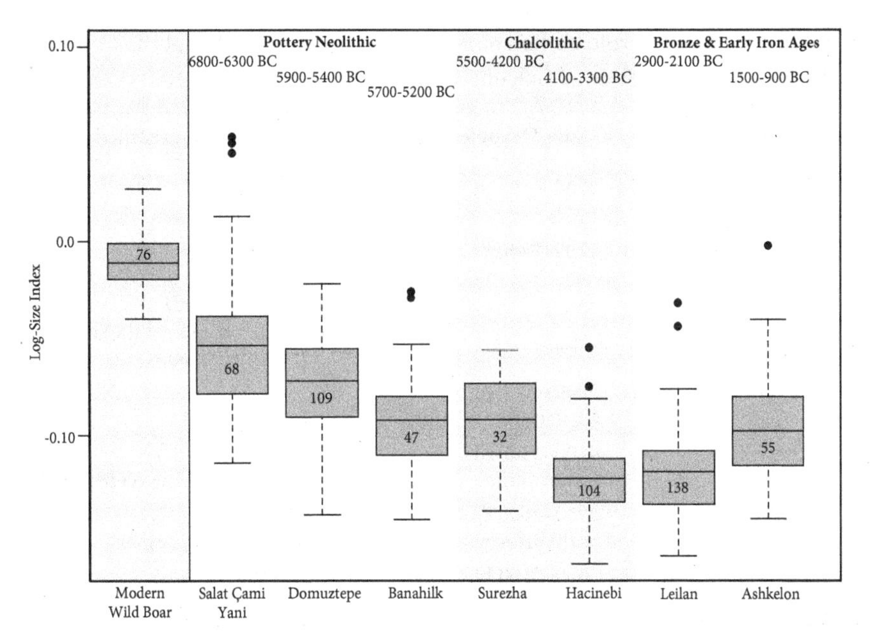

Figure 4.1. Steady and continual reduction in the size of pigs. This figure was drawn with log-size index values, using average modern wild boar as a standard.

were almost 20 percent smaller than the teeth of wild boar.[40] This is symptomatic of a rapid decrease in facial size, something likely related both to more intensive selection for tameness and to reproductive isolation from wild boar. Penning explains both processes. As Figure 4.1 shows, this process of progressive size diminution would continue over the next several millennia, probably as a result of increasingly intensive husbandry regimes.[41]

Second, pig teeth displayed a higher number of hypoplasias. Penning would be expected to increase hypoplasias either by contributing to a higher incidence of disease or by providing an environment in which sick or malnourished individuals (e.g., runts) could receive better care.[42] Third, examination of microscopic plant remains embedded in pigs' dental calculus, or "tartar," at the 6th millennium BC site of Domuztepe in southern Turkey revealed a diet that included processed and cooked cereal grains (Table 4.1). The results suggest that pigs were eating kitchen scraps, a feeding practice common in intensive husbandry systems.[43]

The most direct evidence for intensive pig husbandry derives from a spectacular find at the site of Mezraa-Teleilat in southern Turkey. In a structure labeled Building AY, excavators found the burnt remains of five pigs ranging

Table 4.1 Plant Microfossil Remains (Starch Granules and Phytoliths) Recovered from Pig Dental Calculus

Archaeological Site	Dates Occupied	Plant Remains Identified
Hallan Çemi	9700–9300 BC, PPNA	Tubers Unidentified grasses
Domuztepe	5900–5400 BC, Late Neolithic	Acorns Oats (processed and cooked) Barley (processed) Unidentified cooked starches Unidentified grasses
Tell Ziyadeh	5000–4200 BC, Chalcolithic (Ubaid)	Oat (?)
Hacinebi Tepe	4100–3300 BC, Chalcolithic	Barley (processed?) Unidentified grasses (cereal chaff)
Tell al-Raqa'i	2900–2600 BC, Early Bronze Age	Oat (processed) Unidentified grasses
Tell Leilan	2600–2100 BC, Early Bronze Age	Unidentified grasses

Source: After Weber and Price 2016.

in age from one month to two years old.[44] Building AY was a house dating to the latter half of the 7th millennium BC, and the excavators concluded that the building had burned down in an accidental fire. We can deduce what happened from the contextual evidence and the ages of the animals. A family's herd of pigs, living with their human caretakers, had become trapped inside the house when it went up in flames.[45] Perhaps a few escaped, but the ones that didn't have provided us with the first clear evidence of pigs penned within a domestic structure in a Late Neolithic village.

Feasting on Pork

There are several reasons people may have decided to pen pigs in the late 7th and 6th millennia BC. One is practical. People in the Fertile Crescent during this period relied on mixed farming economies of sheep, goat, cattle, pigs, cereals, and legumes, with intensive manuring and garden cropping designed to increase agricultural yield.[46] With all these different types of food production, space was probably at a premium. People could herd sheep and goats farther away from their settlements, but cattle-herding and crop-growing took place in the fields surrounding the villages. These are not ideal conditions for extensively managed pigs, which can frighten cattle and ravage corn. Moving them into pens would have solved the problem and, as an added benefit, provided nitrogen-rich pig dung for manure.

Another reason people may have decided to pen their pigs was to produce more pork for feasts, a goal to which intensive husbandry was well suited. In fact, there is zooarchaeological evidence for pig feasts in the Late Neolithic. At Domuztepe in southeastern Turkey, excavators uncovered a large assemblage of animal bones in a deposit referred to as "the Ditch," of which pig bones represented around 23 percent of the medium and large mammal remains.[47] Farther south, in Israel, several episodes of pig feasting took place at the 6th millennium BC site of Tel Tsaf. At Tsaf, pig bones were much more abundant in those contexts identified with feasting (33–51 percent) than in those identified as quotidian (28 percent).[48]

The use of pigs for feasts finds many parallels around the world. American readers are probably all familiar with the centrality of pork to barbecues, especially in the cuisine of the US South. Pork is also a traditional food for Christmas feasts throughout the Western world; for example, the fattening and slaughter schedules for traditional swineherds in Spain in part revolve around the Christmastime spike in demand.[49] In the anthropological literature, pig feasts and the politics that surround them are prominent in ethnographies of peoples in New Guinea and the South Pacific.[50]

Feasts, recall from Chapter 3, are celebrations that inspire a sense of community and togetherness through the sharing of food. They can help mend social tensions. Feasts also offer backdrops to political dramas, offering ambitious individuals the opportunity to position themselves as providers to the people and to justify their authority.[51] However, in the fiercely egalitarian Late Neolithic societies, feasts and other rituals may well have served to suppress the ambitions of elites. Regular feasts may have been ritualized and enforced means by which villagers shared food and other resources.[52]

Pigs in Other Rituals

Beyond feasting, we find occasional, if mysterious, uses of pigs in other types of rituals. Many of these rituals involved pig crania and tusks, perhaps related to the masculine associations attached to boar. For example, at 'Ain Ghazal in Jordan in the middle to late 7th millennium BC, skulls and other pig bones were buried alongside humans, and one burial contained a pendant made of out a pig tusk.[53] Similarly, at Domuztepe, excavators found a pig skull buried alongside that of a human,[54] while at Çatalhöyük in central Turkey, an unusual wild boar skull was recovered from a late 7th millennium BC deposit. This skull had been modified—the upper jaw had been removed and some of its lower teeth knocked out, perhaps to transform it into a headdress or a wall decoration.[55] Pig crania were even occasional subjects of art: at the 6th millennium BC site of Choga Mami in Iraq, excavators recovered a finely crafted terracotta pig head from the remains of a house.[56] The meanings of these rituals are obscure, as indeed is the seeming focus on pigs' heads. But they highlight the fact that pigs served more than just an economic role; they were important for the ritual life of Late Neolithic communities.

Pigs in the Chalcolithic: The Great Transformation

The emergence of complex societies in the 5th and 4th millennia BC upended traditional egalitarian values, enforced new economic concepts such as (family-based) private property, and cast individuals into specific social roles. Food played an important part in this process. For one thing, certain food procurement activities took on new meanings. Although hunting was no longer a major means of subsistence in most of the Near East, it became an activity symbolically important for the elite, not only as a form of recreation, but also as an allegory for their power.[57] Beginning in the 4th millennium, Mesopotamian kings commissioned images of themselves hunting to symbolize their masculinity, strength, and dominance over nature. While lions and bulls were the main animals displayed, wild boar were also portrayed. Figure 4.2 shows cylinder seals recovered from late 4th millennium Uruk in southern Iraq and Susa in Iran. The seals depict these cities' rulers hunting wild boar.[58]

Feasting was another way that food was mobilized during the transition to complex societies. The transition to complex societies required the breakdown of mechanisms that had limited social inequality. Emerging elites thus co-opted the feasts that had once served to promote equality.[59] By casting themselves as providers for the people and as instruments of social cohesion, elites found in feasts a way to justify their existence and authority,[60] even as they increasingly used their power to appropriate surplus from the peasantry in the form of taxes, tithes, rents, or involuntary contributions.[61]

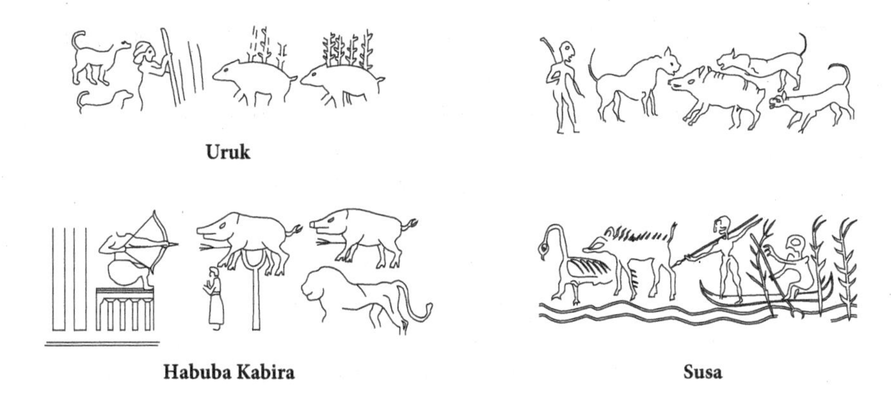

Figure 4.2. 4th millennium BC cylinder seal impressions containing scenes of wild boar hunting.

Feeding the masses, or claiming to do so, was and would remain an important justification of power in the ancient Near East.[62] But feasting served another important role: to create a distinction between the upper classes and the *hoi polloi*.[63] Large-scale celebrations intended only for the entertainment of aristocrats represented a radical new use of food beginning in the Chalcolithic.

Interestingly, pork began to play a less prominent role in feasts in the Near East. With some exceptions, most of the feasting deposits identified from 5th and 4th millennia BC sites contain few or even no pig bones.[64] Chalcolithic feasts focused instead on sheep, goats, and cattle, even when the settlements at which these feasts took place raised pigs in abundance.

One explanation for the diminished presence of pork in Chalcolithic feasts is that pigs became tied up with emerging class identities. That is, people may have begun to associate pigs with lower social classes or with quotidian as opposed to ritual activities. Zooarchaeological data from the site of Arslantepe in Anatolia supports this hypothesis. At Arslantepe, pig bones were concentrated in nonelite household garbage deposits. In contrast, sheep, goat, and cattle bones were found in all contexts, but they were especially abundant in temples, where they represented sacrificial offerings and/or animals slaughtered for feasts[65]—sacrificial animals were frequently feasted upon in the ancient world.[66] Pigs were largely absent from these ritual deposits, representing about 1 percent of the medium and large mammal bones. Pig remains were also more abundant in nonelite households (23 percent) than in elite residences (14 percent).

The data, in other words, show two dichotomies in pork consumption: one between the ritual and the mundane, and one between upper and lower classes. Similar changes occurred in the art produced for the upper classes. By the Chalcolithic, artists depicted pigs much less frequently then other animals, such as goats, bulls, and lions. This does not mean people stopped eating pork or portraying swine (for example, Figure 4.3 shows a ceramic pig vessel from southwestern Iran). Rather, in the Chalcolithic, a certain number of contexts emerged in which pigs were less appropriate than other animals.

The Arslantepe data seem to suggest a downgrading of the status of pork to that of a food eaten more frequently by the lower classes in nonritual settings. But pigs' absence from ritual contexts might have been due not so much to the development of negative attitudes toward pork as to the elevation of the status of mutton and beef. If pigs had developed a negative connotation, we would expect their complete absence from ritual settings. That's not the case; they simply played a minor role. Similarly, elites continued to eat pork, but they ate it less frequently than they did the meat of ruminants. I argue that this reflects a developing association between ruminants and wealth.

Figure 4.3. Proto-Elamite vessel of a pig. Southwestern Iran, ca. 3100–2900 BC.

The association of sheep, goats, and cattle with wealth likely took root in the wake of the secondary products revolution. During this economic transformation, communities in Anatolia and the Fertile Crescent altered the ways in which they managed their sheep, goats, and cattle, shifting from production strategies geared primarily to meet subsistence needs to those aimed at maximizing the production of commodities for exchange. Milk, wool, and traction power became important livestock products in addition to meat.

A new ruminant economy emerged, but the impacts of the secondary products revolution extended beyond livestock keeping, strictly speaking. In particular, grain production received a major boost from cattle traction.[67] Attached to a plow, a pair of oxen could increase the amount of land brought under cultivation about tenfold.[68] Deep plowing, moreover, brought nutrient-rich soils to the surface, enriching the output of fields. Cattle could also be used in threshing and for hauling sheaves of wheat, helping to overcome production bottlenecks. Although the idea to use cattle for these purposes may have occurred to people in the Neolithic, there is only extensive evidence for plowing in the Chalcolithic and later periods.[69] One reason for this is that keeping a pair of oxen alive year-round required a substantial investment in feed and maintenance. That may not have been possible on a large scale until the Chalcolithic. For the same reason, beginning in the Chalcolithic, ownership of oxen probably began to define the differences between the haves and have-nots.[70]

Wool and goat hair represented another set of important commodities in the ancient Near East, ones that later would serve as the main trade exports from Mesopotamia.[71] Instead of slaughtering most males as juveniles to ensure herd growth and propagation, herders waited until these animals were four or five years old to maximize the amount of fiber they could obtain from them. Such strategies came with downsides; herds composed of a high ratio of adult males to females grow at a slower rate and are a less efficient means of achieving subsistence needs. In order to make intensive fiber exploitation an attractive pursuit, herds may have had to be quite large in the first place, something that would benefit from elite-directed coordination and managerial oversight. For that reason, small-scale subsistence-based pastoralists tended to shy away from such a strategy. But elite owners of large herds, pastoralists integrated into market economies, or herders sponsored by an elite patron were better positioned to assume those risks.[72]

As a result of the secondary products revolution, pastoral ruminants—sheep, goats, and cattle—became factors in the development of economic inequality and specialization.[73] Their products became commodities whose exchange financed the elite takeover of agriculture and craft production. Ruminants thus became sources of wealth. But in addition to being the means of production, ruminants also represented wealth in themselves. Controlling herds, and propagating them like the financial assets they were, gave elites another building block with which to construct their power. Although pigs could serve other purposes, such as removing waste from settlements or producing nitrogen-rich manure to fertilize fields, swine were raised primarily for one reason—pork. Their lack of a commodifiable secondary product excluded them as sources of wealth, a point we will take up in the next chapter..

The faunal data in parts of the Near East bear out the impact of the secondary products revolution on pig husbandry. In Mesopotamia in the latter half of the 4th millennium BC, pigs became less common in faunal assemblages as sheep and goats began to dominate. At some sites, pigs were almost or completely absent, especially those associated with the Uruk Expansion in Mesopotamia, Iran, and Anatolia.[74] The decline in pork consumption even occurred in areas where pig husbandry had been—and would again be—a major agricultural activity, such as the marshes of southern Mesopotamia and the Khabur drainage in northern Syria.[75]

Nevertheless, pork remained an important source of food for many people across the Near East. Pig bones typically account for at least 25 percent of medium and large mammals remains—which at this point consist almost entirely of domestic livestock—at Chalcolithic sites in the hilly regions of the southern Levant and the Jordan River Valley,[76] north and central Anatolia,[77] and parts of northern Mesopotamia.[78] In the Nile Delta, pigs made up around 50 percent of the medium and large mammal bones.[79]

To sum up, in the Late Neolithic, farming spread across the Near East. However, people were more hesitant about adopting pig husbandry than other forms of animal husbandry. In places where pig husbandry was adopted, there seems to have been considerable interbreeding with local wild boar. In Europe, local wild boar genes, once introduced into domestic pig populations, were passed on to successive generations and ultimately succeeded over Near Eastern derived genes.

Pig husbandry also became more intensive in the Late Neolithic in northern Mesopotamia. Extensive husbandry did not disappear, but penning pigs became far more commonplace, for two likely reasons: the use of settlements' immediate environments for other types of agricultural activities and the use of pigs in feasts.

The agricultural innovations that developed in the Late Neolithic laid the foundations for the secondary products revolution, which occurred in the 5th and 4th millennia BC. The "revolution" transformed sheep, goats, and cattle into potential sources of wealth. While swine remained a part of the mixed agricultural lifestyle that characterized many Near Eastern settlements, the value of pigs diminished in comparison with that of ruminants. In the next chapter, we will see how Bronze Age societies built upon these patterns and how the role of pigs evolved yet again .

5

Urban Swine and Ritual Pigs in the Bronze Age

Toward the end of the Chalcolithic period, a new type of settlement appeared in the Near East: cities. These communities, consisting of thousands of people, were much larger than Neolithic villages, an order of magnitude larger than even the "megasites" of the Late PPNB. But they were also unique in terms of their social, economic, and political significance. A city is a community of communities, a collection of people inhabiting a diverse number of roles that is organized through a division of labor and socioeconomic and political hierarchy. Cities project their influence into the countryside, controlling production and consumption in satellite villages and enfolding these smaller settlements into coherent political units headed by urban elites. As central and strategic places, cities were more vulnerable to attack than other settlements; people therefore built walls around them of mudbrick and stone.

The first cities appeared in southern Mesopotamia and the Khabur drainage in northern Syria in the early 4th millennium BC. Urbanism then spread, and by 2500 BC, cities dotted the landscape from Iran to Anatolia. In Egypt, the Levant, and western Anatolia urban centers were much smaller than their Mesopotamian counterparts, typically encompassing 10–25 hectares compared with the 100 or more hectares of the behemoths in the Khabur, southern Mesopotamia, and Iran. Though smaller, like the cities in the East, these centers served as seats of highly centralized political and economic power—what archaeologists refer to as "states." By the early 2nd millennium BC, cities and states controlled, or least exerted considerable influence over, the vast majority of people in the Near East.[1]

At the heart of cities' political and economic domination were the so-called institutions—palaces, temples, and elite manorial estates. The institutions controlled vast herds of animals, land, and other resources. They developed elaborate bureaucratic networks and assembled hundreds of workers, free and unfree. The elites who ran these institutions constructed massive houses and religious monuments, many of which are found on the acropolises at the

Evolution of a Taboo. Max D. Price, Oxford University Press (2020). © Oxford University Press.
DOI: 10.1093/oso/9780197543276.001.0001.

center of major archaeological sites. The scale of institutions and their activities inspired new organizational techniques. It was in the need to manage and document economic activity over large areas that led institutional record-keepers to develop the world's first writing systems by the late 4th millennium BC in Egypt and southern Mesopotamia.[2]

Institutions sat at the center of all the states that emerged in the Chalcolithic and Bronze Age, but there were regional differences. Most institutions focused on the agrarian sector of the economy to finance their power. But some, especially in western Anatolia, also relied on trade in precious commodities like metals. Some states, like Egypt, extended over large territories. In southern Mesopotamia, or Sumer, a collection of highly competitive city-states persisted throughout the 3rd millennium BC. Much like their Greek counterparts two millennia later, Sumerian city-states typically controlled only a single city and its surrounding countryside. Each city-state vied with neighboring powers for regional supremacy through warfare and political intrigue. As time went on, just as in ancient Greece, some Sumerian city-states were able to extend their hegemony over others. Ultimately, territorial empires—political units that controlled numerous cities spread over several regions—would supersede city-states. By the Middle (2000–1600 BC) and Late Bronze Ages (1600–1200 BC), empires of increasing size and complexity dominated the Near East from Mesopotamia to Anatolia to Egypt.

While it is perhaps unavoidable to see the Chalcolithic period and Bronze Age as a time of increasing political development, it is important to take stock of the numerous pauses and retreats in the process. Indeed, there was a cyclical nature to Bronze Age urbanism and state power that is reminiscent of the historiographical observations of 14th century AD Arab historian Ibn Khaldun.[3] Much attention has focused on the ends of these cycles, or why civilizations collapse. Answers have included resource overexploitation, climate change, warfare, economic contraction, and popular discontent, among others.[4]

Whatever the specific causes, the Bronze Age Near East has provided some of the most dramatic examples of collapse and rebirth. The end of the Early Bronze Age witnessed a panregional process of deurbanization coinciding with the collapse of the Akkadian Empire, the Old Kingdom in Egypt, and city-states in the Levant. Perhaps not coincidentally, a global climatic downturn referred to as the "4.2 ka event" also occurred around this time.[5] Urban society reorganized in the Middle Bronze Age, but the Hyksos conquest of Egypt around 1650 BC and the Hittite sack of Babylon in 1595 BC led once again to regional power reshuffling and a century-long "Dark Age."[6] The Late Bronze Age, beginning in 1600 BC, was a time of internationalism and imperialism, when great powers—the Hittites in Anatolia, the New Kingdom of Egypt, the Elamites in Iran—controlled the Near East. But this world, too, would fall apart and set the stage for new powers to emerge in the Iron Age.[7]

Plows, Wool, Warhorses (and Wardonkeys), and Wealth

The development of cities and states had a major impact on livestock husbandry, one that brought the secondary products revolution to fruition. So central were animals to early states and empires that it is not too great an exaggeration to say that many of the institutions at the heart of Bronze Age societies, including those in Egypt and Mesopotamia, built their economic power on the backs of sheep, goats, cattle, and later equids. Of course, managing herds for secondary products was nothing new. But during the 3rd millennium BC, institutions sought to produce truly massive volumes of tradable and storable commodities. If the idea that ruminants equaled wealth first appeared in the Chalcolithic, it became a foundational feature of economic thought in the Bronze Age, particularly in Mesopotamia.[8]

Among the most important animals of the Bronze Age were cattle. As agents of traction power, cattle became "the engines of Bronze Age agricultural systems"[9] that were increasingly needed to feed the burgeoning populations of cities. But cattle also contributed to greater inequality. Families that owned cattle could produce more grain than those that did not. Families that did not own cattle often had to rent oxen to prepare their fields, exacerbating existing wealth disparities and leading to situations in which cattle-owning families could hold cattle-borrowing families in debt or social dependence. By the late 3rd millennium and early 2nd millennium, oxen rentals were such a regular feature of daily life that Mesopotamian authorities composed laws regulating transactions and compensations for damages.[10]

Cattle were key elements in large-scale grain production, which helped underwrite state-making in the Bronze Age. The institutions at the top of the socioeconomic ladder, especially in Mesopotamia, produced vast surpluses of grain to distribute to their clients, dependent workers, and slaves as rations. In this way, they conducted "gastro-politics,"[11] employing food and the politics of food-sharing to reinforce social inequalities. Similarly, institutions presented themselves as guarantors of security in the face of famine by operating granaries.[12] In part because of the enormous increase in grain productivity enabled by their traction power, institutions amassed vast herds of cattle. Texts document that Bronze Age institutions in Mesopotamia, Egypt, and Anatolia owned and managed large herds of cattle.[13] Additionally, elites presented these valuable animals as gifts to one another, sought after and captured them in war, and sacrificed them to the gods.[14]

Sheep and goats[15] provided what was, in addition to grain, the single most important agricultural commodity in the Bronze Age: wool. The expansion of the wool economy in Mesopotamia and neighboring regions at the end of the 4th and 3rd millennia BC[16] was so pronounced that some have argued it constituted a veritable "Fiber Revolution."[17] By the 3rd millennium, wool and woolen textiles were the major export of Mesopotamia, driving trade throughout the Near East.[18] Wool also impacted labor. In their attempts to increase the volume of traded textiles, individual institutions controlled tens or even hundreds of thousands of sheep and goats. Institutional authorities contracted specialist herders to manage these assets. Institutions also created textile workshops that employed or enslaved thousands of workers, especially women, to pluck fiber, clean raw wool, spin it, dye woolen cloth, and weave textiles. Thus, the increase in wool production was supported by (and contributed to) the creation of a landless lower class, debt slaves, and captives taken in warfare to bolster institutions' labor forces.[19]

In addition to ruminants, equids (donkeys and horses) were also treated as wealth. These animals were not a part of the original Neolithic package, but were added to the mix in the Bronze Age. First domesticated in Egypt, donkeys were particularly valuable in Early Bronze Age ritual and warfare. Elites routinely sacrificed donkeys and *kunga* (specially bred hybrids of donkeys and wild onagers), burying them alongside their dead. As depicted on the famous "Standard of Ur," donkeys or *kunga* also pulled wheeled carts into battle. Equids thus facilitated conquest and the appropriation of wealth from others.[20] Horses, originally domesticated in Central Asia and introduced to the Near East by the 3rd millennium BC, eventually replaced donkeys and *kunga* in military roles and enabled the development of the fast-moving chariot and, later, cavalry.[21] By the Late Bronze Age, horses were incredibly valuable assets and taken, along with sheep, goats, and cattle, as tribute and as booty in war.[22]

Pigs are conspicuously absent from the list of animals that constituted wealth. For one thing, as mentioned in the previous chapter, pigs did not provide commodifiable and storable secondary products like wool or something that could be translated into a commodity, like traction. While pigs did provide lard, around which a lively trade developed, this product was never as valuable as grain or wool.

For another thing, raising large herds of sheep, goats, cattle, or even equids was more efficient in many Near Eastern environments than raising large numbers of pigs. Large-scale intensive pig husbandry requires vast amounts of grain to fatten hogs before slaughter, something that institutions would have wanted to avoid to keep up their image as the guarantors of grain for the populace.[23] Herding and fattening pigs in nut-bearing hardwood forests would have provided an alternative, one that Roman institutions took advantage of (Chapter 8). But these landscapes were not as common in the ancient Near East as they were in Europe; instead, vast tracts of steppic grasslands covered Syria, central Anatolia, Iran, and other places. These landscapes are excellent for raising ruminants and equids, but not pigs.

There is also a political angle to pigs' exclusion. The anthropologist James Scott[24] has written extensively about how states seek to measure, tally, and quantify the resources under their control. States and their institutions focus their attention on, and encourage the production of, the types of resources that are the most quantifiable and amenable to large-scale coordinated management. Grain, ruminants, and equids fit this description. Pigs do not. The problem would become particularly pronounced as states attempted to derive income from taxes.[25] Any tax collector or tax farmer sent to assess the number of pigs in a peasant village would have to take into account the fact that a sow might farrow anywhere from three to eight piglets every time she gave birth. And she might give birth once, twice, or not at all within a given fiscal year. Of the piglets born, some might die of malnutrition or hypothermia. Given these uncertainties, it would be easy for pig breeders to underreport their gains and cheat the tax collector. Unless under regular surveillance, pig owners could easily sell off or trade their piglets before the tax collector arrived and convincingly claim an underproductive year.[26]

None of this is to say that pigs had no value at all. They remained prominent sources of food in the Bronze Age. Mesopotamian institutions themselves raised pigs, sometimes several hundred at a time. Authorities even meted out punishment to pig thieves, indicating that these animals were valuable. For example, an entry from Hammurabi's famous Law Code (1792–1750 BC) specifies:

> If a man steal ox or sheep, ass or pig, or boat—if it be from a god (temple) or a palace, he shall restore thirtyfold; if it be from a freeman, he shall render tenfold.[27]

A later Middle Assyrian law code (11th century BC) even detailed fines for the theft of different types of pigs—for example, 12 *shekels* of silver for fattened pigs, 6 *shekels* for ordinary pigs, and 6 *shekels* plus 1 *parisu* for pregnant/nursing sows.[28] Other sources document the prices in silver or grain of cuts of pork.[29] However, it is worth mentioning that pigs were evaluated at lower prices (in grain or silver) than other domestic animals.[30]

Although valuable, pigs were never considered worthy of large-scale investment by Bronze Age institutions. Even for middle-class families in Mesopotamia, pigs were not considered wealth. Documents discussing property inheritance among families frequently discuss livestock but only occasionally mention pigs.[31]

Regional Diversity in Pig Husbandry in the Early and Middle Bronze Age

While pigs were not sources of wealth, pork did not disappear. Zooarchaeological work in Mesopotamia and Egypt over the past 30 years has disproved the myth, once popular among researchers, that the Bronze Age witnessed a decline in pork consumption.[32] In fact, during the Early and Middle Bronze Ages, there was a boom in pig husbandry in several key regions of the Near East. But the pattern was uneven. In some places, notably the Levant and western Syria, pig husbandry declined precipitously. And even in those places in which pig husbandry did expand, it did so primarily in lower-class and urban contexts.

Table 5.1 shows three variables that were key to the success or decline of pig husbandry in the 3rd and early 2nd millennia BC: how institutions developed wealth, the degree of urbanism, and preexisting food traditions. During the Bronze Age, pork was an important part of the diet in regions in which cities were large and where a tradition of eating pork had been pervasive in the preceding Chalcolithic. To a lesser degree, pigs tended to be more abundant in regions where institutions financed themselves primarily by accumulating sources of wealth *other than* grain or secondary products (e.g., precious metals).[33] Similarly, pig production declined in regions that

Table 5.1 Relative Abundance of Pig Remains across Regions of the Bronze Age Near East as a Function of Three Variables: Urbanism, Food Traditions, and Sources of Institutional Wealth

Region	Pig Remains in Early–Middle Bronze Age ca. >20 Percent	Degree of Urbanism in Early Bronze Age, Small or Large Cities	Pigs Included in Chalcolithic Food Traditions >20 Percent	Grain/ Secondary Products Major Sources of Institutional Wealth
Egypt	Yes	Small	Yes	Yes
W. Anatolia	Yes	Small	Yes	No
Khabur	Yes	Large	No	Yes
S. Mesopotamia	Yes	Large	Not enough data[a]	Yes
Iran	No	Large	No	Yes
Levant	No	Small	Yes	Yes
W. Syria	No	Small and medium	Yes[b]	Yes

[a] There were few pig remains at 4th millennium Uruk and none at Tell Rubeideh (Payne 1988; Vila 2006:140).

[b] There was a decline in pig husbandry at many sites in western Syria during the Uruk Expansion, but pig husbandry rebounded around 3000 BC (see Price 2016; Price et al. 2017).

lacked large cities and in contexts in which institutions were able to monopolize livestock management and direct it toward the production of secondary products and grain. People from regions lacking a tradition of pork consumption in the Chalcolithic period tended to continue to refrain from eating pork in the Bronze Age. However, which of these variables was the determining factor of pig production varied from case to case. There is no one pig principle we can apply to explain all cases of swine abundance in the Bronze Age.

The (Informal) City Pigs of Early and Middle Bronze Age Mesopotamia

In Mesopotamia during the Late Chalcolithic, the secondary products revolution seemed poised to make pork a thing of the past. Sheep and goats had largely replaced pigs throughout much of Mesopotamia[34] during the Uruk

Expansion in the 4th millennium BC, which was a form of colonialism at least in part connected to expanding wool economies.[35] Its impact significantly curtailed pig husbandry in Anatolia and northern Syria.[36]

The situation changed during the 3rd and early 2nd millennia BC, when pork consumption flourished in the heavily urbanized parts of Mesopotamia. Take southern Mesopotamia. In most of the major Sumerian cities for which we have zooarchaeological evidence, pigs constituted anywhere from 20 to 50 percent of the main livestock animals slaughtered for food.[37]

Pork was even more popular in the cities of the Khabur drainage in northern Syria, where urbanism exploded in the 3rd millennium BC at sites such as Tell Leilan, Tell Mozan, Tell Brak, and Tell Hamoukar. At each of these cities, pig remains constitute at least around 25 percent of the main livestock species and often as much as 50 percent. Concentrations of pig bones were particularly high in areas associated with lower socioeconomic classes. In fact, because archaeologists tend to focus their excavations on elite areas (temples and palaces), zooarchaeological summaries of these sites probably underestimate the overall importance of pork for the general urban diet.

The dependence on pig husbandry survived episodes of urban retreat. Many of the cities of the Khabur were abandoned, perhaps due to a global climate change event (the 4.2ka event) at the end of the 3rd millennium BC. Nevertheless, when cities were once again repopulated in the early 2nd millennium, pig husbandry was a crucial component of the livestock economy. At Middle Bronze Age sites in the Khabur, pig bones typically make up around 30 percent or more of the bones of the main livestock species (see Table A.2 in the appendix).[38]

Urbanism thus appears to have played a key role in supporting pig husbandry in Bronze Age Mesopotamia—and vice versa. Cities offered a new type of environment that was particularly suitable to pig production. As examples from modern-day Cairo demonstrate, pigs thrive in urban spaces as consumers of waste. Indeed, from the pigs' perspective, Bronze Age city life must have been a never-ending feast composed of accumulated human and animal feces, table scraps, and spoiled grain. Food production refuse, such as spent brewery grain or whey from cheesemaking, would have provided another source of calories. From the human perspective, whether penned or allowed to wander city streets in search of food, pigs provided an efficient waste management system, converting garbage into calories.[39]

Pigs served as a crucial source of meat for the burgeoning urban lower classes. Zooarchaeological data from Mesopotamia and Egypt suggest that, although people of all classes ate pork, most pig husbandry took place outside of institutional settings.[40] Textual records of palaces and temples, as Figure 5.1 demonstrates, mention pigs far less frequently than sheep, goats, and cattle. Yet zooarchaeological data indicate that people regularly consumed pork. Zooarchaeological data also indicate that pig bones occur in lower relative frequencies[41] in institutional contexts than in domestic ones, especially those associated with the lower classes.[42]

The class-based patterns of Bronze Age pork consumption reveal how different economic opportunities presented themselves to different status groups. On the one hand, institutions were busy concentrating on forms of livestock production that would increase their wealth and ability to trade with distant parties. Pig husbandry, as we have seen, served neither purpose. In fact, it seems that elites in Mesopotamia continued to raise pigs not in order to develop their wealth, but rather simply because they enjoyed the

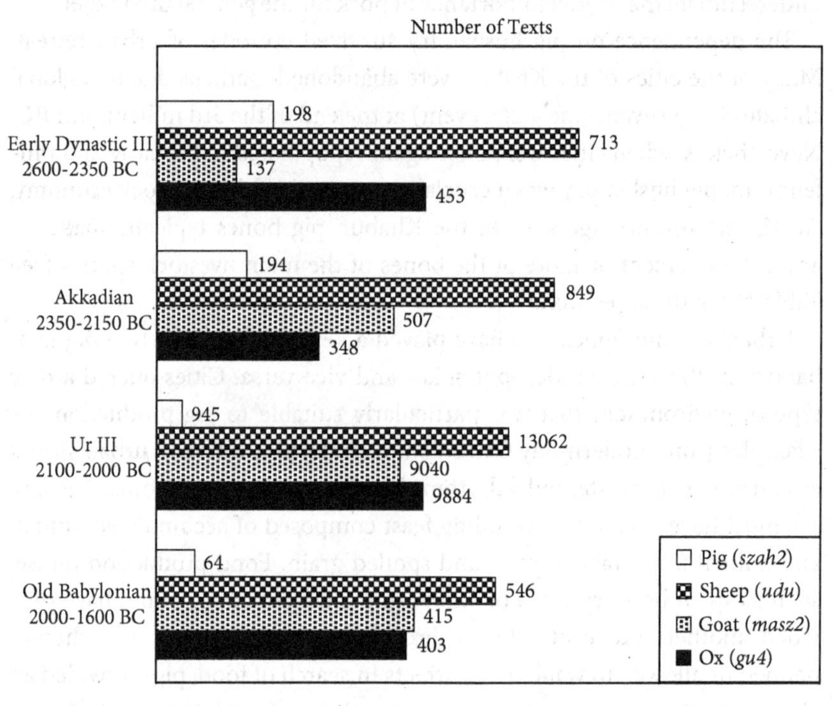

Figure 5.1. Number of administrative texts mentioning pigs and other livestock species. Cuneiform terms for species in parentheses. Texts refer to live animals or animal products.

taste of pork on occasion. But pigs were largely a luxury for the elite, not an asset.

The situation was quite different for the urban poor. Lacking access to institutional herds and frequently landless, members of the lower classes probably turned to the one type of animal they could feasibly raise—pigs.[43] This class divide in pork consumption explains an apparent contradiction in the zooarchaeological pattern: why pig husbandry increased in Mesopotamia during the Bronze Age at the same time that state institutions were attempting to maximize the production of secondary products and grain.

Because institutions focused their attention on other animals, pig husbandry presented an opportunity for the lower classes to pursue a form of food production that fell largely outside institutional interests and beyond their interference. Pig husbandry may therefore have acted as a sort of Bronze Age "informal economy."[44] Today, the informal economy exists in the shadows of global capitalism, beyond the reach of state authorities to tax and tabulate. It includes all off-the-books transactions, from working "under the table" to dealing in narcotics.[45] In the Bronze Age, the informal economy was that which took place beyond institutional oversight. It would constitute, therefore, the goods and services that were not recorded in textual records. Pig bones appear to be a fossil of one of these activities. They occur in large numbers in the zooarchaeological record, but not in texts. Indeed, difficult to tax, simple to raise, and largely ignored by institutions, pigs were ideally suited to the informal economy of Bronze Age cities.

The Inheritance of Tradition in Western Anatolia, Egypt, and Iran

The connection between urbanism and pig husbandry, mediated through class relations, is clear in southern Mesopotamia and the Khabur. But pork was important in locations without large cities, like western Anatolia and Egypt. Conversely, in Iran, large cities did not correspond to a rise in pig husbandry. Although there must have been a number of cultural variables at play, it is notable that these contexts show continuity in animal husbandry patterns from the Neolithic to the Bronze Age. This suggests that, tradition, even in the face of radical socioeconomic change, played a determinative role in some contexts.

Western Anatolian economies in the 3rd millennium were unique in two ways. First, cities were small, rarely covering more than 10 hectares. Second, the main economic focus of institutions was trade and metal production, in contrast to the agrarian focus of Mesopotamian institutions.[46] The animal economy was based on mixed livestock production, with high levels of rainfall supporting abundant cattle and pig husbandry. Pig bones typically represent 20–30 percent of the main species at 3rd and early 2nd millennia BC sites, including the iconic city of Troy.[47] These percentages are in fact equivalent to those found at western Anatolian sites dating to the Late Neolithic and at Chalcolithic sites.[48] Because pig husbandry neither served as a distraction from the accumulation of institutional wealth nor found a welcome niche in large urban environments, there was no incentive to increase pig production or limit it. People produced pork and other meat in roughly the proportions as their ancestors had done and in accordance with local environmental conditions.

Though a powerful and highly centralized state, Egypt also lacked large cities—at least ones of comparable size to the cities of southern Mesopotamia and the Khabur.[49] Yet in Egypt, too, pigs often dominated the faunal assemblages of the Old Kingdom (ca. 2630–2130 BC) and Middle Kingdom (ca. 2050–1650 BC), often representing 50 percent or more of the livestock slaughtered for meat.[50] The incredible preservation of Egyptian sites has provided additional insights into pig husbandry. For example, excavators uncovered intact pig pens, which included preserved hoofprints, at the Old Kingdom site of Dendara in Upper Egypt.[51] Nevertheless, the role of pork in the Bronze Age Egyptian diet was not significantly greater than its role in the Neolithic and Predynastic periods. Again, the data suggest a continuation of inherited food traditions rather than an increase as in Mesopotamia.[52]

Egyptian elites held large agrarian estates and developed their wealth via control over secondary products and grain. As in Mesopotamia, it is possible that this led to a situation in which pigs were a more important source of food for the poor than the rich. In fact, the infrequency with which swine are mentioned in texts and depicted in tomb art has suggested to some that pigs may have lost status in the early 3rd millennium BC.[53] This may be a reflection, again, of the fact that pigs, while eaten with great frequency, were no more considered wealth in Egypt than they were in Mesopotamia. Yet as in Mesopotamia and the Khabur, pork probably played a vital role in livestock-keeping among the lower classes.

Finally, Bronze Age communities in Iran produced very few pigs (5 percent or less), even though the region contained cities as large as those in Mesopotamia.[54] On the one hand, the climate of the Iranian Plateau is hot and dry—not necessarily ideal for raising pigs.[55] Indeed, this is likely a major reason pig husbandry failed to take off in the Neolithic and Chalcolithic.[56] On the other hand, climate fails to explain why pig husbandry would not increase in cities, which are typically located close to major bodies of water and which represent prime environments for raising swine. Again, it is hard to shake the idea that communities in Bronze Age Iran were simply continuing to follow food traditions inherited from earlier periods, ones in which local natural environments would have placed limits on the scope of pig husbandry.

The Erosion of Pig Husbandry in the Levant and Western Syria

Western Syria and the Levant extends from the Negev Desert in the south to the Upper Euphrates River Valley in the north. In this region, swine had been a common part of the agricultural tableau prior to the Bronze Age. But beginning in the 3rd millennium BC, pig relative abundances declined. With a few exceptions, both regions saw pork progressively disappear from the diet from the Early through the Late Bronze Age.

In the Levant, especially in the river valleys and oak-covered hills of the southern Levant, archaeologists frequently find that pig bones represent 25–40 percent of the main livestock remains from Chalcolithic sites. However, as Figure 5.2 shows, pigs made up only 10 percent or less of livestock remains from sties dating to the 3rd and 2nd millennia BC. An excellent example comes from the hilly Galilee region of northern Israel. At Chalcolithic Marj Rabba, pigs made up 32 percent of the livestock remains found in settlement debris dating to just before 4000 BC.[57] By the middle to late 4th millennium BC, a period referred to in the Levant as the "Early Bronze I" (EB I, 3600–3000 BC), pig relative abundances remained the same or slightly lower. At the sites of Qiryat Ata II and Yiftahel II, pigs represented 30 percent and 16 percent of the livestock remains.[58] Around 3000 BC, however, the numbers began to decline more precipitously. Pigs composed 21 percent of the remains of livestock taxa from the Early Bronze II phase (3000–2800 BC) at

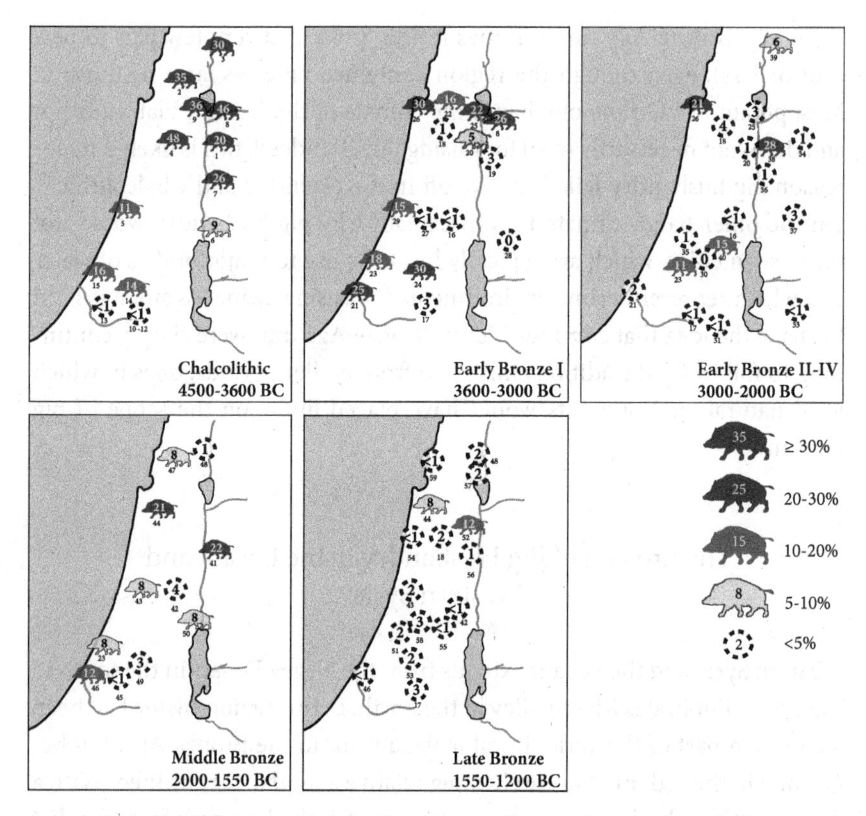

Figure 5.2. Relative abundances of pigs in southern Levant. Numbers inside pig silhouettes show percentage of pigs. Numbers below indicate site.

Key to sites: 1, Pella; 2, Marj Rabba; 3, Tel Teo; 4, Meser; 5, Tel Aviv Jabotinsky St.; 6, Teleilat al Ghassul; 7, Abu Hamid; 8, Tel esh Shuna; 9, Tel Ali; 10, Bir es Safadi; 11, Horvat Beter; 12, Abu Hamid; 13, Shiqmim; 14, Gilat; 15, Grar; 16, Ai et Tell; 17, Tel Halif; 18, Megiddo; 19, Tell Abu al Kharaz; 20, Tel Yaqush; 21, Tel es Sakan; 22, Yiftahel; 23, Ashkelon; 24, Tel Hartuv; 25, Tel Bet Yerah; 26, Qiryat Ata I III; 27, Tel Dalit; 28, Tell Madaba; 29, Tel Lod; 30, Tel Yarmouth; 31, Tel Arad; 32, Tel Erani; 33, Khirbet al Minsahlat; 34, Kh. ez Zeraqon; 35, Tel es Safi; 36, Tell Handaquq; 37, Tell al ʻUmayri; 38, Tell Abu en Niaj; 39, Tel Dan; 40, Refaim Valley; 41, Tell el Hayyat; 42, Shiloh; 43, Tel Aphek; 44, Tel Yoqne'am; 45, Tel Haror; 46, Tell Jemmeh; 47, Tel Kabri; 48, Tel Hazor; 49, Tel Nagila; 50, Jericho; 51, Tel Harasim; 52, Beth Shean; 53, Lachish; 54, Tel Dor; 55, Beth Shemesh; 56, Tel Rehov; 57, Tel Kinrot; 58, Miqne Ekron; 59, Nahariya.

Qiryat Ata I[59] and only 4 percent from EB II and III phases (3000–2500 BC) at Tel Bet Yerah on the southern shore of the Sea of Galilee.[60]

A roughly similar story unfolded in western Syria, the area extending from Damascus to the Middle-Upper Euphrates. There, the abundance of pigs had declined somewhat toward the end of the Chalcolithic period. This was in large part due to the influx of southern Mesopotamian colonists during the

Uruk Expansion.[61] Although the percentage of pigs rebounded impressively during the first half of the Early Bronze Age (3000–2600 BC), pork consumption plunged again after 2600 BC.[62]

There are two components to the decline of pig husbandry in the Levant and western Syria in the 3rd millennium BC. The first is economic. Communities would have felt drawn to the production of valuable commodities like wool and grain, which lent themselves to intensive secondary product exploitation. Several lines of evidence indicate that livestock keepers concentrated on raising sheep for wool. Sites dating to the 3rd millennium BC show much higher ratios of sheep to goats than in previous periods. Moreover, the ages at which both of these animals were slaughtered increased, a signature of intensive fiber exploitation.[63] Finally, mid-3rd millennium BC palace archives from Ebla in western Syria demonstrate institutional officials' almost singular focus on sheepherding and wool production.[64]

The second component of the decline in pig husbandry relates to the scale of urbanism in the Levant and western Syria. As in Mesopotamia and the Khabur, the production of animal fiber textiles, livestock, and cereals—and their collection through taxation—were key ingredients of institutional wealth in the Levant and western Syria. But in Mesopotamia and the Khabur, the massive size of cities and larger numbers of urban poor seeking alternative means of subsistence created a niche for swine husbandry. Such factors were not at play in the Levant, where cities were typically less than 30 hectares in size. And while larger than those in the Levant, western Syrian cities were only about half the size of their counterparts in the Khabur and southern Mesopotamia.[65] In other words, the urban environments that had lent themselves to pig husbandry in Mesopotamia and the Khabur were far more limited in size in the Levant and western Syria, and both regions were less densely settled.[66]

The situation in the Levant and western Syria contrasts with that in Egypt. In both regions, societies were marked by low-level urbanism, a tradition of pork consumption inherited from the Chalcolithic, and an institutional focus on grain and other secondary products. One can speculate that differences in the structure of political and economic power may have had something to do with the stark disparities in pork consumption. Egypt was a highly centralized state with a deified monarch. Class differences were pronounced. The Levant, on the other hand, contained small-scale city-states, while western Syria contained regionally expansive states that were nevertheless much weaker than Egypt. In these regions prevailed a more tribal ethos based on extended patriarchal family units and, at times, more collective forms of governance.[67] One wonders if the greater degree of economic inequality in Egypt inspired the development of a more robust informal economy or a distinct lower-class cuisine.

There is zooarchaeological evidence to suggest that a weak informal economy in pork may have operated in the Levant and western Syria. But disparities in pork consumption are apparent only in the urban-rural dichotomy rather than within cities themselves. Pig bones are found in much higher proportions (15–30 percent) in a few of the rural villages that served as satellite communities for the urban centers.[68] This suggests that people living in the urban centers were able to pursue forms of livestock production that would enhance their wealth and prestige, while rural communities were excluded from those lifestyles—or possibly opted out of them. Instead, the people on the periphery pursued a strategy of mixed agriculture that sometimes included a healthy dose of pig husbandry. This informal sector, if indeed it was one, was spotty and ultimately unable to offset the decline of pig husbandry on a regional scale. By the end of the 2nd millennium BC, even in rural areas pig bones rarely represented more than 5–10 percent of livestock remains at sites in the region.

Pigs and the Changing Environment

During the Holocene, the Near East experienced periodic aridification and steady deforestation.[69] For decades, archaeologists have hypothesized that these environmental changes may have led to the regional decline in pig husbandry.[70] Pigs certainly require more water than sheep and goats, and they do not thrive in grasslands like the ruminants. In the Levant and western Syria, sheep and goats claimed most of the percentage points lost by pigs in the Early Bronze Age.[71] This might be an indication that herders were adapting their livestock choices to changing environmental conditions.

However, close inspection of the climatological and zooarchaeological data reveals that environmental change had less impact on pig husbandry than often suggested. The first line of evidence comes from the Early Bronze Age settlements of the Khabur. The communities living there practiced extensive cereal agriculture and livestock-herding that led to a significant depletion of local forests.[72] Despite deforestation, pigs were extremely common, often the dominant form of livestock. This was because the urban niche provided a ready-made ecosystem for pigs. Pig husbandry also weathered the 4.2ka climatic downturn, which brought cooler and drier conditions to much of the Near East. In fact, the percentages of pigs remained more or less the same at sites with layers predating and postdating 2200 BC.[73]

Data from the Levant and western Syria also problematize an environmental explanation for the decline in pigs. At first glance, there appears to be a connection between aridification and pig husbandry. Beginning around 3000 BC, an episode of climate change caused rainfall to be less abundant in the region.[74] This was precisely when people began raising fewer pigs in the Levant. However, between 3000 and 2600 BC, pig husbandry was *more* popular in western Syria and the Euphrates Valley than it had been in the late 4th millennium BC; in that region, proportions of swine fell only after 2600 BC and with the emergence of urban-based political systems.[75] Additionally, in the 3rd and 2nd millennia BC, communities in the Levant and western Syria raised large numbers of water-guzzling cattle (10–30 percent)[76] and grew flax and grapes, both of which require a lot of water.[77] These are hardly strategies adopted by people facing water shortages.

This is not to say that environmental changes had no impact on pig husbandry—they certainly did. But how they did so was neither uniform nor straightforward. Pig husbandry is flexible, adaptable to urban, forested, or other types of environments. Archaeologists must seek more nuanced ways to explain how environment, culture, and social pressures combine to present people with opportunities for altering agricultural practices, rather than attempting to explain pork consumption simplistically in terms of aridity. One avenue for future research is to investigate how pig-keepers shifted their husbandry practices in the face of environmental change. For example, no one has explored how the depletion of oak woodlands in parts of the Levant and Syria in the Bronze Age[78] specifically affected extensive pig husbandry systems that relied on seasonal crops of acorns and other nuts. Asking more tailored research questions will enable archaeologists to better understand the interactions between climate change, anthropogenic impacts on the landscape, and agricultural choices.

Pigs in Texts

Texts written in Sumerian, Akkadian, Egyptian, and other languages provide glimpses of day-to-day life. These perspectives are, however, not unbiased.

They reflect the interests and concerns of those wealthy enough to be literate or employ scribes. As such, much of the textual corpus details the management of livestock and the people dependent on institutions. These administrative texts are heavily skewed toward sheep, goats, and cattle,[79], as Figure 5.1 shows. The relative paucity of texts detailing pig management has contributed to scholars' general dismissal of the importance of swine across the ancient Near East. The zooarchaeological data demonstrate how mistaken this impression is.[80]

Texts are not entirely silent on pigs, however. Although institutions did not focus on pig production, written records establish that temples, palaces, and elite manorial estates owned and managed swine. Swine appear in some of the earliest Sumerian texts. For example, a tablet recovered from Uruk and dating to the late 4th millennium BC discusses a herd of 95 "grain-fed pigs" belonging to two temples.[81] Even more intriguing is an archive dating to the 2300s BC that concerns the exploits of a Sumerian swineherd named Lugal-Pa'e from the city of Lagash. Lugal-Pa'e worked for the household of the governor's wife, and the roughly 200 pigs put in his charge were referred to as "reed-thicket pigs," a moniker that seems to denote swine raised under extensive husbandry in the marshlands around the city.[82] The archive records herd statistics for seven consecutive years. During that time, Lugal-Pa'e increased the number of reproductive two- and three-year-old sows from 49 to 70 and oversaw three birth cycles per year. He also managed a small number of male "wild pigs," which he used as breeding stock.[83] Why he wanted to produce hybrids is unclear. Perhaps he wished to increase the size of his animals; zooarchaeological data make it clear that 3rd millennium pigs were quite small, about 60 cm tall at the shoulder.[84]

Egyptian texts of the Old and Middle Kingdoms also mention swine, although, again, not as frequently as they do other animals.[85] Temples sometimes accepted pigs as sacrifices,[86] and Egyptian institutions occasionally kept large numbers of swine. One 16th century regional elite (or "nomarch")[87] even claimed to possess 1,500 pigs—more than all of his sheep, goats, and cattle combined.[88] As in Mesopotamia, specialist swineherds raised these animals for the institutions. In fact, swineherds are mentioned in what might be the oldest biography in the world on the tomb of an Old Kingdom official named Methen.[89]

Swine also make an appearance in one of the oldest parables of legal justice, *The Eloquent Peasant*, an Egyptian story composed in the late 3rd millennium BC. In the story, an estate overseer tricks a well-spoken peasant into trespassing onto a nobleman's field with his donkey. Bronze Age justice ensues: The overseer confiscates the donkey and beats the peasant, who then pleads his case before the overseer's boss, the nobleman. Following a lengthy trial and several more undeserved beatings, the nobleman comes to learn of his employee's dishonesty. He returns the donkey and then compensates the peasant for his pain and suffering with some of his overseer's livestock—including pigs.[90]

The Eloquent Peasant is a work of fiction, but bookkeeping documents corroborate that institutions gave pigs to members of the lower classes, although not for the purposes of justice. Instead, institutions sometimes distributed pork as rations to temple workers and soldiers in both Mesopotamia and Egypt.[91] Institutions also provided their menial workers with rations of lard.[92] While the size of these rations paled in comparison with that of grain rations, lard was something of a valuable commodity in the Bronze Age. In fact, there was a lively trade in lard in Egypt, Anatolia, and Mesopotamia.[93] People, especially those of the lower classes, ate lard and probably used it for cooking. Lard was also applied as a lubricant to farm equipment—literally greasing the wheels of agriculture—and when mixed with lye (which can be leached from ash) was a key ingredient in soap.[94] And while soap can be made from beef tallow and other types of animal fat, lard could be obtained without killing a high-value animal, such as a sheep or a steer.

Here, then, was an important product that pigs could provide, one that would be valued by institutions and members of the elite. Soap's utility extended beyond personal hygiene; it was essential for woolen textile production. Anyone who has gotten close to a sheep knows that raw wool is a filthy mass of dirt, oil (lanolin), and feces. Washing raw wool was therefore critical to the institutional textile industry. In fact, Mesopotamian institutions regularly issued animal fat (often lard) to textile workers.

A Sumerian tablet dating to around 2200 BC[95] reveals the scope of lard production. The tablet records the receipt or distribution of 22 jars of pig fat, each containing 18 liters for a total of 396 liters, which were probably intended for making soap. How many pigs went into making this much fat? At a density of 0.86 kg/liter, 396 liters equals 341 kg. A modern 100-kg hog can yield about 10 kg of lard. Assuming that a fattened pig could reach 75 kg—which is a generous assumption given the Bronze Age pig's small size—then each pig could yield about 7.5 kg of lard. The 22 jars therefore represent *at least* 45 fattened adult pigs. To put that into perspective, Lugal-Pa'e's herd of swine totaled around 80 adult animals, but most of those would be needed as breeding stock for the next generation. The 22 jars thus represented the off-take from several large herds of pigs.

In sum, while a small component of their overall economic activities, texts reveal that 3rd and 2nd millennia institutions in Mesopotamia and beyond did raise pigs. Curiously, however, there are very few documents relating to institutional pig-keeping after around 1600 BC, suggesting that institutions largely stopped owning pigs at that time.[96] The reason for the abandonment is unclear, but it would ultimately put greater distance between the members of the elite and pigs in the Late Bronze Age.

Pork Consumption in the Late Bronze Age

The Late Bronze Age ushered in a new era defined by international connectedness and powerful empires. The tenor of this period is perhaps best captured by the Amarna Letters, a series of clay tablets composed in the 1360s–1330s BC that consist of correspondences between the Egyptian pharaoh Akhenaten and his contemporaries in the Levant, Cyprus, Assyria, the Hittite Empire, and other places.[97] The Amarna Letters showcase the political machinations and the dance of Late Bronze Age empires. They do not discuss pigs.

While Akhenaten's scribes were sending and reading letters, the workers of Amarna were, in addition to fulfilling their duties as artisans and laborers, raising swine. In fact, pork was the main source of meat consumed by Amarna's nonelite.[98] Excavations in the 1980s uncovered what have proved to be the best examples of ancient pig pens, complete with traces of pig bristles and coprolites (preserved dung). The Amarna Workmen's Village contained a number of pens grouped into larger compounds.[99] An example is what excavators labeled Building 400, a cluster of six pens and adjoining courtyards, shown in Figure 5.3. The pens themselves consisted of low mudbrick/stone walls about 1.5 meters in diameter—just enough room for a sow and her piglets—and contained small doorways through which the swine could enter and exit into common courtyards. One of these courtyards even contained a stone trough.[100] Analysis of coprolite samples collected from the Amarna pens revealed that many swine were infected with tapeworm (*Taenia solium*) and roundworm (*Ascaris suum*) parasites, both of which are common in pigs and transmissible to humans. The dung also

Figure 5.3. Pig sties in Building 400 at Amarna, Egypt.

contained the remains of what these pigs ate: rye, emmer, and wheat seeds as well as small mammal bones.[101]

The Amarna pens vividly document that pork remained a staple in the New Kingdom, the apogee of Egyptian imperial hegemony in the Near East.[102] But the importance of pork in Egypt contrasts sharply with its status in other parts of the Near East. Although zooarchaeological data for the Late Bronze Age are unfortunately limited in many places, they suggest a general pattern of decline in pig husbandry. However, one must offer a caveat to this pattern: archaeologists have focused their excavations on elite areas of sites. With the exception of the Amarna Workmen's Village and a handful of other excavations, the nonelite of the Late Bronze Age have been little explored. It is possible that pigs continued to be consumed in high proportions by the lower classes in the Late Bronze Age in certain places.

The available data from Mesopotamia and the Khabur, regions where pigs had been abundant in previous periods, show that swine typically represented less than 20 percent of the main livestock animals consumed at most settlements in the Late Bronze Age.[103] In the Levant and western Syria, where many people had already largely given up on pig husbandry as a major means of food production, relative abundances of swine slid even further. Pig bones make up less than 5 percent of the main livestock taxa at almost every site for which there are published data.[104] Only in Anatolia did levels of pig production hold steady. Pig remains typically make up around 20 percent of the remains of livestock recovered from the towns and cities of that region.[105]

Although people continued to eat pork and raise pigs, zooarchaeological data suggest that pigs had become less common in every major region of the Near East except Egypt and Anatolia. Part of the reason for this may have been that, despite the rise in imperial power throughout the Near East, urbanism contracted. By the Late Bronze Age, population numbers in much of Mesopotamia and the Levant were at their lowest since the Chalcolithic. Additionally, a greater proportion of the population resided in villages and small towns than in previous periods.[106] The urban niche, which pigs had so successfully colonized in the Early Bronze Age, had shrunk. But the reduction of swine paralleled another important change: in some Late Bronze Age religious contexts, pigs had begun to develop a reputation as being ritually impure.

Pigs and the Gods

The combination of textual evidence and zooarchaeology provides considerable insight into the uses of pigs in celebrations, rituals, and religion in the Bronze Age, especially after around 2000 BC, when writing was more frequently used for purposes beyond bookkeeping. Together, the datasets shed light on how the ritual roles of swine evolved. We will focus on four key areas: the role of pork in feasts, pigs as sacrificial animals, their symbolic significance, and their connection to certain deities.

Although the most convincing zooarchaeological evidence for large-scale pig feasts dates to the Neolithic and Chalcolithic (Chapter 4), texts clearly indicate that Bronze Age people of all social classes feasted on pork. In fact, pork played a central role in some festivals. For example, early 2nd millennium BC Babylonian *elunum* celebrations took place around the time of the summer solstice and featured roasted piglets.[107] Akkadian documents (2350–2150 BC) also attest to pork being eaten at wedding feasts and even in high-society gatherings meant to impress foreign and local dignitaries.[108] In these Akkadian feasts, whole pigs were typically roasted over an open fire made from bundles of reeds.[109]

Nevertheless, there is some indication that pork's status as a ritual animal or feast food eroded over the course of the Bronze Age. In Mesopotamia, despite pork's role in Akkadian social gatherings, descriptions of public and royal feasts rarely mention swine; pork was generally left out of recipes prepared by institutional cooks.[110] This suggests the exclusion of pork from Mesopotamian *haute cuisine*. Perhaps because of its associations with the *hoi polloi*, pork gradually lost its luster as a food worthy of the menus of elite feasts. Pigs were also infrequently depicted in art. In Egypt, pigs rarely appeared in tomb engravings and, when they did, tended to be shown in association with lower-class individuals.[111] An intriguing example is found in a scene from the Old Kingdom Tomb of Kagemni at Saqqara. In it, a kneeling man, thought to be a peasant, places his lips on those of what appears to be a piglet (others have argued it is a strange-looking puppy), while another man offers him a jar.[112] The meaning of the image is unclear: Is he orally providing his animal water or milk? Is this some sort of ritual? Is he *kissing* the animal? . . . Perhaps some mysteries are best left unexplored.

The value attached to pigs as livestock animals—their exclusion from the group of animals bearing wealth—influenced their limited role as animals sacrificed to the gods and in burial rituals. Ample zooarchaeological and textual evidence shows a clear bias against swine, especially in temple and mortuary contexts, in favor of the more valuable sheep, goats, cattle, and equids.[113]. Nevertheless, pigs were sacrificed on rare occasion with the dead. Pig bones, for example, were found in rock-hewn tombs at Saqqara in Egypt, dating to the 26th century BC,[114] and a few other Early and Middle Bronze Age burials throughout the Near East.[115]

While uncommon in temple and mortuary contexts, pig sacrifice was common in rituals connected to fertility rites and magic. For example, archaeologists have uncovered pig or piglet sacrifices beneath buildings at a number of sites across the Bronze Age Near East.[116] These sacrifices were probably foundation deposits intended to bless the buildings and their inhabitants. By offering a piglet, which magically conveyed fertility, families hoped to ensure health and reproductive success.

The association between pigs and fertility probably drew upon the fact that these animals are prodigious breeders, able to give birth to many offspring at once—something they hold in common with dogs.[117] Examples of this connection can be found in literature and in art. A passage in *The Benedictions of Labarna*, a Hittite text dating to the 17th or 16th century BC, contains the words for bestowing a blessing on a vineyard in order for it to produce grapes in such great numbers "as a single pig gives birth to many piglets."[118] Similarly, Figure 5.4 shows a fist-sized clay amulet recovered from Nippur in Mesopotamia, probably dating to the 2nd millennium BC. On it, a boar mounts a sow, who is simultaneously nursing a litter of piglets.

Pigs (along with dogs) were associated with lust and erotic excitement.[119] But in addition to sows' fecundity, swine's connotations of lustiness may derive from the fact that copulating boars climax with a lengthy and voluminous ejaculation; over the course of several minutes, a breeding boar will release up to half a liter of semen. Akkadian and Babylonian sexual potency incantations for both men and women invoke pigs. A particularly vivid example comes from an early 2nd millennium BC tablet found at Isin in southern Mesopotamia:

> Place your(m.) mind with my mind!
> I hold you(m.) back just like Ištar held back Dumuzi,
> (Just like) Seraš binds her drinkers,
> (so) I have bound you(m.) with my hairy mouth,
> with my urinating vulva,
> with my drooling mouth,

Figure 5.4. Amulet or clay plaque from Nippur showing a boar mounting a nursing sow (probably Old Babylonian period, early 2nd millennium BC). 10.5 × 7 cm.

> with my urinating vulva.
> May the enemy-woman not come to you!
> The dog is lying, the boar is lying—
> you lie forever in between my thighs.[120]

Additional evidence for the sacrifice of piglets in connection with fertility comes from what is perhaps the most exciting example of pig sacrifice in the Near East. In 1999, archaeologists digging at Tell Mozan in northern Syria uncovered a massive stone-lined pit near the southern wall of a palace.[121] Dating to around 2300 BC, the pit was around 5 meters in diameter and over 6 meters deep. Within it were the remains of at least 60 piglets, 20 puppies, 60 sheep/goats, and 20 donkeys, in addition to a number of other offerings that included a ceramic jar in the shape of a pig's head and another depicting a nude woman.[122]

The remains found in the Mozan pit bear similarities to deposits associated with chthonic rituals (those relating to the underworld). While it may seem counter-intuitive, chthonic and fertility rites are often connected through dialectical opposition—life and death give meaning to one another; they are frequently perceived as operating in a cyclical manner. In fact, rituals of the Hurrians, who are thought to have inhabited Tell Mozan, mention special ritual pits called *abi*, or channels to the underworld.[123] Another parallel to the Mozan sacrifices derives from ancient Greece: the Thesmophoria festival.[124] This festival took place in autumn to honor Demeter, the goddess of agriculture and fertility, whose daughter, Persephone, was abducted by Hades. The sacrifice of piglets, offspring of a highly reproductive animal, figured prominently in the rituals as a symbol of the seasonal cycles and Persephone's sojourn in the land of the dead. Of course, Mozan was occupied almost 2,000 years before the earliest records of the Thesmophoria festival. Nevertheless, the parallels between the two suggest long-standing commonalities in the religious thought of Anatolian and Mediterranean peoples.[125]

Beyond fertility, and its connection to sexuality and death, the Mozan offerings might relate to another characteristic of pigs in ancient Near Eastern thought: pigs served as *substitutes* for humans in purification rituals. In other words, they could be sacrificed in place of a human being. They could also take on a curse or illness afflicting a person. In this way, pigs acted as a sort of ritual sponge that absorbed the burden of one's sins or the evil eye. For example, in Bronze Age Mesopotamia and Anatolia, doctors used swine to expel illnesses from patients.[126] Oftentimes, these rituals of substitution were connected to pigs' chthonic association in magical rites designed to placate the gods. One example derives from a ritual prescribed by a Hittite priestess named Hanitassu to exonerate an individual following a moral transgression:

> When night falls, the petitioner digs a hole in the ground and slits the throat of a piglet, letting its blood flow into the pit. Grains and liquids are offered into the pit as well. The doors to the Underworld are symbolically opened and the divine images of the Underworld deities are set around the pit to draw the deities up from the earth. Finally, they are invoked to plead with the Sun Goddess of the Earth, Queen of the Underworld, on behalf of the petitioner so that his offense may be forgiven.[127]

Pig sacrifice could even be prophylactic if malicious intent was suspected, as seen in one Babylonian royal ritual:

> Then a white pig is slaughtered and the king spills its blood to the four cardinal directions [. . .] Both the figurine [of an enemy] and the dagger are enclosed in the pig's skin, which makes a perfect container for impurity and evil, being pure and white from outside, while holding all the contagious materials inside. [It] is carefully sealed with a clay bulla, then the king puts his hand on the sealed package and orders the evil to depart.[128]

Swine's feeding behavior may explain their usefulness as substitutes. Bronze Age city dwellers would have been well accustomed to seeing pigs consuming waste and filth, neutralizing it, and transforming something defiling (garbage, feces) into something innocuous (pork). In the same way, pigs could metaphorically collect and neutralize curses or bodily afflictions.[129]

The symbolic qualities that swine acquired during or even before the Bronze Age fostered connections with certain deities. Our evidence primarily derives from the Middle and Late Bronze Ages. For example, their service as substitutes probably drew pigs into affiliation with the Lamashtu, a demon-goddess who was believed to enter houses and murder newborn babies.[130] Amulets and spells meant to appease, terrify, outmaneuver, or otherwise thwart Lamashtu are found throughout Mesopotamia, Anatolia, and the Levant in the 2nd millennium BC.[131] While taking a variety of forms, the amulets often depict the demon-goddess suckling a puppy and piglet on her poisonous breasts. In other images, these animals accompany Lamashtu, as for example in Figure 5.5. One can interpret these amulets as a plea: take a piglet; leave the baby.[132]

Pigs or wild boar were associated with two other deities, Baal and Seth—the Levantine and Egyptian gods of storms, disorder, and fertility/male sexual potency. Baal's connection to wild boar derives from a series of stories, known as the *Baal Cycle*, written in the middle to late 2nd millennium BC in the Canaanite city of Ugarit. The connection is brief, but in one passage, Baal hunts animals described as "the voracious ones" that inhabit oak forests and marshes and that possess armor and horns (tusks).[133] Prestige objects from Ugarit, including boar-head-shaped bronze spear points found near the Temple of Baal, also depict these animals.[134] Why Baal was associated with wild boar is not clear, but it perhaps symbolized the deity's ferocity. Wild boar and boar hunting, symbols of masculinity and power in the Mediterranean and Anatolia, fit neatly with Baal's persona.[135]

Figure 5.5. Obsidian amulet of Lamashtu with dog and pig. Early 1st millennium BC. 5.7 × 4.7 × 0.9 cm.

The Egyptian god Seth filled many of the same roles as Baal—in fact, the two were often syncretized. But the connection between pigs and Seth is better defined. In one myth, Seth, having murdered his brother Osiris, attacks Osiris's son Horus in the form of a pig.[136] This initiated a cosmic grudge against pigs, as described in the New Kingdom text *The Book of the Dead*:

> So Re said to the gods: "Put him [Horus] on his bed, that he may recover." It was Seth, who had assumed his form of black boar. Then he had struck him in the eye. So Re said to the gods: "Abominate the pig for Horus' sake,

> so that he may recover." Thus came about the pig-abomination for Horus' sake by (the gods), his Train. (But) when Horus was (in) his childhood his sacrifices used to consist of his beef cattle and his pigs. (Now) his Train abominates (them).[137]

Egyptologist Richard Lobban[138] has argued that the struggle between Seth and Horus mirrors that between Lower and Upper Egypt, especially after the Hyksos conquest of—and, later, expulsion from—Lower Egypt. Lobban even claims that the hatred of Seth (= Lower Egypt = the Hyksos) initiated a taboo on pork toward the end of the 2nd millennium BC. This is a neat hypothesis, but unfortunately one for which there is little evidence beyond vague textual allusions and the effacement of images of Seth in the 1st millennium BC.[139] In fact, zooarchaeologists have clearly demonstrated that pigs remained a food source for the vast majority of Egyptians long after the expulsion of the Hyksos.[140]

Nevertheless, it is hard to escape the conclusion that pigs lost some of their status after 1600 BC in many regions of the Near East, at least among the elite. The shift was subtle and the evidence is far from overwhelming, yet the recording of unambiguously negative attitudes toward pigs (e.g., in *The Book of the Dead*), the zooarchaeological evidence for a decline in pig production in much of the Near East, and the abandonment of pig husbandry by institutions all point to a shift in attitudes toward swine in the Late Bronze Age. While pigs were associated with certain gods and occasionally depicted in art, it is noteworthy that most of these associations are negative—pigs are associated with demons and gods of chaos. Their roles in fertility rituals or as grave/house foundation deposits gradually eroded.[141] In fact, beyond the folk magic/medical rites among the Hittites and in Mesopotamia, pigs seem to have disappeared from rituals by the Late Bronze Age. By the end of the 2nd millennium BC, they were no longer sacrificed to chthonic deities (as they may have been at Tell Mozan) or even eaten near temples.[142]

People also started to associate pigs with filth to a greater degree in the Late Bronze Age. The association had probably existed for a long time; swine's connection to urban filth probably drove it home on a daily basis.[143] But it took on a new intensity beginning in the Late Bronze Age and extending into the Iron Age. At this time, texts explicitly banned pigs—and their scavenging counterparts, dogs—from temples and other sacred areas in Anatolia and Mesopotamia because they were impure.[144] Even the Hittites, who sacrificed pigs in magical rites, banned them from temples for fear they would pollute the sacred spaces and paraphernalia.[145] In later Assyrian texts (8th–7th century BC), not only would pigs and dogs be labeled profane, but they would also called upon to defile and mutilate the corpses of enemy combatants.[146]

It is not hard to see why people might have assigned ritual impurity to pigs and dogs—those lustful, overfertile scavengers. But what is interesting is how the very same features that had made these animals ritually important had been flipped on their heads. Where once pigs could neutralize foulness, now they were carriers of impurity. Where once they symbolized fertility, holding in the balance death and life, now they were portents of chaos. People had recast pigs' symbolic power in an entirely negative light, transforming them from (occasionally) sacred to profane. In this way, they were made taboo in certain—and quite restricted—contexts.

In terms of social, economic, and political changes, the Bronze Age was perhaps the most turbulent period in Near Eastern history. The evolving relationship between people and pigs was likewise dynamic. The first and perhaps most significant change was that pigs were excluded as sources of wealth. On the one hand, this meant that institutions were less interested in raising pigs. On the other hand, it mean that pig husbandry found a place within an informal economy, especially in Bronze Age cities in Mesopotamia. At the same time, cities provided ideal environments for pig husbandry—ones full of pooling wastewater, garbage, animal carcasses, and mud. Just as in the Pre-Pottery Neolithic, burgeoning human settlements offered new opportunities for pigs to adapt and thrive.

However, one can argue that it was the initial exclusion of pigs as sources of wealth—something that evolved from the secondary products revolution in the Chalcolithic (and ultimately had roots in the Late Neolithic)—that set the wheels in motion for the incredible evolution of swine's cultural significance in the Bronze Age. The fact that equids and ruminants equaled wealth generated a strong pull toward those forms of livestock production. Except in places where large urban environments made pig husbandry attractive and where members of the lower classes sought to opt out of ruminant economies increasingly controlled by institutions, that pull severely reduced the popularity of pig husbandry. This led to conspicuous losses of pig husbandry in some places in the Early Bronze Age, such as Levant and western Syria. I have argued that the virtual abandonment of pig husbandry had more to do with economic than other factors—for example, environmental ones. Whatever the case, Iron Age societies in these regions found themselves with an inherited food tradition largely lacking in pork. While there is no evidence to indicate that a taboo existed among the general populace, communities in the Early Bronze Age Levant developed culturally specific foodways that did not include much pork. Not eating pork, in other words, developed in the Bronze Age into a passive tradition passed on from generation to generation.

The transition toward more active forms of pork avoidance in the Iron Age Levant would largely be the result of the continued development of—and clashes between—culturally specific foodways, themselves articulating, as we have seen, with political and economic patterns. But the evolution of pigs' ritual and religious roles also contributed to the changes that would take place in the Iron Age Levant. Pigs began the Bronze Age on the wrong foot, so to speak. As animals not conveying wealth, they were excluded as sacrifices fit for the temples. While they maintained important roles in rituals connected to fertility, sexual potency, and death, as well as in those requiring a substitute, by the Late Bronze Age the ritual meanings associated with pigs had shifted. Filth and pollution were emphasized. While one could easily find inspiration for these new attitudes toward pigs in the animals' feeding and wallowing habits, these perceptional changes probably represent a case of symbolic inversion. In any case, by the end of the Bronze Age, pigs were banned from some temples, perceived as less valuable livestock, rarely useful in ritual, and often not even eaten in parts of the Near East. The stage was set for a more general and all-encompassing taboo—what we can call *the* pig taboo.

6
Theorizing the Taboo

Up to this point, we have dealt with the evolution of pigs from wild boar and their integration into the economies of the Near East up to around 1200 BC. We have seen how pigs were enmeshed in wide-reaching developments such as urbanization and the formation of social classes, and how pigs developed new ritual meanings over the course of the Bronze Age. In the next three chapters, we will look closer at the formation and evolution of the pig taboo over the past 3,200 years. Before doing so, we need to examine what it means for something to be taboo and to investigate some of the existing theories about why the pig taboo emerged in the Iron Age.

The Anthropology of Taboo

Although, arguably, taboos exist in every culture and have been around for thousands of years, the English word "taboo" is much more recent. The word *tapu* (or *tabu*) originated in Pacific Island languages as a way to signify something or someone imbued with special spiritual power.[1] "Taboo" entered into Western vocabularies only in the decades after James Cook's voyages.[2] A century later, the anthropologist James Frazer offered a systematic discussion of taboos in his 1886 *Encyclopedia Britannica* article on the topic as well as in his magnum opus, *The Golden Bough*.[3] Soon after, scholars across a variety of disciplines began writing about taboos, including the anthropologist A. R. Radcliffe-Brown,[4] the sociologist Émile Durkheim,[5] and the psychologist Sigmund Freud.[6]

What does it mean for something to be taboo? Each person's concept of taboo is based on his or her experiences with specific types of avoidance behavior. In a general sense, *a taboo is a form of culturally prescribed avoidance of a thing or an activity that is surrounded by a high degree of social energy. This avoidance is bound by a moralized cosmological order—an idea of what is right and wrong based on fundamental principles that appear to people as immutable.*[7] Taboos are frequently enforced not through law or commandment,

Evolution of a Taboo. Max D. Price, Oxford University Press (2020). © Oxford University Press.
DOI: 10.1093/oso/9780197543276.001.0001.

but on an emotional level through the sentiment of disgust.[8] Disgust is a powerful emotion that children readily pick up on from their parents; it provides a "culture's most effective means to enforce a prohibition."[9]

Note that the term "social energy" in the definition of taboo is neither positive nor negative. In fact, it is often ambiguous whether the object of a taboo is avoided because it is too sacred or because it is too profane. In any event, breaking a taboo is perceived as dangerous for the individual and the community because the taboo essentially keeps powerful forces separate from everyday life. Like radioactive material, that which is tabooed is not only inherently powerful but also dangerous if not contained properly.[10] But unlike the effects of unleashing radioactive substances, the specific real-world consequences of breaking a taboo (transgression) are rarely clear. They may include illness, impotence, bad luck, or, as in Leviticus and Deuteronomy, becoming "impure." The reason for the ambiguity is that taboos are not, in essence, about health or physical well-being, but about maintaining a cosmological order ("the way things ought to be") for a person and the society in which he or she lives.

Taboos may apply to an entire community or ethnic group. Or they may apply only to certain of its members, for example, those of various age groups. Among the Gidra people of New Guinea, men can eat the flesh of pigs only once they reach marriageable age, and only old men are allowed to eat pig fetuses.[11] Cross-culturally, priests/shamans, menstruating/pregnant women,[12] and ceremonial participants are frequently subject to taboos.[13] Such taboos protect *liminal* individuals—people inhabiting the "betwixt and between" spaces. For example, ceremonial initiates reside betwixt and between social positions; pregnant women between life and death/nonbeing; and priests between the sacred and the mundane. Taboos separate these liminal characters from forces that might upset their ambiguous and thus perilous situation. They also protect the rest of society from the unusual social energy surrounding such individuals.

When taboos apply to a whole group of people, they often serve as markers of, on the one hand, identity and solidarity and, on the other hand, difference from "others."[14] In this way, taboos play as important a role as cuisine or dress in creating and maintaining ethnic or religious boundaries.[15] The Jewish and Muslim pig taboos, as we will see, have functioned in this manner and, in fact, many anthropologists have suggested that taboos derive their strength and staying power from the fact that they reveal differences between people.[16]

While taboos can apply to just about anything, they often surround three types of behaviors: eating, sex, and speaking. The reason that these three behaviors are subject to taboos is perhaps not surprising. Food, sex, and language are among the most basic elements of the human experience. They sustain people, bind them together, and enable them to create the next generation. They fix our position within the cosmos as creative and social beings. As a result, even seemingly minor transgressions in the proper ways of eating, having sex, or speaking can upset the cosmological balance—with dire consequences. For example, anthropologist Signe Howell discusses a taboo on mocking animals among the Chewong hunter-gatherers in Malaysia. She refers to an incident in which a few members of the group laughed at some millipedes that had entered their dwelling. This transgression resulted, according to Howell's informants, in a severe a thunderstorm that knocked over a tree and killed three people.[17]

At this point, it is important to remind readers that taboos are universal, even among people who don't think of themselves as superstitious or religious. Most Westerners, secular or religious, avoid eating dog meat, feel uncomfortable watching or even thinking about their family members having sex, and wince if a friend uses a racial slur. Catholic priests are not supposed to marry or have sex. Pregnant women should not drink even a sip of wine. Teenage siblings of the opposite sex should not sleep in the same bed even if there is no possibility of sexual behavior. On the day of their wedding, a bride and groom are not supposed to see each other before they are married. These are just a few of the everyday taboos many "rational" people in the West follow. While one can suggest any number of practical explanations for these types of avoidance behavior, when push comes to shove, the reason we follow these rules is that, deep down, we feel that not doing so isn't the way things ought to be. Violations of the rules threaten a deep-seated order of being.

Because this book is about pigs and pork, food taboos will play a central part in the discussion. In fact, most food taboos involve meat.[18] On the one hand, meat spoils easily and frequently carries diseases. An innate distrust of meat, translatable into taboos in certain cultural contexts, might be adaptive.[19] But what makes meat good to taboo largely reflects the symbolic and social power of animals in human cultures.[20] Animals are metaphors for human behavior; they reflect our humanity back at us. Killing them, while necessary to provide the meat that most humans crave, is an emotion-laden, guilt-ridden task. There is no culture that does not recognize the "intimate bond" between humans and animals.[21] Breaking that bond, even out of necessity, feels wrong. Every culture thus struggles with the moral dilemma of eating our fellow creatures of the earth. The powerful emotions surrounding meat give animal flesh its power,[22] a power that is potentially dangerous and must be controlled.

The significance of taboos is inherently social. Transgression threatens the order of things on a level that is deeper than that of the individual. But such threats are nebulous; large-scale social or cosmological consequences are difficult to imagine concretely. For that reason, the justifications people cite for following taboos are often personal. Thus, many believe that transgression will cause physical harm or bad luck. For example, Amazonian Nukak hunters fear that eating monkey heads will cause them to have unsuccessful hunts in the future.[23] In other cases, breaking a taboo is thought to cause illness, impotence, or death.[24] People seek rules to make sense of an unknowable world and to keep them safe within it.[25] Similarly, humans have a tendency to moralize misfortune. We seek to explain "bad luck" as a product of behavior—perhaps for no better reason than to give us the sense that we have a modicum of control over our fates.

But break taboos people will. People frequently challenge, renegotiate, and eliminate taboos. A century ago, Sigmund Freud[26] suggested that, despite our fears of tabooed things, we harbor an inherent desire (or urge) to transgress. The forbidden is, after all, titillating. But a "transgression fetish"[27] is not the only reason for breaking a taboo. It can also be done out of necessity—and in fact, both the Quran (e.g., Quran 5:3) and Jewish rabbinic law (*pikuach nefesh*) expressly allow the consumption of pork in cases of survival. Taboos can also fade from popularity seemingly on their own. For example, taboos on beef in Late Imperial China[28] and fava beans in parts of Classical Greece[29] are no longer practiced.

Beyond pleasure, survival, and fading customs, the power surrounding forbidden behaviors makes them potential sites of self-conscious political struggle. Breaking a taboo often sends a powerful message of resistance, a declaration of one's unwillingness to participate in an oppressive society.[30] In some cases, the taboo itself may be a form of oppression, and in such cases, social activism may precipitate the elimination of a taboo. In recent years, many segments of American culture have witnessed the rapid, if not total, lifting of taboos on homosexuality, largely as a result of LGBT activism. As we will see, the elimination of the pork taboo in Christianity was, in part, a radical gesture toward broad social reform.

Theories of the Pig Taboo

Perhaps no taboo has received as much scholarly scrutiny as the pig taboo in Judaism and in Islam. This is due in part to the fact that these versions of the taboo, historically, were some of the most visible in Europe. But theories about the pig taboo have a much longer history. For over 2,000 years, various thinkers have penned theories about why the pig taboo came into being. Fewer, as we'll see, have thought about why it may have persisted for such a long time and how it evolved over the millennia. Nevertheless, modern zooarchaeologists have inherited a rich tradition of speculation, informed or otherwise, about pig consumption and reasons for its avoidance. Much of this inherited wisdom is found in the pig principles mentioned at the end of Chapter 2.

Biblical Explanations

The authors of Leviticus and Deuteronomy, as we will see in the next chapter, inherited a tradition of pork avoidance, elaborated upon it, and embedded it into a set of food taboos. Perhaps for largely political reasons, they sought to crystallize it and other taboos into a code of behavior for the Israelites (later, Jews). To do so, they provided a classification of animals that were good to eat (e.g., ruminating bovids like sheep, goats, and cattle) and those that were not (e.g., certain types of fish, birds of prey, and swine). They explained the taboos on pigs and other forms of meat primarily in terms of animal physiology. For example, fish that possessed scales and fins were good to eat, but fish lacking these appendages (e.g., catfish) were not (Leviticus 11:9–10).

While a traditional reading of the biblical texts would see these explanations as generative of the taboo, in fact, they represent some of the oldest attempts to understand why people avoided pig meat. That is, they were post hoc explanations; the biblical writers sought to provide a framework for understanding food proscriptions that already existed (although the question of how frequently they were followed will be addressed in the next chapter). The focus on animal physiology is perhaps telling of the logic followed by the biblical authors. Genesis 1–2 explained how the god Yahweh had created the universe and all animals. The rules for purity and impurity, therefore, should be grounded in the immutable characteristics of animals themselves, the features that Yahweh gave them at Creation.

A close reading of the biblical text is thus warranted. The taboo on swine is found in Leviticus and Deuteronomy. Pigs are included in a list of animals that are prohibited, as well as a much shorter list of animals that are good to eat. The taboo on swine is often translated as follows:[31]

> And the swine—although it has true hoofs, with the hoofs cleft through, it does not chew the cud: it is *tame* [impure] for you. (Leviticus 11:7)

> Also the swine—for although it has true hoofs, it does not bring up the cud —is *tame* for you. You shall not eat of their flesh or touch their carcasses. (Deuteronomy 14:8)

There are two important features of this passage. First, the Hebrew word *tame*, which is often translated as "unclean" in many English-version Bibles, is best translated as "impure." The biblical taboos have nothing to do with cleanliness or health[32]—although this is a common misconception. This is where the metaphor of radioactivity is useful to remember. Breaking a taboo creates a state of impurity because it brings into contact things that should remain separate, subjecting a person or thing imbued with less power to high doses of cosmological/social energy. Rituals are needed to purge the person or thing of that energy. The rabbinical scholars writing several centuries after the Torah certainly understood this principle. Pork can make a person impure (*tame*), but so too can something holy. The rabbis use the same terminology to describe how (mundane) hands become impure (*tame*) after touching a (sacred) Torah scroll (Mishnah Yadayim 3:5). By using the Hebrew word *tame*, the biblical authors were not declaring that pigs were unclean but rather that they harbored a certain energy that was dangerous, somehow, for the people of Israel.

Second, the biblical authors are at pains to point out that the pig, though it bears a resemblance to animals that are good to eat (in that it has cloven hoofs), nevertheless has an important natural property that differentiates it from them (it does not ruminate). One can read between the lines the biblical authors' struggle to comprehend the taboo on pigs that they had inherited and that they likely believed was divinely inspired. The pig presented some problems; it was an animal they knew was eaten by other peoples in the Near East, and it seemed similar to other livestock species. Why they settled on the lack of cud chewing—as opposed to, for example, the pig's omnivory, its unique ability to grow tusks, or any other distinguishing feature—remains unclear. Nevertheless, the authors emphasized a natural and essential characteristic of this animal that clearly separated it from sheep, goats, and cattle, the animals the biblical authors identified as good to eat (e.g., Leviticus 11:3).

Classical Writers on Pig Taboos

The earliest writers about the pig taboo from an external cultural perspective were Greek and Roman historians. Though their depictions of other cultures must be taken with a grain of salt, they were nevertheless some of the first to struggle to understand why people in the Near East avoided pork. Earliest among the Classical historians was Herodotus (lived ca. 484–425 BC), who wrote about the Near Eastern cultures of his time and attempted to parse out some of their histories. Interestingly, he does not mention the taboo in Judaism, but rather the one in Egypt, claiming (*Histories*, II.47) that Egyptians considered pigs unclean and that if an Egyptian man were to touch a pig, he would jump into the river to cleanse himself.

Herodotus and other Classical writers, however, make it clear that the pig taboo they observed in Egypt applied only to certain members of society—probably priests. Several centuries later, Sextus Empiricus (ca. AD 160–210) spelled this out, commenting that Jews or Egyptian *priests* would rather die than eat pork (*Outlines of Pyrrhonism* 24.223, emphasis my own), as did the Jewish historian Josephus (ca. AD 37–100) (*Against Apion* 2.14). In fact, Herodotus, in his rambling manner, also made it clear that the pig taboo applied only to certain segments of Egyptian society or perhaps in certain contexts. Other passages in his *Histories* count swineherds as one of the main social classes—one whose members were so low in status that those of other classes would not marry them (*Histories*, 2.164). Herodotus also claimed that Egyptians ate pork during certain festivals (2.47–48), that pigs were used to tread seeds into the ground during planting (2.14), and that they served as bait for hunting crocodiles (2.70). This is a depiction not of a society observing a universal taboo on pigs and pork, but of a society in which a taboo applied only to certain contexts, probably those related to ritual activity.[33]

Herodotus aside, much of the literature on food taboos in antiquity derives from Roman sources. As a general rule, the Roman writers were bemused by the Egyptian pig taboo and other taboos they encountered among the peoples they conquered. The Jewish pig taboo, on the other hand, incensed them (Rosenblum 2010b; Schäfer 1997:70–81). In part, this was due to the simmering resentment between Greco-Roman and Jewish cultures that frequently broke out into violence. For example, Tacitus (AD 56–117) wrote, "Jews regard as profane all that we hold sacred [. . .] they permit all that we abhor" (*Histories* 5.4) and depicted Moses as a vindictive miscreant who designed Jewish laws specifically to offend other peoples. He went on to denigrate the pig taboo in particular, claiming that it originated from Jews afflicted with a skin disease carried by swine (*Histories* 5.4).

Roman satirists mocked the Jewish pig taboo and occasionally provided theories of its origins. Petronius (AD 27–66), for example, mused that the Jews abstained from pork because they worshipped a pig-god.[34] While intended to be derogatory, the joke played off the ambiguous nature of food taboos. It is not always clear whether an animal whose flesh is not eaten is too sacred or too profane. In fact, confusion surrounding the origins of the pig taboo has propagated a long tradition of anti-Semitic humor and folklore identifying Jews as pig worshippers or even pigs themselves.[35]

Jewish scholars trained in the Classical tradition also turned their attention to the pig taboo. Philo (ca. 25 BC–AD 50), a Hellenistic Jew who lived in Roman Alexandria, echoed Leviticus (11:7) in identifying the taboo in terms of pigs' unique physiology (*On Husbandry* 32). He argued that the pig's inability to ruminate (in the digestive sense) was analogous to a person's inability to ruminate (in the mental sense). Swine's flesh, therefore, inspired idleness and epicurean impotence. Jews should avoid pork not because it caused illness or defilement, but because it was so sweet as to make the mind and spirit weak.

Health-Related Explanations

One of the most commonly accepted explanations for the pig taboo is that it was intended to prevent people from becoming sick. Poor translations of the word *tame* are partly to blame for this misconception, but this problematic explanation has a longer history. In fact, Classical writers cited Egyptian/Jewish fears of "leprosy" and other unspecified skin diseases.[36] Later, in the 12th century, the Sephardic Jewish polymath Moshe ben Maimon, or Maimonides, proposed two main health-related explanations for the taboo on swine in Leviticus: (1) pigs were dirty and (2) pork contained an overabundance of "moisture" (*Guide of the Perplexed* 3.48). For Maimonides,

Yahweh gave the Jews laws in order to improve them bodily and mentally. Avoiding pork and other foods that were not *kosher* (fit) must, he reasoned, achieve these goals.[37]

The discovery in 1835 of *Trichinella spiralis* larvae in a human cadaver and their subsequent connection to the consumption of raw or undercooked pork was a watershed moment for health-related explanations.[38] *Trichinella spiralis* is a roundworm parasite that causes trichinosis, a disease characterized by diarrhea, vomiting, fever, muscle pain, and inflammation of the eyes. In the decades that followed its discovery, medical authorities began speculating whether the Jewish taboo on pork was a hygienic measure intended to reduce trichinosis and other infections by parasites carried by pigs, such as tapeworms.[39] Biologists and physicians today still peddle these stories, typically with a focus on trichinosis.[40]

The problem is that there is no evidence in the Bible or, for that matter, the Quran that the food laws were health measures. And as pork was not widely eaten in the Levant in the Bronze and Iron Ages, pigs were almost certainly a very minor vector of disease for the ancient Israelites—although diseases carried by swine probably affected Egyptians with some frequency. More importantly, if pork was so harmful, why did the civilizations of premodern pork-loving peoples thrive around the world, from Polynesia to New Guinea to China to Europe? And why, if ancient Israelite priests had somehow cracked the epidemiological code for trichinosis, would they keep such a tight lid on their discovery, opting to shroud a perfectly logical reason for not eating something ("Thou shall not eat the pig, for it will give you diarrhea") in a religious commandment justified on the grounds that the pig has a cloven hoof but does not chew its cud (Leviticus 11:7)?[41]

Poor translations and overzealous epidemiologists may explain the modern acceptance of health-related explanations, but these accounts achieve some of their greatest traction among those who follow taboos. The emotion of disgust shields taboos from transgression. When a taboo is broken, disgust can easily manifest as a feeling of illness. When shown a photograph of a lowland ringtail possum, a Seltaman elder in New Guinea responded viscerally, exclaiming "Ach . . . we don't eat that."[42] Similar reactions can be found among many vegetarians or vegans at the sight of meat,[43] and many Americans cannot stand the thought of eating dogs.[44] Fans of the film *Blade Runner* may recall that measuring the disgust felt upon imagining a scenario in which one ate canine flesh was part of the Voight-Kampff test to detect androids masquerading as humans.

To those who follow a taboo, transgression makes one vulnerable to illness or reveals something unhealthy (or just *wrong*) about an individual. Fear of illness and anxiety about other physical consequences are cross-culturally typical reactions. But these reactions are psychological in origin, not epidemiological. They stem from the intense emotions that surround taboos, not their genesis.

In the final analysis, the health-related explanations attempt to force-fit the pig taboo into a functionalist framework rather than pursue a scientific interpretation of the available historical information. A number of anthropologists have made this point already.[45] But despite their work, health-related theories remain popular, probably because they offer a just-so story that rationalizes the irrational, grounding taboos in everyday experience—as opposed to cosmological and social life.

Religious Explanations

Another class of scholars have identified the pig taboo exclusively as a religious phenomenon, namely as a transition from sacred to abominable. James Frazer provided one of the earliest anthropological accounts of this explanation in his wide-reaching and heavily interpretive book, *The Golden Bough*.[46] Frazer did not have access to much archaeological data, and his account does not include any concrete dates or specific cultural settings. Rather, it is based on comparative readings of different mythologies, especially in Classical and early European literature, and his rather nebulous reconstructions of the past.

Frazer argued that the pig was once associated with the "corn goddess," whose domains included the harvest, fertility, and rebirth. He noted, for example, the sacrifice of piglets during the Thesmophoria festivals.[47] He also argued that, paradoxically, pigs were *chthonic* symbols (i.e., symbols of death and decay), citing the Greek deity Attis, who was associated with death, resurrection, and pigs and whose priests allegedly abstained from eating pork.[48] In fact, the duality of death and life, the notion that death and birth are linked processes, may very well have been a feature of Mediterranean and Near Eastern ideology. For example, the Sumerian/Babylonian goddess Inanna (Ishtar) was said to be the sister of the goddess of the underworld, Ereshkigal. But Frazer went on to posit an evolution in the symbolism of pigs. Pigs, he claimed, were once sacred to the Israelites[49] and Egyptians,[50] who avoided pork either out of respect for the deities associated with birth/death and fertility/decay or because of the power imparted to the pig by its association with these essential themes of human life. But over time, the priesthood forgot the pig's sacred nature, even while they continued to avoid eating its flesh. Eventually, they replaced worship with disgust. In other words, the priesthood continued to perceive pigs as powerful animals, but they recast the basis of their conceptualization of this power as repugnance rather than awe.[51]

Archaeology and text-based research have borne out the idea that pigs were ritually associated with fertility, death, rebirth, and decay. The 3rd millennium BC sacrifices at Tell Mozan provide one example.[52] The biblical scholar Jacob Milgrom[53] has also resurrected the hypothesis that the taboo in Judaism was connected to pigs' associations with chthonic deities. Milgrom[54] mused that the taboo may have been a proscription put in place to reject or ban rituals invoking the spirits of the underworld (e.g., "the Witch of Endor" in 1 Samuel 28). In other words, the biblical writers may have found these rituals antithetical to the worship of the single god Yahweh. Or the taboo might simply have emerged from fear of the considerable energy associated with death and its ability to render impure a person offering sacrifices in Yahweh's presence. In fact, Milgrom[55] identified many of the forms of ritual impurity in Leviticus with death. Not to be eaten, he argued, were those animals and things most redolent of death and decay: birds of prey, blood, carrion, and, ostensibly, pigs.

The religious hypotheses ultimately harbor a kernel of truth. There were ritual uses of pigs, especially in chthonic rites. This may well have contributed to the unease with which the biblical authors approached pigs. But there is little evidence that pigs were sacred to the extent that Frazer hypothesized or as strongly connected with death as Milgrom suggested. Although pigs were ritually powerful in some contexts, they were hardly sacred; pigs were relatively uncommon among animals sacrificed to the gods in the Bronze Age. Although they were used in chthonic rituals, so too were other animals that were not forbidden; for example, at least 60 sheep and goats were found in the Mozan pit alongside the piglets.[56]

On the other hand, one of the major changes that occurred in the Late Bronze Age was pigs' apparent loss of ritual status. The evidence is admittedly thin, but texts do seem to suggest an increased focus on the impure or defiling nature of pigs, a symptom of their changing religious significance in multiple cultures of the 2nd millennium BC Near East. The biblical writers were probably aware of these negative connotations and drew upon them in their codification of the pig taboo.

Douglas's Physiological Explanation

The biblical books of Leviticus (11:7) and Deuteronomy (14:8) specify a concrete physiological reason for not eating pigs: they have cloven hoofs but do not chew their cud. Similar explanations rooted in empirical observations of anatomy are given for not eating other animals—for example, the camel (because it does not have hooves) and certain fish (if they lack scales or fins). The fact that the Bible provides its own justifications has made some scholars question why we should search for other explanations. Instead, they argue, we should try to understand the reasons for the justifications more fully. No modern scholar has done this as prominently as the anthropologist Mary Douglas.

In her book *Purity and Danger*, Douglas[57] set out to develop a general theory of ritual pollution in human cultures, but one of her most memorable chapters focused Leviticus. Relying strictly on the biblical wording, she argued that the "abominable" animals, such as pigs and camels, were those that did not fit neatly into Israelites' taxonomic categories, but were rather positioned in liminal categories between archetypal animal forms. Among the mammals, the pure animals were those most useful to pastoralists.[58] The tabooed animals were those whose characteristics were did not neatly match those of the pure animals. Their impurity made them dangerous.

There were several problems with Douglas's thesis. One was her incomplete knowledge of Hebrew at the time of writing, something for which biblical scholars lampooned her.[59] Second, as several anthropologists noted shortly after *Purity and Danger* was published, it is just as likely—or even more likely—that the taxonomic categories were composed as a response to extant food taboos, and not the other way around.[60] Douglas assumed that either the authors of Leviticus invented the food laws themselves or they had access to the knowledge of why certain foods were taboo in the first place. Imagine, however, if the authors were themselves befuddled by what they did and did not eat. Just as we struggle to understand the meaning behind the taboos, so too did they. If so, the physiological explanations in Leviticus are nothing but post facto attempts to justify existing and mysterious ethnoreligious traditions. Given the late date for the codification of the food laws in Leviticus—probably the late 8th century BC at the earliest,[61] but certainly centuries after the Israelites appeared in the archaeological record and were generally avoiding pork—the case for post facto justifications is quite convincing. In other words, the biblical authors wrote the taboos, and in the process imbued them with new meaning, but they did not invent them.

Like any good scholar, Douglas accepted the critiques of her work and adapted her thesis. Within a decade, she was advancing a radically new argument, one vaguely reminiscent of that of Tacitus (see below). The taboos, she argued, were laid down by priests during the Babylonian Exile (6th century BC) in order solidify the difference between themselves and their conquerors.[62] Not quite satisfied, Douglas continued to rework her theory. In her late work, *Leviticus as Literature*, Douglas referred to pigs as "sacred contagions,"[63] arguing, à la Frazer, that pigs were not reviled but rather revered. As she explained in the preface to the 2002 edition of *Purity and Danger*, it was wrong to assume

> that the rational, just, compassionate God of the Bible would ever have been so inconsistent as to make abominable creatures [. . .] I now question that they are abominable at all, and suggest rather that it is abominable to harm them.[64]

Ecological Explanations

Since the middle of the 20th century, many anthropologists have embraced ecological explanations for the pig taboo. Their work was part of a larger trend in the discipline to frame culture as essentially the way that people adapt to their local environments.[65] Carleton Coon[66] was an early proponent of this approach, hypothesizing that the pig taboo emerged because regional deforestation made swine management unfeasible.

Marvin Harris developed these ideas further as part of a larger effort to demonstrate that food taboos increased ecological fitness around the world. His argument boils down to two main points.[67] First, much of the Near East is hot and dry, especially in the summer, and pigs are more susceptible to heat stress than sheep and goats. Second, deforestation—which Harris conveniently dated to the Iron Age[68]—removed the few habitats that made pig husbandry ecologically viable. Once forests were depleted and extensive husbandry became impossible, raising pigs would have meant competing for water with cereal agriculture and other forms of livestock rearing.

There are some serious flaws in Harris's argument. For one, he made some highlight-worthy bloopers when it came to geography. For example, he claimed that the spread of Islam was checked at places where pig husbandry was viable; in regions where ecological conditions were suitable to suid physiology, few converted to the faith.[69] Harris apparently neglected to consider Indonesia, the world's largest Muslim country and the evolutionary birthplace of the genus *Sus*.

Harris also exaggerated the effects of high air temperatures on pigs. True, pigs do not thrive in hot and dry places. But if they can wallow in mud or water, pigs cool off even when temperatures rise to 40°C or higher.[70] Large rivers, like the Nile, Jordan, Tigris, and Euphrates offer refuge for pigs, even in the broiling summer heat. Finally, intensive pen-based husbandry is often *complementary to*, not competitive with, other types of agriculture. Pigs can subsist on organic waste, including spent brewery grains, rotting vegetables, table scraps, and the feces of other animals, while providing nitrogen-rich manure.[71] Arguably, keeping pigs in sties or allowing them to wander around settlements is one of the most ecologically sound types of animal husbandry.

Despite these problems, Harris's argument gained considerable traction in archaeology, where ecological adaptation remains a powerful explanatory framework. However, in addition to the specific issues mentioned in the preceding paragraph, ecological explanations have one more serious shortcoming: they do not explain why the *taboo* developed. That is, ecological explanations indicate why people might stop raising pigs (something that is technically not, in fact, prohibited by Leviticus) or how prohibitions of pig husbandry might make sense. One could go a step further and claim that, once people stopped raising pigs, the groundwork was laid for a taboo to emerge. One could push the argument even further and suggest that societies with a pig taboo had an ecological advantage over those without one. But in addition to stretching the logical limits of the argument, none of these claims account specifically for why people began *reviling* both swine and those who ate pork. Not eating pork because it is not available or because raising pigs does not make ecological sense is quite different from not eating it because doing so is taboo.

A final problem with ecological explanations is that they don't address why the writers of Leviticus—or the Quran, for that matter—formalized the prohibition without ever citing ecological factors. As we saw in the case of health-related explanations, there is something missing from the argument that the pig taboo was intended to address a real-world problem. Just who was making these decisions? Why did they keep the real reason a secret? One is left to imagine a rather ridiculous scenario in which a group of priests conspired to trick their followers into doing something that was, ostensibly, already in the people's best interests.

Political-Economic Explanations

Not everyone was pleased with Harris's ecological explanation. Anthropologists Paul Diener and Eugene Robkin[72] provided an alternative explanation that focused specifically on the pig taboo in Islam. They argued that pigs were an ecologically viable part of agricultural economies. In fact, they were so ecologically viable that they posed risks to urban elites' control over the lower classes, especially the peasantry. Pigs, they suggested, "render peasant villages dangerously rich and autonomous."[73]

In other words, Diener and Robkin hypothesized that the pig prohibition was a political move. They maintained that in the post-Neolithic world, the "appropriation processes" by which people were linked into large-scale political economic structures were more important in dictating people's decisions than considerations of local ecological conditions.[74] Support for this idea is not hard to find. Whether by drilling for small amounts of oil in fragile ecosystems, precipitating climate change by burning fossil fuels, or overharvesting forests for timber and fuel, people living in state societies often prioritize short-term political and economic gains over long-term ecological ones. The pig prohibition, Diener and Robkin argued, was an example of this sort of prioritization. Early Islamic authorities, who exhibited a strong "mercantile focus,"[75] banned ecologically viable pigs to ensure that villages produced large numbers of flock animals that could be herded back to urban markets.

Like ecological explanations, political-economic explanations are useful for understanding patterns of pig husbandry in the Bronze and Iron Ages. Diener and Robkin's article had great influence on zooarchaeologists.[76] Their work shed light on the ways in which pigs were excluded from institutional economies and how pork became an important food source for the lower classes. It is possible that, in certain contexts, pork was considered a low-status food whose very ingestion signified poverty and was therefore eaten less frequently by the upper class.[77] Such sentiments might have contributed to the taboo on pork observed among Egyptian priests by the Classical authors.

However, as an explanation for the Islamic or Jewish pig taboo, the political-economic argument falls short. For one thing, Diener and Robkin's depiction of early Islam is problematic. Islamic states certainly became mercantile, urban-based, and elite-driven. But when Muhammad pronounced the pig *haram*, the Muslim community was none of these things.[78] In fact, the teachings of the Quran, like those of the Bible, stress economic egalitarianism.[79] Second, Diener and Robkin exaggerated the degree to which pig production necessarily resists integration into urban markets. While it is true, as discussed in Chapter 5, that Near Eastern economies historically excluded pigs as wealth, others did not, including the Romans, who had well-developed markets for pork in Italy and other parts of their empire (Chapter 8). Finally, once again, we are faced with the image of holy people conspiring to trick the masses via a taboo. The image perhaps makes a bit more sense in the case of political-economic explanations. If the urban elite were trying to manipulate the rural peasantry against their economic interests, then a bit of chicanery might have been in order. Nevertheless, without evidence to back it up, the political-economic argument remains an unfounded conspiracy theory.

Ethnic-Political Explanations

The anthropologist Frederick Simoons's *Eat Not This Flesh*[80] is a go-to manual for anyone interested in food taboos. In an 89-page chapter devoted to the pig, Simoons summarized a vast amount of ethnographic and archaeological data related to pork consumption in Europe and Asia and developed his own hypothesis, which was that the pig taboo was essentially rooted in ethnic differentiation.[81]

The idea that the pig taboo was a marker of ethnic distinction had already been around for some time. Tacitus stated that Moses invented his laws to set the Israelites apart (*Histories* 5.4), and post–*Purity and Danger* Mary Douglas[82] argued that the pig taboo was invented during the Babylonian Exile to distinguish Jews from their conquerors. The historian and religious philosopher Jean Soler[83] went a step further, arguing that priests during the Exile created the food taboos to separate the Jews not only from other people, but also from Yahweh, the creator and owner of all animals. Simoons[84] drew upon these ideas but reframed them in a less conspiratorial way. Rather than hypothesizing that Moses or exiled priests invented the taboos, he based his argument on the organic development of disgust for the food habits of others.

To understand Simoons's theory, one has to know something about what is referred to as the *instrumentalist theory of ethnogenesis*. Ethnogenesis is the process by which a group of people create and, as time goes on, reproduce an awareness of an identity that is separate from that of other peoples. This tends to be accomplished through the development and employment of a unique language, mythology, sense of shared heritage (real or imagined), material culture (e.g., dress), and diet. Food plays a particularly significant role, not only because of its importance for the well-being of the individual and the community, but also because it is an element of daily life and thus something regularly encountered. As a result, food choices, cooking styles, and eating habits are all common means by which ethnic groups differentiate between "us" and "them."[85] In this way, food taboos are particularly useful boundary markers between groups.[86]

The instrumentalist theory of ethnogenesis began as a reaction against the "primordialist" theory, which held that ethnic groups are essentially eternal and arise because of inherent (i.e., biological) differences between peoples. Primordialism not only ran counter to ethnographic observation but also had the unpleasant side effect of perpetuating racist stereotypes. In seeking a better explanation for how distinct ethnic practices emerge and persist, many anthropologists[87] argue that unique ethnic practices such as those related to food arise largely in opposition to the practices of another group. That is, people pick up on arbitrary and often subtle differences in daily life between themselves and the members of another group. This perception of differences may begin organically, and the initial stages of differentiation may be largely unconscious. But over time, people in different groups begin to see their dissimilarities in dress or food or language as salient markers of group membership and, in doing so, often entrench themselves further in this view. They may begin to perceive the practices of "the other" as alien and even disgusting, especially in times of conflict.[88] By the same token, groups drawn together may adopt the practices of one another via acculturation. Or they may "hybridize" practices by blending elements together.[89] The result is a process of identity construction and reconstruction dictated by interactions among groups of people. Ethnicity is not immutable, but a cultural construct in continual motion.

In line with the instrumentalist theory, Simoons argued that certain peoples began to associate pigs with their enemies. Simoons suggested that the Egyptians' rivalry with the Hyksos, who conquered Lower Egypt in the 17th century BC and who held the pig-associated god Seth in high esteem, inspired disgust for pork.[90] He hypothesized that the Israelite taboo arose from the ethnic rivalry that played out between mobile pastoralists and sedentary farmers. In doing so, he claimed that pastoralists like the ancient Israelites often develop a sense of disgust for the sedentary way of life, a disgust that sometimes focuses on pigs.[91] These sentiments, he argued, evolved into a taboo that was eventually codified in Leviticus.

Simoons's argument makes good anthropological sense, and it dispenses with the conspiracy theory nature of some of the other explanations. There are problems with it, however. First, it does not work well in the Egyptian case. There is no solid evidence for an Egyptian taboo on pork before Herodotus's writings in the 5th century BC, and even then it appears to have been confined to specific social classes, not an entire ethnic group. Moreover, the Hyksos probably invaded Egypt from the Levant, which zooarchaeological data show was relatively pig-free in the Bronze Age. If anyone were going to be associated with pork in the Hyksos-Egyptian rivalry, it would be the Egyptians!

The identification of the ancient Israelites as mobile pastoralists is also a controversial idea, one that archaeologists have debated for decades. The archaeological data generally paint a picture of the earliest Israelites living in small but sedentary hilltop villages, not campsites as imagined by the biblical authors (Chapter 7). Aware of this issue, Simoons echoes some archaeologists by hypothesizing that although most Israelites may have been sedentary farmers, a small and currently archaeologically invisible mobile pastoral component was dominant in generating the cultural and religious ideology.[92] This is a convenient idea, but one for which there remains little evidence.

Despite these problems, ethnic explanations hit upon some important points. First, ethnic difference is an important, perhaps the *most* important, factor driving the persistence of many taboos today.[93] Taboos and ethnic difference feed off one another: taboos maintain boundaries between people, while a sense of difference gives taboos more significance. Indeed, one of the most powerful mechanisms for reinforcing a taboo is learning of a foreigner's transgressions and feeling disgust.[94] Second, Simoons builds on an interesting connection between pork avoidance and pastoralists. He may have taken this argument too far—mobile pastoralists are not inherently averse to pork[95] and pig husbandry need not be entirely sedentary.[96] Mobile pastoralists are also not always at odds with sedentary farmers; in fact, the two often intermarry and forge alliances.[97] Nevertheless, it is fair to say that, in the majority of cases, a mobile pastoral way of life, especially in the Near East, generally does not include raising pigs. And even if some have oversimplified the relationship as necessarily antagonistic, pastoralism has historically contrasted with settled life—the so-called steppe versus sown dichotomy.[98]

The relationships between pastoralists, sedentary peoples, and pigs were and are highly diverse. Yet, at least in some cases, it is reasonable to assume that certain pastoralists' negative attitudes toward sedentary peoples became focused on the association between pigs and peasants. I argue in the next chapter that the connections between mobile pastoralism and pig avoidance were, in fact, important components of the crystallization of the taboo, albeit indirectly. That is, Israelites' pig taboo in part reflected the fact that they *imagined* themselves as pastoralists, or at least as having descended from pastoralists, and that they connected pastoralism with eating sheep, goats, and cattle, but not pigs.

The Chicken Explanation

Zooarchaeologist Richard Redding has published a novel theory explaining the origins of the pig taboo.[99] He started with the premise that pigs are excellent sources of food for small-scale farmers, especially those in urban areas where waste can be converted into pork.[100] They were too valuable to be made taboo unless a more cost-effective animal came along to fit this role. Such an animal was the chicken:

> With the introduction of the chicken in the Near East, we have two taxa, the pig and the chicken, functioning in almost the exact same role in the human subsistence system and probably competing with each other for food and labor [. . .] I suggest that the chicken, once introduced, was favored by humans and largely replaced the pig in most village and poor urban contexts [. . .] First, chickens are a more efficient source of protein than pigs. Second, chickens produce a secondary product, the egg, which is also a more efficient source of protein than the pig. Third, the chicken

is a smaller package than the pig, and a household can consume a chicken within 24 hours.[101]

Redding's is the most innovative explanation to come out in years. But it has two flaws. First, as with the ecological and other explanations, it technically does not explain why pigs would be made taboo, but only why that taboo might make sense from an economic perspective. The second, and more fatal, flaw is that while chickens arrived in the Near East from South Asia in the Early or Middle Bronze Age, there is no evidence for a generalized pig taboo operating at the level of an entire nation or group until the Iron Age.[102] Moreover, judging from the bones that archaeologists have collected,[103] chickens were relatively rare in the Levant until the Hellenistic period—centuries after Leviticus was penned.[104] The arrival of chickens did not precipitate the end of pig husbandry, and they became popular only after the taboo developed. To his credit, Redding[105] admits that the chronology poses a problem. But he does not float the theory that chickens filled a gap in peasant economies left by the pig taboo. Given the chronological gaps, this scenario seems more likely.

Making Sense of the Pig Taboo

Taboos are culturally prescribed practices of avoidance, "negative rituals" as anthropologist Roy Rappaport[106] once described them. The tabooed thing tends to elicit a sense of disgust, although the intensity of the emotional response varies. Some taboos are associated with more social energy than others. For example, many Jews won't eat pork but will wear *shatnez* (mixed-fiber clothing), which is also banned (Leviticus 19:19). For reasons discussed in Chapter 8, the pig taboo became surrounded by a greater amount of social energy over time. Particularly in the case of "higher-energy" taboos, transgression is an offense to the social order, and for that reason taboos are often battlegrounds in times of intercultural conflict and social revolution.

Taboos sometimes exist at the level of entire ethnic or religious groups, as in the case of the pig taboo in Judaism and Islam. When they do, they help maintain boundaries between groups through the strong emotional responses associated with transgression. It is simply more difficult to intermarry, dine with, or even interact with someone whose cultural habits disgust you and threaten your sense of cosmological order. As a result, taboos often become more deeply entrenched, especially when groups have other reasons to maintain distance from one another. In many ways, Simoons's explanation based on ethnic politics is an extrapolation of this process; if taboos become more intense with adversarial culture contact, then one can imagine taboos beginning as slight differences between groups, which develop greater significance as people become aware of these differences and begin to view the habits of "the other" with disgust.

But taboos also separate people within the same ethnic or religious group. Some taboos, for example, separate people who are in temporary or liminal states, like pregnant women. Class- or caste-based taboos, such as the taboo on eating meat among Brahmins, are more permanent. Working from the logic of instrumentalism, it is not hard to imagine how class-based taboos emerge. High-status individuals frequently develop cuisines to distinguish themselves from the low-status members of their societies.[107] These are conscious (or semiconscious) attempts to separate status groups. But on a more unconscious level, food practices or habits observed in another social class often become identified with that class (think, for example, of someone raising his or her pinky finger while drinking tea).

Status-based habits often become entrenched. The nonelite may seek to emulate the elite in order to join their ranks, leading to a pattern in which elite habits become commonplace. But it is also common for these habits to divide societies further. Across social fault lines, they provide status groups ammunition for enforcing their distrust and dislike of the other. This is because, as classes are organized hierarchically and reproduced through oppression, they are inherently at odds with one another to the point of mutual hatred—to borrow a phrase from the feminist Simone de Beauvoir, "[O]ppression creates a state of war."[108] On this battlefield, transgressing class-based food practices, sexual mores, or speech patterns can feel like a betrayal of one's identity and sense of self. It is an act redolent of the "other," even if, on a deeper level, that "other" is recognized as one of "us."

A class- or caste-based taboo likely explains the origins of the Egyptian pork taboo described by Classical writers. In the Bronze Age, members of the lower classes ate pigs more often than those higher up on the social ladder. Perhaps this influenced the pig taboo—first as a divisive food habit separating the rich from the poor, then later as a more democratic rejection of pork as lower-status groups emulated the elite. Diener and Robkin's political economic explanation does not directly mention this possibility. But it is interesting to consider that, rather than a conspiracy by the elite, some versions of the pork taboo, such as that in Egypt, may have arisen organically from the "state of war" between the rich and poor. Although I suggest that pigs were excluded from temple life because they did not embody wealth in the Bronze Age, the initial identification of "pork = poor" may also have helped inspire a reconceptualization of the power of pigs, leading to their banishment from temples in Mesopotamia and Anatolia and to their abomination by the Egyptian priestly class.

However, class antagonism does not explain the taboo that developed among the Israelites and, later, Jews. That taboo emerged in a context in which there were no major differences in the consumption of pork between classes—in fact, there was very little pork consumption at all in the Levant except among the Israelites' neighbors, the Philistines. While an explanation of the Israelite pig taboo must focus on the Iron Age, the roots of a pork-free diet extend to the Early Bronze Age. The simplest and probably most accurate reason that people in the Levant did not eat much pork in the Late Bronze and Iron Ages was that they had inherited a tradition of not eating it. It was the contact with pork-eating groups, such as the Philistines and, later, Greeks and Romans, as well as the romanticization of an imagined past that galvanized the sentiments surrounding pigs and transformed passive pork avoidance into more active abomination and taboo. This process drew strength from, and itself helped forge, a unique identity for Israelites, one derived from an idealized pastoral origin lacking in pork and in contradistinction to pork-eating others. The pig taboo written in the Bible, in other words, is most closely associated with taboos that develop through ethnogenesis. In that sense, Simoons and those who followed his line of thinking are closest to the truth.

Ethnic politics, however, was not the only factor involved in bringing about and reproducing the pig taboo. In fact, there is a kernel of truth in each of the explanations. For example, there *was* a major change in the perception of pigs in Near Eastern religious thought during the Bronze Age: swine came to be regarded as capable of defiling Hittite or Mesopotamian temples. Israelites may well have found that these and other negative attitudes toward pigs, including class-based ones, supported their justification of pork avoidance. Similarly, although they did not inspire the origins of the taboo, both health-related and physiological explanations were important for defending its existence later on—the first propagating a sense of visceral disgust, the second grounding the taboo in the logic of a creator-god. Although the chicken probably didn't drive the prohibition on pork consumption, its success in the

Near East probably did stem from the fact that their meat could serve as a substitute for pork (in the sense of being an inexpensive form of meat) after the taboo had emerged. Although pigs were not banned because they were ecologically unfit in the Near East, their absence from certain highly arid regions (such as the Arabian Peninsula) certainly facilitated the spread of the taboo, making it more readily adoptable by desert-dwelling messianic figures and their followers in the centuries to come. In short, many of the factors hypothesized by theorists in the past have played *some* role in the evolution of the pig taboo.

The Pig Taboo as an Evolving Cultural Element

Our review of the various explanations for the pig taboo reveals a logical flaw common to all of them. In addition to the specific factual problems associated with each explanation, many scholars have failed to recognize and differentiate between (1) the reason the taboo developed in the first place and the conditions that made it possible; (2) the reason it endured over the short term and, in some cases, spread to other cultural contexts; and (3) the reason it succeeded over the long term. While all of the explanations claim to address (1), they often end up focusing instead on (2) or (3). This shortcoming, which amounts to a conflation of temporal scales, has caused considerable confusion and led to inaccurate conclusions. In fact, there is every reason to believe that the specific sentiments underlying taboos mutate and evolve. Why a taboo persists today may have nothing at all to do with why it began in the first place. A satisfactory explanation for a taboo ultimately must account for the conditions and cultural context(s) that set up its emergence, its first appearance and early history, and its change over the long term.

Thinking in terms of how cultural elements evolve is the bread and butter of archaeology, even if it is uncommon in the scholarly writings about taboos. A refreshing perspective has been offered by zooarchaeologist Naomi Sykes. In her book, *Beastly Questions*, Sykes investigates the 1,500-year history of the swan in England. Her discussion clearly delineates the unexpected ways that food meanings can evolve. In the fifth century, the swan was imbued with religious significance and eaten primarily in monastic settings. Later, in the medieval period, rural elites began to raise and eat swans, perhaps in an attempt to claim piety. Following the Black Death, swan consumption increased as upwardly mobile urban families attempted to mimic aristocratic habits. Such democratization elicited a backlash from the established elites, who passed legislation to prevent people of lower social status from eating the bird. The Act of Swans in 1482–1483 declared that all swans belonged to the English monarch. This law effectively prohibited swan consumption. As time went on, law became tradition and tradition was used to justify the law. Negative sentiments associated with eating swans evolved. Sykes ends her story in 2006 with a judge sentencing a man to 57 days in prison for attempting to eat a swan and thus committing a "taboo act."[109]

Like the English swan, the Near Eastern pig had a long history before and after it became taboo. The meanings associated with it evolved. An archaeological perspective allows us to investigate the pig's many transitions. In doing so, we have to abandon the search for simplistic explanations and instead pursue a scientific analysis of pigs' tortuous journey in the Near East, from the first wild boar hunts until today. The result is less elegant and satisfying than the theories put forward by previous scholars. Then again, the histories of human cultures are rarely elegant or satisfying.

7

The Coming of the Taboo: Pigs in the Iron Age

The Iron Age and Israelite Ethnogenesis

We come now to the Iron Age, a dynamic and tumultuous period in which the first versions of the pig taboo (at least those for which we have solid evidence) took root. In the early part of the 12th century BC, dozens of cities along the Levantine and Anatolian coasts of the Mediterranean were sacked, and the great imperial powers of the Late Bronze Age lost control of the Near East. Scholars continue to debate the causes and consequences of this collapse, but likely contributing factors include drought, migration and warfare, the disruption of trade routes, and internal rebellions.[1]

In the Levant and parts of Syria and Anatolia, the power vacuum left by the retreat of the Late Bronze Age empires created room for a regionwide process of ethnogenesis. This process was inspired in part by the migration to the Near East of people from the Aegean and other parts of the Mediterranean, especially those known in Egyptian inscriptions as "Sea Peoples." One of these groups, the Peleset (or Philistines), came to rule several city-states in modern-day Gaza and southern Israel. These people, perhaps unsettled by the calamity at the end of the Bronze Age, brought with them their customs and food habits.[2] In some cases, these traditions contradicted those of local peoples.

At the same time, local Near Eastern peoples reconfigured themselves into new groups. One group, the Hebrew-speaking Israelites, coalesced as a distinct ethnic group in the 12th–11th centuries BC in the southern Levant, the region between the Aramaean kingdoms to the north, the Jordan River to the east, and the Philistine city-states to the south. By the 10th–9th centuries BC, the Israelites were organized into two distinct kingdoms—Israel in the north and Judah in the south.[3] It was in this setting that the most long-standing and most widely applicable pig taboo would emerge.

Evolution of a Taboo. Max D. Price, Oxford University Press (2020). © Oxford University Press.
DOI: 10.1093/oso/9780197543276.001.0001.

The first five books of the Bible (the Torah; literally "instruction") and the books of Joshua, Judges, 1 and 2 Samuel, and 1 and 2 Kings provide a narrative of these time periods, beginning with the creation of the universe by the Israelites' god Yahweh (sometimes "Elohim") and ending with the conquest of the kingdom of Israel by the Assyrians in the 8th century BC and that of Judah by the Babylonians in the 6th century BC. This story, perhaps first composed in the late 8th or 7th century BC in the kingdom of Judah, contains a mixture of facts, myths, and heavily spun accounts of historical events. However, extrabiblical sources provide independent verification of some of the events described in the Hebrew Bible. The Merneptah Stele (dating to 1207 BC), found in Egypt, provides the first mention of a people called "Israel" living in the Levant. Other texts and monuments, such as the Mesha Stele found in Jordan, records a war between the Moabites and the kingdom of Israel in the mid-9th century BC (mentioned in 2 Kings 3).[4] In general, while the events described in the Torah are largely mythological (including the exodus from Egypt, for which there is no concrete archaeological or historical evidence)[5], the historical accuracy of the Hebrew Bible appears to increase with time up to the 8th–7th centuries BC.

The small territorial kingdoms of the Israelites and other groups would ultimately fall to the major empires that reappeared in the Near East beginning in the 9th century BC. The greatest powers were based in Mesopotamia, the Neo-Assyrian Empire (ca. 900–609 BC), and its short-lived successor the Neo-Babylonian Empire (609–539 BC). These empires developed a clear imperial ideology that justified conquest as a divine mission.[6] As part of this imperial project, the Assyrians and Babylonians conducted mass resettlements of conquered or rebellious peoples. Many of the inhabitants of Judah, whose kingdom the Babylonian Nebuchadnezzar II destroyed in 586 BC, spent a roughly 50-year "exile" in and around the city of Babylon. During that time, they built upon a new form of theology that had begun to emerge in Jerusalem in the century prior. Central to this theology—Judaism—was a special attachment to their god (Yahweh), their holy texts, and the land of Israel and its people.

The Iron Age came to a close with the rise of the Persian Empire. Persia began to emerge as a major power in southern Iran in the 6th century BC. The empire expanded rapidly beginning in 559 BC with the ascension of Cyrus, who captured Babylon in 539 BC. The Persians' policy of imperial domination, while remaining focused on conquest, differed from that of their Neo-Assyrian and Babylonian predecessors in that they adopted a more tolerant stance toward local customs and traditions. In this spirit, Cyrus issued a decree allowing the Jews in Babylon to return to Jerusalem, where they would soon rebuild the temple to Yahweh that had been destroyed by the Babylonians.[7]

The Writing of the Torah and the Pork Taboo

The partial historical validity of the Bible brings up a question about when it was written. The Hebrew Bible contains three main section—Torah ("instruction") or the five books of Moses, Prophets (in Hebrew, *Nevi'im*), and Writings (*Ketuvim*)—each of which is composed of texts from several sources that were later joined together. The dating of the sources varies. Much of the latter half of the Prophets and the Writings deal explicitly with events that occurred during or after the Babylonian Exile. They are late 1st millennium BC compositions. The Torah and the first few books of the Prophets are earlier in date, but even these books consist of texts written centuries apart that editors in the late 1st millennium BC compiled and modified. Dating these texts is a complex issue. We need not get too bogged down in the debate. But it is important to establish when, where, and how those parts of the Torah related to the pork taboo were composed in order to understand the context in which it was codified.

Since the 18th century, scholars have adopted a skeptical stance toward the Torah and its authorship. Rejecting the idea that Moses composed it shortly before his death, scholars have identified a number of clues suggesting that the Torah was a living document, edited and reconfigured until well into the Persian period.[8] At least five major authors (or groups of authors) have been identified: the Yahwist (J), Elohist (E), Priestly (P), Holiness (H), and Deuteronomist (D), the last of whom composed many of the historical books of the Prophets, the so-called Deuteronomistic History.[9] Using place names, word choices, and references to historical events, scholars have attempted to provide dates for these authors, ranging from the 10th to the 5th century BC.[10] In the biblical timeframe, this encompasses the time of David to the post-Exilic period.

Particularly important for the pig taboo are the books of Leviticus and Deuteronomy, especially Leviticus 11:7 and Deuteronomy 14:8, the passages that explicitly ban the consumption of pork. These portions of the text relate to the P, H, and D sources.[11] Scholars have proposed various dates for these sources, especially P. Many argue that they were composed by, or at least have their origins in, small groups of priests writing in the 8th and 7th centuries BC. This time corresponds to the reigns of Hezekiah, Josiah, and other later kings of Judah. Other scholars argue for a composition in 6th or 5th centuries BC, corresponding to the Exilic/post-Exilic period, when the remnants of the Judahite priesthood reconfigured their religion into something we would now recognize as Judaism.[12] We can be reasonably sure that priests of Judah, and not Israel, wrote these texts, as they exhibit a clear bias in favor of the southern kingdom and its monarchs.[13]

Israel Finkelstein and Neil Asher Silberman have proposed one particularly convincing argument, and one that is based on historical and archaeological data.[14] These authors argue for a pre-Exilic date for many of the sources, and they identify the D source as living under the reign of Josiah (639–609 BC), with the P source living somewhat later. Other sources, like H, may have lived a bit earlier. Assuming Finkelstein and Silberman are correct—and I stress again that dating the Torah is a complex and heavily debated topic—we can date the *textual codification* of the pig taboo to the late 8th or 7th century BC in the kingdom of Judah.

The Archaeology of the Israelites and the First Jews

Archaeology tells a somewhat different story of the origins of the Israelite people than the Bible. Early archaeologists largely accepted the biblical account of an exodus from Egypt, nomadic wandering, and the conquest of southern Levantine cities.[15] But a century of research has painted a very different picture of Israelite ethnogenesis. The archaeological data strongly suggest that the Israelites coalesced as a distinct group a bit before 1200 BC from local southern Levantine ("Canaanite") groups living in the hilly regions west of the Jordan River.[16] In this sparsely inhabited territory, several dozen new villages appeared in the Iron I period (1200–950 BC) that were characterized by their location on hilltops, lack of defensive walls or fortifications, densely packed houses with a four-room layout, agricultural terraces, an absence of socioeconomic differentiation, and very few or even no pig bones.[17]

Relying on these facts and drawing on parts of the Bible, several modern theories have sought to explain the origins of the Israelites. Some scholars have argued that the Israelites descended from people who revolted against the Egyptian-backed Canaanite city-states and their presumably oppressive social institutions.[18] Others, not necessarily accepting the revolt hypothesis, have argued that the Israelites descended largely from people who left Levantine cities at the end of the Bronze Age.[19] Another group of scholars have argued that the Israelites descended from nomadic pastoralists, either native to the region or arriving from the east, who had settled down into agricultural villages in the relatively empty hill country.[20] Seeking consensus, some recent publications have argued that the Israelites were a mixture of groups, including nomadic pastoralists, fleeing urbanites, disenchanted peasants, and even the ʿApiru bandits sometimes mentioned in Bronze Age texts in Mesopotamia and Egypt.[21]

Whatever its origins, several features defined Iron I Israelite society. First, Israelites were not monotheistic, but worshipped a number of gods in addition to Yahweh. These included Baal, the Canaanite storm god mentioned in Chapter 5. Also important were Astarte (or Ashtoreth), the goddess of erotic love whom Solomon allegedly worshipped (1 Kings 11:5), and Asherah, a fertility goddess who was probably initially the consort of Yahweh.[22] Eventually, these gods would become targets of the biblical writers' ire as they attempted to elevate Yahweh to the status of the one and only god.

Israelite society was also characterized by a strong tribal ideal. This contained three components. First, it prized a family-centered social organization built on paternalism and filial obligation.[23] Even though powerful and independent female figures populate the biblical texts (e.g., Deborah and Yael in the book of Judges 4–5; Naomi and Ruth in the book of Ruth), male heads of households were perceived as the building blocks of society. Second, the tribal ideal promoted a fiercely egalitarian ethos, at least among adult Israelite males. This is detectable archaeologically at early Israelite sites by the lack of prestige objects and the overall uniformity of house sizes.[24] Third, it encouraged pastoralism of sheep, goats, cattle, and, later in the Iron Age, camels.[25] This pastoral archetype, especially one based on highly mobile or "nomadic" pastoralism, is abundantly clear from the Bible, whose writers projected—and exaggerated—the image of their ancestors living largely by herding animals.

These ideals, however, did not always match reality. Most Israelites by the Iron I, and perhaps at any time in their history, were not nomadic pastoralists. By the Iron II, Israelites would form monarchic societies that would set aside social equality. But the tribal ideal remained intact. Though contradicting daily realities, the tribal ideal served as a rallying point for revitalization movements later on. Jewish, Christian, and Muslim thinkers alike have drawn on its archetypal principles for centuries.[26]

Beginning around 950 BC, in the Iron II period, the Israelites abandoned their small hilltop villages and congregated in larger, walled settlements.[27] A more rigid social hierarchy was adopted, and political power was centralized in the hands of kings. This might have occurred in response to military confrontations with other groups in the southern Levant, such as the Philistines.[28] Whether these developments coincided with a united monarchy (under Saul, David, and Solomon) or not remains the subject of considerable debate.[29] Regardless, by the 9th century BC, two Israelite kingdoms existed: a larger and more prosperous one to the north (Israel) and a smaller one in the south (Judah).

Archaeological and extrabiblical textual evidence corroborates much of the history of the two kingdoms—their wars, accomplishments, and ultimate downfalls—recorded in the books 1 and 2 Kings. The events correspond to a date beginning around the 10th century BC and ending with the fall of Jerusalem in 586 BC. However, it must be stressed that the biblical authors' agenda was not to write an accurate history. Rather, it was to understand and glorify their god Yahweh, explain folktales and traditions in terms of Yahweh's relationship to his people, and sanctify the authority of the kings under which the priesthood worked. We must therefore examine critically the biblical presentation of history. We will focus especially on the developments in Judah after the Assyrian conquest of the kingdom of Israel in 722 BC. The Bible depicts Judah being ruled in the late 8th and 7th century by kings who are described as righteous (Hezekiah and Josiah) or wicked (Manasseh and Amon). The righteous kings reformed religious practice, gained Yahweh's favor, and led Judah to new prosperity. Indeed, from the late 8th century until 586 BC, Judah's kings and priests initiated a series of policy and religious reforms. These changes must be understood in their political and social contexts, which can be inferred from archaeological and historical data.

After the fall of Israel, Judah found itself the center of Israelite culture. Over the next few decades, as Assyrian power waxed and waned in the Levant, Judah's kings embarked on a series of military expeditions to conquer territory in the north and incorporate it into their kingdom. Judah's elevation to a position of prominence was bolstered by the fact that populations in and around Jerusalem had swelled to unprecedented levels beginning in the mid-8th century BC, turning the Judahite capital into a much more important civil and religious center.[30]

In this context of, on the one hand, newfound glory and, on the other, a real threat of foreign imperial conquest, the kings and priests in Judah sought a new identity for themselves and their subjects. They pursued a strategy, so common among states in the ancient and modern world, of promoting ethnogenesis within their territory—akin to what we would call nationalism today.[31] And like so many other kings and politicians, they did so by concentrating political power in the hands of a single person (one approved of by the priesthood) and painting an image of a glorified version of the past to justify their ambitions.

This process of glorifying and unifying Israelite identity began in the late 8th century BC when, according to the biblical narrative, Hezekiah centralized political leadership, in part by banning all religious sanctuaries except the Temple in Jerusalem. He also may have commissioned a history of the Davidic kingly lineage and a code of laws.[32] These reforms served to transform Jerusalem into the center of Judahite religious and cultural expression.[33] Reforms continued into the 7th century BC, reaching a critical moment during the reign of Josiah, which lasted from 639 to 609 BC. Josiah initiated a religious-political revolution that constituted, according to Israel Finkelstein and Neil Asher Silberman, "the most intense puritan reform in the history of Judah."[34] His reforms tolerated only one type of religious expression—the worship of Yahweh—and he commissioned his priests to write down codes of moral behavior. These codes, perhaps in combination with ones written in Hezekiah's time, probably included at least the nucleus of Leviticus.[35] Over the next several centuries, priests continued to redact Leviticus and added Deuteronomy (which was allegedly discovered in 622 BC by Josiah's priests).

The codes written down in Judah would serve as the basis of Jewish Law (*halakha*) and included taboos on pork and many other foods, as well as numerous other commandments (*mitzvot*), such as laws dictating sexual behavior and dress. The writing down of these *mitzvot*, inscribing them in holy documents ostensibly of divine inspiration, was revolutionary.[36] It cemented the moral prescriptions of the Judahite priesthood as law, bestowed upon these practices a permanency that was integrated into the core of Judahite and later Jewish life, and prevented any "cultural drift" that could mute or alter them over time. Henceforth, any violation of this Iron Age code of moral behavior would represent a self-conscious rejection of *halakha* and all its associations with Judahite and Jewish identity.

Nebuchadnezzar II's sack of Jerusalem in 586 BC brought an end to Judahite autonomy. According to the Bible—and the brutality of Babylonian kings leaves no reason to doubt the biblical account—Nebuchadnezzar forced the last king of Judah, Zedekiah, to watch as his sons were executed. Nebuchadnezzar then had Zedekiah blinded, shackled, and sent into exile (2 Kings 25:6–7). Along with the king, Nebuchadnezzar deported a portion of the Judahite population to Babylon, several thousand people that included most of the elite. This deportation unexpectedly laid the foundations for the development of a uniquely Jewish identity. It is at this time that we identify those following the Judahite religion as "Jews."

During the so-called Babylonian Exile, Judaism crystallized around the religion and laws promulgated by the Judahite kings and priests. The Jews in Babylon made heavy redactions to existing biblical texts and composed new ones, fortifying *halakha*. The laws served as a covenant that bound Jews to a set of daily practices. These practices set Jews apart from their conquerors and inspired a sense of dislocation. This perception of being "a stranger in a strange land" (Exodus 2:22) inspired Jews to direct their gaze toward Jerusalem, a city to which they could not return but which became the center of their shared ethnoreligious experience.[37] It was also at this time that the Jews adopted a philosophy that can be summarized as being "*the* counterculture of the oppressed."[38] It would serve as a rallying point for future revitalization movements, including Christianity. Accordingly, although the Jews might have experienced political and social hardships, they perceived themselves as loftier in the eyes of their (one and only) god, a position they maintained through adherence to his commandments.[39]

The Persian king Cyrus conquered Babylon in 539 BC and issued an edict allowing Jews (or Yehudim) to return to Jerusalem. Many did, bringing with them a new form of religious and ethnic identity refashioned in the crucible of the Babylonian Exile. Plans were soon laid to rebuild the Temple, and Jews repopulated the southern Levant with particular zeal. They continued to redact parts of the Torah and to write down new material that would be included in the Bible.[40] But the Persian period also witnessed the first Diaspora communities. Ironically, at the same time that Judaism emerged firmly focused on Jerusalem, it also became a pan–Near Eastern ethnoreligious entity. Thus, some Jews chose to stay in Babylon, which would remain a center of Jewish thought for centuries to come. Others traveled as far as Elephantine in Upper Egypt, where a group of Jewish mercenaries guarded Egypt's southern borders and constructed their own temple to Yahweh sometime in the 6th century BC.[41] Although spread across the Near East, Jews maintained their identity based on monotheistic worship of Yahweh and observing *halakha*.

Pig Husbandry in the Iron Age

Before examining the pig taboo and its evolution, it is important to establish the patterns of pig consumption around the Near East. Unfortunately, zooarchaeological data from many key regions are sparse. Nevertheless, the available data largely show that, continuing traditions begun in the Late Bronze Age, pork contributed less to the average Near Eastern diet than it had prior to about 1600 BC.[42] Pork consumption even declined in Egypt, where pig bones make up only around 10–25 percent of the remains of livestock recovered from sites spanning the 11th–4th centuries BC—a far cry from the predominance of pork consumption in earlier periods.[43] In fact, the only places where pig husbandry remained the leading type of meat production in Egypt were settlements with a strong presence of foreigners, such as the 5th-century Greek emporium at Naukratis.[44]

Nevertheless, patterns of pig husbandry were quite variable, judging from several Iron Age faunal assemblages in Anatolia, Syria, and the Levant. At some settlements, pigs were almost or completely absent.[45] At others, faunal assemblages yielded pig bones at a relative abundance of 20 percent or more of the main livestock species.[46] This variability in pig relative abundances is striking, but not well understood. However, it is interesting that pork consumption did not correlate with social class in the Iron Age. At two cities, the Neo-Assyrian city of Ziyaret Tepe (ancient Tushan) and the Phrygian capital at Gordion, pig bones were just as common in upper-class residential areas as in lower-class ones.[47] This might suggest that pig consumption had less to do with social status than with other facets of social identity—at least in certain contexts.

Pig Husbandry and Avoidance in the Southern Levant

Researchers have spilled much ink in the past few decades over the ethnic significance of pig bones in the Iron Age Levant. The general understanding is that Israelites did not eat (much) pork, while their frequent rivals, the

Philistines, did. While that is essentially true, the reality was more complicated than this simplistic identification between pigs and people.

Zooarchaeologists remain divided over their approaches to Levantine pig bones.[48] Complicating matters is the difficulty of identifying ethnicity in the archaeological record.[49] The material record is not always a reliable indicator of people's self-ascribed identities. For one thing, people frequently adopt "hybrid" identities or practices. People also borrow materials, techniques, and traditions from other groups. Additionally, communities can be composed of two or more ethnic groups, which may deposit artifacts in the same archaeological contexts.[50] These problems notwithstanding, several patterns emerge from the zooarchaeological data.

The Philistines occupied the Levant beginning around 1200 BC. Their presence is typically deduced from ceramic styles (including those of cooking vessels), architectural features, and unique linear script.[51] Their settlements also generally included higher proportions of pig bones.[52] Pork consumption also increased over the course of the Iron I period at many Philistine sites.[53] As a result, many have emphasized the importance of pork and other foods, as well as certain cooking styles, to the Philistine identity.[54] That is, there appears to have been a unique Philistine cuisine or foodway in which pork was one component. However, it is important to recognize that pigs at most constituted around 20 percent of the main livestock animals slaughtered for meat in Philistine centers—a far cry, for example, from the pork-dominated diets of Bronze Age Mesopotamian and Egyptian cities (see Table A.3 in the appendix).

Philistine settlements exhibited considerable variability in their pig husbandry. For example, while pork consumption was common in Philistine cities, it appears to have been rare in rural villages, such as Qasile, where pig bones constituted around 1 percent of the livestock remains.[55] Additionally, although Philistines consumed more pigs over the course of the Iron I, they ate considerably fewer after about 950 BC.[56] By the Iron II and Iron III periods, the percentage of pig bones dropped at several key sites—for example, 4 percent at Tel Miqne (ancient Ekron)[57] and less than 1 percent at Ashkelon[58]—although they remained stable at 13 percent at Tell es-Safi (ancient Gath).[59] Some have argued that the Philistines underwent an "acculturation" to local southern Levantine food traditions at this time, adopting many of the local practices. Indeed, the Philistines stopped using certain types of cooking ware in addition to reducing the amount of pork in their diets.[60] Others have argued that this variability over time and space weakens the link between Philistine identity and pigs.[61] If so, archaeologists must reconsider the importance of pork to Philistine identity.

A contentious issue is the impact that Philistines' foodways had on their rivals, the Israelites. Many archaeologists have argued that Philistine pork consumption inspired a pig taboo among the Israelites.[62] While I essentially agree with that assessment, the matter is complicated. First, as I stressed in Chapter 6, we have to think of the taboo as an evolving cultural element. It did not emerge fully formed. The backward projection of the taboo as it exists in modern Judaism and Islam is anachronistic. Second, identifying pig consumption solely in terms of ethnic identity ignores the other reasons that pork may or may not be eaten (the pig principles discussed in Chapter 2), such as the practical benefits of swine husbandry in urban environments. Third, detecting a taboo in the archaeological record is by no means a straightforward endeavor.

The difficulties inherent in the archaeological identification of taboos[63] are perhaps best illustrated by a hypothetical example. If a team of archaeologists in the future were excavating garbage dumps from a modern Midwestern American town, they would not find many dog bones. The reason for the absence would, of course, be that most 21st century Midwestern Americans harbor a taboo on eating dogs. But the excavators would also not find many bones of other locally available animals, like beavers or cranes, to which no specific taboos are attached but nevertheless are not eaten. They might not find many bones of goats, animals that are eaten, but infrequently. To make matters worse, assume a team member found a single dog bone and that the bone displayed cut marks suggesting it had been butchered and eaten. How should the archaeologists interpret that find? Did someone break the taboo? If so, under what circumstances? Perhaps the community was culturally heterogeneous and included a minority population that occasionally ate dogs. Archaeologists would have trouble evaluating these possibilities and sorting taboos from other forms of meat avoidance or nonconsumption. Unfortunately, when it comes to the pig taboo in the Iron Age, many scholars fall back on preconceived and potentially anachronistic notions of what they imagine pork meant to Israelites and Philistines.

The zooarchaeological detection of a taboo is possible, however, through inspection of the data for unusual spatial or temporal patterns.[64] To detect a taboo that was applicable to an entire ethnic group, one should expect the presence of absence of certain species to match up against other potential archaeological signatures (e.g., architectural, ceramic) of ethnic groups.

One need not expect a total absence—rituals can nullify taboos,[65] or certain members in the community may chose not to follow them, even at the risk of social isolation. What is important is a stark and "conspicuous absence"[66] of a food source—one that is eaten at nearby sites. In this sense, when viewed against the backdrop of the Philistine faunal data, there is evidence for a "conspicuous absence" of pig bones at settlements identified with Israelite occupation. At the vast majority of sites, pigs represent 1 percent or less of the livestock remains in the Iron I.[67] This is obviously not a complete absence, and it represents only a slight decrease in patterns already present in the southern Levant in previous periods.[68] But both the extreme infrequency of pig bones and the contrast to Philistine settlements just a few dozen kilometers away are nonetheless striking.[69]

Interestingly, however, like the Philistine faunal data, Israelite pig husbandry patterns changed over time (see Table A.4 in the appendix). In fact, there was an *increase* in pork consumption in the Iron II period at some sites in the northern kingdom of Israel. Namely, at Megiddo and Beth Shean, pig bones represent around 8 percent of the admittedly small assemblages of livestock remains.[70] And on the acropolis of Iron II Tel Hazor,[71] archaeologists found the cranium and vertebral column of a domestic pig—the remains of an animal that had been butchered, the limbs and ribs presumably taken elsewhere for consumption.[72]

One can interpret the uptick in pig husbandry in the Iron II in a number of ways. It might reflect the presence of people who originated elsewhere and had been resettled in the Levant by the Assyrians in the late 8th century BC. Alternatively, the pig bones might be an indication that some people of Israelite ancestry were adopting new traditions. Perhaps the increase in city size in the 8th century BC inspired some Israelites to take up pig husbandry, a form of livestock production ideal for urban environments, despite an existing pig taboo or traditional rejection of pork.[73] One could even read this in light of the biblical authors' railings against the people of northern Israel for their transgressions in 1 and 2 Kings.[74] If so, perhaps we should imagine the pig taboo as a more negotiable feature of Israelite identity, at least until the religious reforms initiated in Judah in the 8th and 7th centuries BC.

Working under an instrumentalist understanding of ethnicity, it is in fact reasonable to suspect that the meanings attached to pigs evolved within Israelite communities over the Iron Age. Philistine pork consumption probably initially inspired a taboo among the Israelites, which evolved out a passive nonconsumption of pork to a more self-conscious avoidance of it in the

Iron I. But pork avoidance would have played a minimal role in identity construction after the 10th century BC, when Philistines themselves largely gave up eating it. At that point, the significance of the pig taboo probably began to wane. To understand why the taboo became codified in Leviticus centuries later, we have to search for factors not only in the Iron I, but also in the cultural and political situation of the 8th and 7th centuries BC.

The Evolution of the Israelite Pig Taboo

Scholars have posited different timelines for the origins of the pig taboo, from the earliest phase of the Iron I through the Babylonian Exile.[75] The various authors in this debate, however, tend to treat the taboo as something that emerged fully formed rather than as something that evolved slowly over the course of the Iron Age, growing like a tree from a sapling until it eventually became enshrined in Leviticus and Deuteronomy as one part of *halakha*. In fact, treating the taboo in this manner can be a way of reconciling many of these previous arguments. Above, I have alluded to this evolution in my review of the arguments that exist among archaeologists and biblical scholars. Here I spell it out more concretely and offer a hypothetical reconstruction of the pig taboo from the 12th through 5th centuries BC.

The nonconsumption of pork was a part of Israelite food practices from the earliest moments of their ethnogenesis. The Israelites' Iron I hilltop villages generally—if not entirely—lacked pig bones beginning in the 12th century BC. By themselves, these data are unsurprising. The reason pork was such a rare feature of the traditional Levantine diet by 1200 BC was probably that the people who settled in the hill country west of the Jordan River did not think to bring swine with them. At least initially, the extreme paucity of pig bones at Iron I Israelite settlements most likely did not reflect an intentional rejection of pork so much as the mostly unconscious continuation of food traditions.[76] On some level, the tribal ideal and the romanticization of sheep- and goatherding may have inspired a glorification of eating ruminant products. But privileging certain types of food need not entail reviling others.

Pork avoidance likely became more active as Israelites came into contact with other peoples who ate pork—namely, the Philistines. The Philistines were originally an Aegean or Cypriot people. Zooarchaeological data from Late Bronze Age Aegean and Cyprus indicate that people ate a significant amount of pork—pig bones comprise typically 20–40 percent of the livestock remains.[77] The Philistines who colonized the southern coast of the southern Levant thus brought with them a food tradition quite distinct from those of their new neighbors.[78]

Food is a potent marker of social identity and the boundaries that groups of people (whether social classes, ethnicities, or gender and age groupings) construct between each other. Food helps shape how we conceive of ourselves and the people we are closest to. Unique food traditions therefore helped Philistines define themselves in a new land among foreign peoples. This might explain why pork consumption increased at some Philistine cities over the course of the Iron I period.[79]

On the other hand, boundaries between people are not static; they are constantly under negotiation. Israelites and other non-Philistine peoples in the Levant adopted some of the Philistine food traditions; for example, ceramic styles crossed ethnic boundaries.[80] Similarly, pigs were not prominent features of daily life in the Philistine countryside, where the mingling of groups may have been more common and the pressure to adopt local Levantine foodways more pronounced. Pork also became increasingly rare in the Philistine diet after the Iron II.[81] This may well be an example of the process of Philistine "acculturation."[82] But it could also indicate that pork was not as crucial to Philistines' self-definition as some scholars have assumed. Pork consumption, remember, was only a small component of the overall tableau of ethnic-based practices among the Philistines. Indeed, centuries later, biblical authors focused their revulsion on Philistine foreskin, not even mentioning pork (e.g., 1 Samuel 18:25–27, Judges 14:3, Judges 16:8).

Whether or not pig husbandry defined Philistine identity to the Philistines themselves, it created the opportunity for Israelites to reflect on their own traditions and markers of ethnicity. It is reasonable to suspect that, as enemies living in close proximity to the Philistines, the Israelites of the Iron I defined themselves in part against this Philistine "other." In all likelihood, they drew on male circumcision, language, dress, religion, and food to distinguish "us" from "them."[83] In this context, the inherited tradition of pork nonconsumption became a more active form of pork avoidance—a taboo.

While Philistines stopped eating much pork in the Iron II, the pig taboo evolved in a new direction. The uptick in pig remains at Iron IIB (ca. 780–680 BC) sites located within the political boundaries of the kingdom of Israel might reflect a growing tendency among city dwellers in the north to abandon the pig taboo, which may no longer have been relevant, in favor of food production techniques suitable for urban environments.[84] Or perhaps the remains indicate the presence of ethnic mixing in these cities.[85] In any case, a change occurred following the dismemberment of the northern kingdom of Israel. Judah, once second fiddle, now found itself the sole independent political entity of the Hebrew-speaking peoples.[86] As their political ambitions to control northern cities grew, the kings and priestly class in Judah may have sought to abolish pig husbandry via religious decree.[87] This was part of their larger political-religious project designed to unite the Israelite peoples and resuscitate the lost glory of an imagined past.

Writing the Taboo

As noted above, Judah in the 8th–7th centuries BC represented a unique political context. To recap: Cultural, religious, and economic friction between the two Israelite kingdoms, as well as the threat of foreign invasion, created a sense of urgency in consolidating power in Judah.[88] Additionally, Judah's expansion into the former territories of the kingdom of Israel after the retreat of Assyrian power in the 7th century inspired the political elite in Judah to forge a nationalist pan-Israelite narrative. The kings and priests of Judah achieved this by cementing Israelite identity around a set of core beliefs and practices that ultimately served to enhance their positions at the head of religious and secular life. Their reforms emphasized "One God, worshipped in one Temple, located in the one and only capital [Jerusalem], under one king of the Davidic dynasty."[89]

To forge a new identity that would serve as the foundation of an expansive Judahite state capable of resisting external threats, the biblical authors needed an origin story that was both believable and sufficiently glorious. They felt the need to stress that their ancestors' might ultimately derived from the power of their god, to whose cult the priesthood was devoted. While embellishing truths and, perhaps, inventing others, the biblical authors probably relied on reframing existing folk stories and traditions. Using existing traditions provided an air of legitimacy to the authors' claims. Weaving them together, they depicted their ancestors as paternalistic, pastoral, pious, and, ultimately, pigless. These traditions formed the core of an ideal life, one that moored the people to the will and power of Yahweh. At the same time, they adorned their ancestors with a melodramatic degree of heroism. In addition to identifying the Israelite people as precious to the most special (and later, *only*) god, the biblical authors connected their patriarchs to great ancient cities, such as Harran and Ur in northern and southern Mesopotamia. They scripted a drama of rebellion and escape from one of the greatest Near Eastern powers (Egypt). These stories, perhaps inspired by the tumultuous events at the end of the Late Bronze Age, established the magnificence of the Israelite people. But the biblical authors were at pains to show that this past glory was dependent on the proper behavior of a people that, almost comically, kept giving in to the temptation to flaunt the rules of tradition.

Food represented an important set of behaviors. The biblical authors spent much energy detailing food laws and taboos, which they believed were crucial to reestablishing the Israelites' past glory. Since the Bronze Age, ruminant meat and milk were the main forms of animal protein in the Levant. This traditional Levantine diet fit well with the cultivated nostalgia for a nomadic pastoral ancestry. But if food connected the Israelite people to their god and his plan for his people, food traditions would have to be written as absolute laws and not simply celebrated as accomplished facts. Thus the authors decreed that, among the mammals, the only animals fit to eat were ruminating ungulates, the animals owned and exploited by pastoral nomads and representing a category in which pigs did not fit.[90]

The existing, but by now fading pig taboo lent itself to this project. The taboo was another piece of tradition that the authors could use to support their claims about the past, the power of their god, and the legitimacy of the monarchy. It is unclear exactly what the taboo meant to biblical authors and to the people of Judah in the 8th and 7th centuries. They may have perceived the lingering memory of pork's association with an ancient enemy, which perhaps persisted in oral traditions. If so, the pig taboo still possessed power, even if transgressions against it were becoming more commonplace. In any case, the authors were able to rely on the fact that most Israelites probably retained an inkling that pork was not part of their traditional diet—that there was something *wrong* about it.

The authors may also have drawn upon existing cultic or religious anxieties about pigs—perhaps pigs' association with chthonic rituals or magic[91] or as potential pollutants of sacred spaces. In any case, the biblical authors likely mixed the two pork taboos together—one derived from ancient Israelite-Philistine interaction and one derived from the ritual associations that pigs developed in the Late Bronze Age. Both added a sense of credibility and power to the newly-codified taboo.

While labeling pigs abominable was nothing new in the Iron Age, the priests of Judah made a revolutionary move by applying the taboo not only sacred places and people, but to *all* the children of Israel at all times (e.g., Leviticus 11:2). Only food fit for the altar was now acceptable on the table.[92] This democratization of ritual purity and its extension into everyday life facilitated the transition to monotheism by providing a constant ritual connection between Judahites and their one and only god.[93] Ultimately, this connection, while focused on Jerusalem, could be forged anywhere. The *mitzvot* made it possible for the Jewish religion to flourish in a Diaspora setting.

We should pause here and note that, among the hundreds of *mitzvot*, the avoidance of pork was just one (Leviticus 11:7, Deuteronomy 14:8). The texts also prohibit the consumption of reptiles, fish lacking scales and fins, several (all?) birds of prey, camels, rock hyraxes, and other mammals that do not possess both hooves and a ruminating stomach. Like many other animals, pigs were described as impure (*tame*),[94] and one is instructed not to eat them or even touch their dead carcasses—although Leviticus stops short of banning the raising or handling of pigs or other impure animals.

In the end, in their desire to resuscitate a glorious pastoral past, the biblical authors inspired a revitalization of foodways. It is perhaps not a coincidence that food represented a way for the authors to draw upon the traditional tribal ideal in a way that did not directly confront or contradict kingly power and social hierarchy. While many passages of the Bible articulate an egalitarian ideal, the authors were careful to avoid undermining their own positions of power and the institution of the monarchy. Instead, much like neoliberals today, they focused on moralizing personal behaviors. An erosion of values, and not the political machinations of those in power, was to blame for any suffering that came upon the people of Israel, including conquest by foreign armies. This habit-based revitalization movement created an opportunity to breathe new life and meaning into the pig taboo.

It would be a mistake to conclude that the biblical authors conspired to invent a tradition and use it to trick the populace of Judah. Rather, it is important to recognize how self-deception is a powerful force in mythmaking and political projects. The biblical writers probably did not fully understand the pig taboo and other traditional food habits they turned into law. They simply believed, or convinced themselves to believe, that their ancestors followed a nomadic pastoral way of life and, through the foods they ate, were connected to a special god who had a special plan for them.

The biblical authors attempted to understand food in terms of the relationship it forged between people and a god. If eating was a sacred act, and

Yahweh an all-powerful creator-god, the food rules must be legible in creation itself. Thus the authors sought justification for the food taboos on physiological grounds. Essentially, they attempted to explain the meaning of these taboos in animals' essences, the unique forms that Yahweh gave different beings at the creation of the universe. The explanations for the taboos the biblical authors penned were therefore reflections of the truths to which they aspired.

When initially written, the pig taboo played a minor role in the consolidation of the people of Judah's identity. The connection between an existentially threating "otherness" and pork had waned with the changing Philistine diet, and even the Israelites in the northern cities who ate pork did so infrequently. But the pig taboo would have certainly resonated in the Diaspora. Jews living in Babylon, Egypt, and other places during the Babylonian Exile and Persian period daily confronted other people eating pork; indeed, it is likely that this animal was the most frequently consumed of all those banned by Leviticus and Deuteronomy. The stark contradiction between biblical commandment and the food habits of Jews' host communities likely amplified the anxieties of forging a Jewish identity, of sensing themselves as a separate and superior people.[95] It is perhaps for this reason that texts dating to the Persian period supply some of the only biblical passages that specifically condemn people consuming pork (Isaiah 66:3–17).

Thus, by the Persian period, the Jewish people embraced a taboo on pork consumption that they self-consciously connected to their ethnic identity and that was written down in unambiguous terms. Whatever initial associations it may have had, by the 4th century BC avoiding pigs was a part of how Jews reproduced their own sense of self and connection to their deity. Yet, at least on paper, the pig taboo held no special status relative to the other *mitzvot*. This situation would change in the Hellenistic period, when Jewish people in the southern Levant were once again faced with a pork-loving political enemy and rival ethnic group (Chapter 8).

Pig Taboos in Other Parts of the Near East

Beyond the Israelites and Jews, other cultures across the Near East persistently held negative attitudes toward pigs. In Chapter 5, we saw that texts dating to the Late Bronze Age indicated that pigs, as well as dogs, were capable of polluting temples. These specifically religious taboos helped separate the sacred from the profane. While they continued into the Iron Age, there is no evidence that they were ever applied to an entire ethnic group.

Mesopotamian texts clearly indicate the persistence of injunctions against pigs in temple contexts. The references are mostly found in popular sayings and aphorisms, the so-called Babylonian wisdom literature. One tablet dating to 716 BC proclaims:

> The pig [.]. has no sense;
> lying [in. .] . . he eats his food
> They do not [say,] "Pig, what respect have I?"
> He says [to] himself "The pig is my support!"
> The pig himself has no sense;
> . [. . .] corn [. .] in the oil pot.
> When at leisure [. . .] he mocked his master,
> His master left him [. . .] the butcher slaughtered him.
> The pig is unholy [. . .] bespattering his backside.
> Making the streets smell. polluting houses.
> The pig is not fit for a temple, lacks sense, is not allowed to tread on pavements.
> An abomination to all the gods, an abhorrence [to (his) god,] accursed by Šamaš.[96]

This passage provides some interesting if rather ambiguous explanations for why pigs were considered polluting—they apparently have "no sense," smell bad, and are reviled by Šamaš, the sun god. However, the passage also demonstrates that, while people considered pigs abominable, they continued to raise and eat them. Thus, it reveals that pigs were encountered on streets and near houses, probably scavenging food and urban waste. The passage also indicates that people brought their pigs to the butcher to be slaughtered.

Other textual and iconographic data indicate that pork remained on the menu and continued to play some roles in ritual life. While there is little evidence for institutions raising pigs, Assyrian kings at least occasionally provisioned their armies with pork.[97] Meanwhile, boar-hunting scenes meant to extol the masculine prowess of princes were depicted on seals in the Persian period.[98] Pigs also remained associated with the demon-goddess Lamashtu,[99] and an Assyrian text,[100] probably dating to the 7th century BC, describes the sacrifice of a pig to the "Mistress of Babylon" during the spring Akitu festival.[101]

Beyond Mesopotamia, textual evidence reveals that other Iron Age societies continued to practice, or adopted, taboos on pigs in certain religious settings. The Egyptian pig taboo, introduced in Chapter 6, provides a good example.[102] And while we must remember that our primary source of evidence for this taboo is Herodotus, a 5th century Greek historian who probably did not understand the nuances of Egyptian culture, later works corroborated its existence and defined it more explicitly as one applying only to priests (e.g., Sextus Empiricus, *Outlines of Pyrrhonism* 24.223).

But like the Babylonians, Egyptians continued to sacrifice pigs in certain contexts. According to Herodotus, pigs were sacrificed once a year on a full moon to the lunar deity (*Histories* 2.47). The 2nd–3rd century AD writer Aelian backed up this claim (*On Animals* 10.16), citing the now lost works of the 3rd century BC Egyptian historian Manetho. Additionally, tomb drawings dating to the Late Period (664–332 BC) occasionally depicted pigs being ferried away on boats on Judgment Day, indicating the removal of sins for the purification of the soul and reflecting the long-standing tradition of pigs as ritual substitutes for humans.[103]

It is something of a pastime in Egyptology to speculate on whether the Jewish pig taboo derived from the Egyptian one.[104] For this there is no evidence. Not only is there no evidence to indicate that an ethnic-based taboo ever applied to ancient Egyptians, but also the date of its first reference is late. Herodotus, the first to unequivocally identify a pig taboo in Egypt, wrote his *Histories* around two centuries after the composition of Leviticus. In fact, one has to wonder if Diaspora Jewish communities may have inspired a pig taboo in Egyptian religion. We know that a thriving Jewish community resided at Elephantine in the 6th century BC and that Egyptian and other Near Eastern religions were certainly not averse to syncretism. Moreover, although Diaspora Jews and Egyptians were often at odds in the Classical period,[105] there is evidence that Jewish rituals percolated into Egyptian magical rituals.[106] The so-called Greek Magical Papyri are a collection of charms and curses dating to the 2nd century BC through the 5th century AD that reflect an amalgamation of Jewish, Egyptian, Greek, and Roman beliefs.[107] They contain a number of incantations in Hebrew or citing Jewish traditions, including one that prohibits the person on whom the spell is cast from eating pork.[108]

Another possible example of a religious pig taboo—one that may also have been influenced by Jewish tradition—was reported at the city of Comana in the Cappadocia region of Anatolia. The evidence derives from a passage written by Strabo (63 BC–AD 25) in his *Geography*. Strabo stated that the people of Comana had banned pork and pigs not only from the sacred precinct, but also from the whole city—that is, until a certain 1st century BC warlord named Cleon of Gordiucome attacked Comana and, to humiliate its citizens, committed sacrilege by eating pork within its walls (*Geography* 12.8.9). While Strabo's story is intriguing, there are, unfortunately, no additional archaeological or textual data to corroborate it.

A final, and quite problematic, example of a pig taboo in the Iron Age concerns the Phoenicians, coastal traders living in modern-day Lebanon. Ostensible evidence for the Phoenician pig taboo derives mainly from a very late source: the 3rd century AD philosopher Porphyry of Tyre.[109] Porphyry advocated vegetarianism and the ethical treatment of animals, but he also contemplated taboos on meat. When he mentioned pigs, he wrote that "Phoenicians, however, and Jews, abstain from [pork], because, in short, it is not produced in those places" (*On Abstinence* 1.14). This seems to suggest that pork avoidance among the Phoenicians and Jews was a passive custom, not an active proscription. We know that was not the case for Jews, especially by the time of Porphyry's writing, but perhaps he was projecting his own experiences as a Phoenician. In fact, in a later passage, he writes:

> [T]he Syrians indeed will not taste fish, nor the Hebrews swine, nor most of the Phoenicians and Egyptians cows; and though many kings have endeavoured to change these customs, yet those that adopt them would rather suffer death, than a transgression of the law (*On Abstinence* 2.61)

If an injunction against pork existed in Phoenician tradition, it would be unusual for Porphyry not to mention it in this passage. It is more likely that the Phoenicians avoided eating pork, not because it was taboo, but for the sake of passively maintaining a tradition.

Zooarchaeological data also offer a perspective on a possible Phoenician taboo on pigs. As in other parts of the Levant, Phoenician sites generally contained very small numbers of pig bones.[110] But pigs were far from absent at Phoenician colonies around the Mediterranean. At 10th–9th century BC Utica in modern-day Tunisia, excavators found a large pit with the remains of feasting debris that included the bones of pigs,[111] and at Carthage, pig remains increased from around 5 percent in early phases to around 40 percent in later phases of the city's history.[112] Both of these lines of evidence cast doubt on the existence of a Phoenician pig taboo.

In sum, there is good evidence for pig taboos in religious contexts in Mesopotamia and Egypt in the Iron Age. This suggests that many Near Eastern peoples were in agreement that pigs had certain properties that made them dangerous or powerful and therefore unfit for temples. However, in none of these cases is there clear evidence for a taboo outside of strictly religious contexts. These other pig taboos had nothing to do with ethnic identity, and people remained content to raise pigs and eat pork. The Israelite/Jewish pig taboo was different. Building on an earlier ethnic taboo, it served the remarkable function of democratizing ritual purity for all Israelites at all times as part of a covenant binding a people to their god.

The Genetic Turnover

Taboos have monopolized the bulk of scholarly interest in pigs in the Iron Age, but other important changes were occurring. Archaeogeneticist Greger Larson and colleagues[113] published evidence indicating that Near Eastern pigs, which initially descended from wild boar domesticated in the Neolithic, were replaced by ones whose ancestors were European wild boar. Recall that in Chapter 4, Anatolian farmers brought swine into Europe in the 7th and 6th millennia BC, where those animals bred with local wild boar. By the 4th millennium BC, most domestic pigs in Europe could trace their ancestry to European, and not Near Eastern, wild boar.[114] Unexpectedly, Larson and colleagues[115] found evidence that a similar genetic replacement took place in the Near East sometime before or during the Iron Age. By the later part of the Iron Age, as depicted in Figure 7.1, the bulk of Near Eastern domestic pigs' ancestry derived from European wild boar.[116]

Studies conducted since Larson and his team published their findings have added new details to the picture. Prior to 2017, researchers were working almost exclusively with mitochondrial DNA, genetic material that is inherited only through the maternal line. Recent studies, however, have largely corroborated the pattern observed in mitochondrial DNA with nuclear DNA, which is inherited from both sets of parents and is therefore a more reliable indicator of ancestry.[117]

A more comprehensive treatment of mitochondrial DNA in pigs in the Near East, utilizing genetic data from 192 pig bone specimens from the Neolithic to the Medieval period, identified four prehistoric lineages: Y1, Y2, Arm1T, and Arm2T.[118] These lineages clustered geographically, with

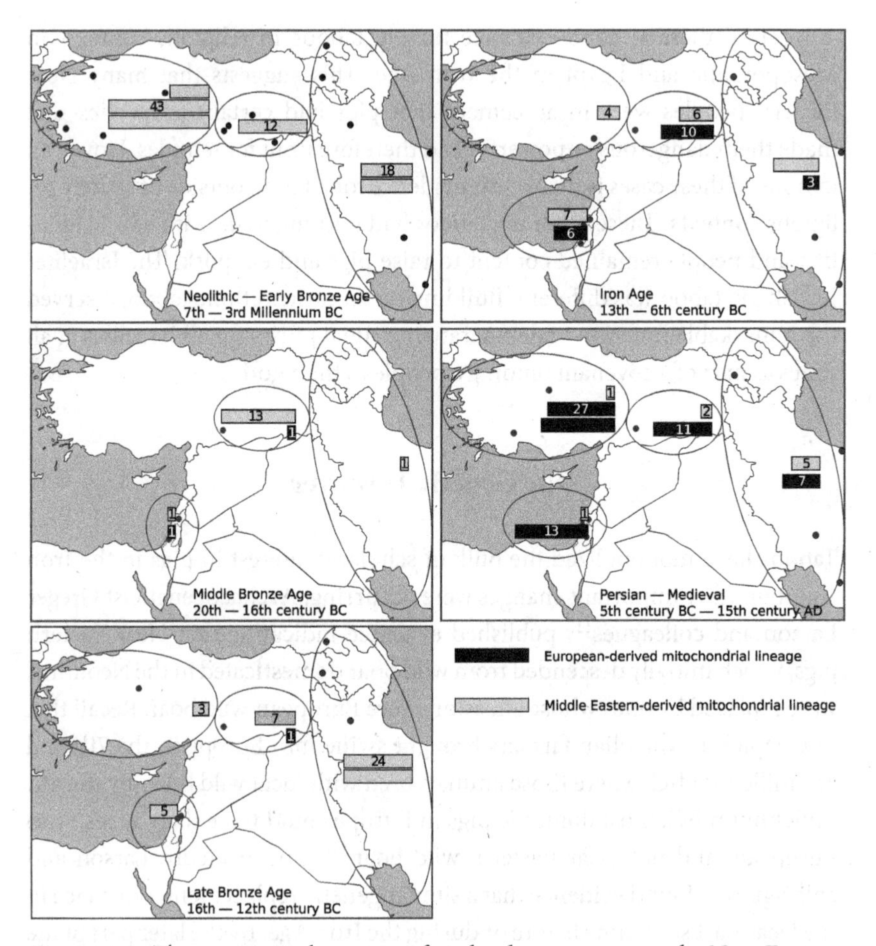

Figure 7.1. The genetic replacement of pig haplogroups across the Near East. Four regions compared: Anatolia, Levant, N. Mesopotamia/S. Anatolia, and Iran/Caucasus. Points indicate locations of sites.

Y1 and Y2 having been more common in western and central Anatolia and Arm1T and Arm2T more common in eastern Syria, the Caucasus, Iran, and Iraq. All domestic pigs and wild boar sampled fell into these four groups until the Middle and Late Bronze Ages, when a small number of European-derived individuals appeared at the site of Lidar Höyük in southeastern Anatolia.[119]

Similarly, in a study focused on the Levant, Meirav Meiri and colleagues[120] detected a rapid replacement of local Near Eastern pig stocks in the Iron Age. In addition, they found that all of the modern wild boar in the southern

Levant that they sampled had European ancestry,[121] which suggests that feral or extensively managed pigs carried European genes into the wild. Meiri and colleagues[122] connected the introduction of European haplotypes to the arrival of the Sea Peoples, chief among them the Philistines.

The connection between Philistines and new pig genes is, however, problematic. For one thing, Meiri and colleagues could not actually find a single pig specimen with European-derived mitochondrial DNA at an Iron Age I Philistine site.[123] Even more problematic was the fact that the researchers found a specimen at Ashkelon with European ancestry in *Middle Bronze* layers. That is, the earliest European pigs appeared to predate the arrival of the Philistines by centuries. Meiri and her team,[124] however, suggested that the specimen could be intrusive from later levels. While possible, offering a post hoc explanation for a piece of data that contradicts one's hypothesis is a convenient and quite problematic approach. Indeed, the presence of European-derived pigs in the Middle and Late Bronze Age levels at Lidar Höyük[125] gives support to the hypothesis that a small number of European-derived pigs began to infiltrate Near Eastern swine stocks well before the beginning of the Iron Age.

European-derived pigs probably began trickling into the Near East in the early 2nd millennium BC or perhaps earlier. But Meiri and colleagues are correct in identifying the Iron Age as a key period. Something happened in the Iron Age that affected how pigs in the Near East passed on their genetic material to succeeding generations, accelerating the genetic turnover to the point that Near Eastern-derived lineages were uncommon by the beginning of the Classical period.

What exactly caused the rapid turnover in swine ancestry is unclear. One possibility is that European-derived pigs were better suited as livestock animals. A study of modern wild boar from Italy might hold the answer.[126] It compared two European genetic lineages that Larson and colleagues[127] had labeled "A-side" and "C-side," with A-side individuals becoming more prevalent over time than C-side ones. Wild boar belonging to the A-side lineage exhibited faster growth rates and had average adult body weights about 7.63 kg larger than C-side individuals. It is not hard to imagine that the bigger and faster-growing pigs were more successful as livestock over the long term. Might something similar have played out in the Near East? Perhaps, but no comparable comparison has been made, as yet, between Near Eastern lineages and their phenotypes and those of European lineages. The answer will have to await further research.

The Iron Age thus saw major changes in pigs in the Near East. On a genetic level, for reasons that remain unclear, the local pig lineages that had dominated the Near East since their domestication in the Neolithic were replaced by European ones. On a cultural level, people continued to raise pigs and eat pork, but there were notable declines in pig husbandry in many parts of the Near East, especially in Egypt. In some religious contexts, people across the Near East perceived pigs as impure, although people continued to sacrifice pigs in other ritual contexts .

The Torah's taboo on pigs was unique. It drew upon a taboo that probably had its origins in the ethnic conflict between Israelites and Philistines. While this taboo waned over the 10th–8th centuries BC, the biblical authors revitalized it during a period of political expansion and state-inspired ethnogenesis. In building an idealized history, the biblical authors found existing southern Levantine food traditions, including the pork taboo, particularly relevant. These already-at-hand traditions exemplified a glorious pastoral ancestry and were mobilized in the writing of the biblical tale.

The biblical authors sought explanations for the taboos they wrote down within the logic of their creator-god, Yahweh. Accordingly, they emphasized animal physiology. In doing so, the biblical authors naturalized existing traditions and used them to construct a wall of taboos around the foodways most redolent of a glorious pastoral ancestry. The food laws in Leviticus and Deuteronomy thereby underscored two of the most important themes of Genesis: the creation of the universe according to a divine and perfect plan and the special place of the Israelite people within that universe.

The puritanical reforms that took place in Judah in the late 8th and 7th centuries BC, especially under Josiah, represent a watershed moment. The writing down of the taboos and other *mitzvot* in the Torah, whose texts were considered sacred, was critical. It made them resistant to change. In the Exilic and post-Exilic periods, regular reading and recitation of the Torah created among Jews a state of perpetual ritual awareness, one in which pigs were a defiling element. Thus, the injunctions against eating pork and other foods became part of a central code of behavior applicable to each individual and necessary for the reinforcement of his or her Jewish identity.[128] Even today, the *mitzvot* make Judaism a religion and ethnic identity defined by daily practices, what religious scholars call *orthopraxy*.

Yet the abstention from pork was but one of hundreds of practices that defined a Jew. It held no special place. This would change when Jews came into contact with Hellenistic and Roman empire builders, who inadvertently helped transform the pig taboo into one of the strongest in the world, while at the same time laying the groundwork for very divergent thoughts on pork in Christianity.

8
Clash of Cultures in the Classical Period

Already in the Iron Age, Greek and Near Eastern cultures had begun to exert considerable influence on one another. Spurred on by the burgeoning pan-Mediterranean trade network, Greek colonists, mercenaries, and merchants began settling in the Near East and initiated a blending of Greek and Near Eastern cultures.[1] Cultural hybridity grew even more pronounced after Alexander's campaigns in the 4th century BC. Greek religious ideas were combined with local Near Eastern theologies, Greek artistic and architectural styles were hybridized with local ones across the region, and people adopted Greek as the *lingua franca*, largely replacing Aramaic.[2]

After Alexander's death in 323 BC, his generals squabbled over the empire and broke it into three parts. In the Near East, the Seleucids controlled Persia, Mesopotamia, and Anatolia. They vied for control over the Levant with the Ptolemies, who were based in Egypt. But Rome's growing power spelled the end of these dynasties. The Romans took control of Egypt, Syria, the Levant, and Anatolia by the 1st century BC. Referred to as the Byzantine Empire after the 4th century AD, Roman imperial power dominated the western half of the Near East from Constantinople (modern-day Istanbul) even after Rome itself was sacked.[3] Yet Roman power in the Near East did not go unchallenged. In eastern Anatolia and northern Syria, Rome contended with powerful Armenian kings. In Iran, the Parthians (ca. 238 BC–AD 224), who had wrested control of the region from the Seleucids, presented a continual military threat. In fact, the centuries of war between the Romans and Parthians, continued by their successor empires, the Byzantines and the Sassanians (AD 224–AD 661), would exhaust both states and lay the groundwork for the Arab Conquest in the 7th century AD.[4]

The Classical period also witnessed profound changes in the sphere of religion. Of great influence was Zoroastrianism, a uniquely monotheist religion native to Iran that predominated under the Parthian and Sassanian Empires. Syncretic offshoots, such as Manichaeism and Mithraism were

Evolution of a Taboo. Max D. Price, Oxford University Press (2020). © Oxford University Press.
DOI: 10.1093/oso/9780197543276.001.0001.

popular, if frequently persecuted, throughout the ancient world. In this context of religious profusion and revitalization, Judaism also flourished. Building on earlier dispersals (diasporas) to Egypt and Mesopotamia, large Jewish communities thrived throughout the Mediterranean as well as in Mesopotamia, Anatolia, Upper Egypt/Ethiopia, and Arabia.[5] While Jerusalem remained the spiritual hub, synagogues at archaeological sites like Dura-Europos in northern Syria bear witness to the local character of these communities. Similarly, the intellectual centers in Babylonia and Jewish-led political states such as the 5th century AD kingdom of Himyar in modern-day Yemen attest to the regional diversity of Judaism in the Classical period.[6]

The success of a more globalized Judaism fed on its unique religious, cultural, and political ideology, while articulating with and incorporating elements of Greco-Roman culture and philosophy.[7] The Hebrew Bible was translated into Greek in the 3rd century BC (the Septuagint), and several Jewish scholars, perhaps most notably Philo of Alexandria (20 BC–AD 50), wrote in Greek and attempted to merge Greek and Jewish philosophy.[8] But Judaism also came into conflict with Greco-Roman imperial ideology, as well as the Greek and, especially, Roman love of pork.[9] The antagonism of some of the more radical Jews toward Greek and Roman cultural and political hegemony set off a series of major revolts in the Levant and beyond. These included the Maccabean Revolt (ca. 167–160 BC), the Jewish-Roman War (AD 66–73), the Diaspora Revolt (AD 115–117), and the Bar-Kochba Rebellion (AD 132–135).[10] These conflicts cost hundreds of thousands, perhaps millions, of lives. They ultimately resulted in the destruction of the Second Temple in Jerusalem and the decimation of Jewish communities throughout the Roman Empire.

Judaism in the Classical world also inspired several revitalization movements. Christianity arose in the midst of Roman-Jewish conflict in the southern Levant, but it quickly spread throughout the empire. Although heavily repressed at first, it eventually became the official religion of Rome under Constantine (AD 306–327).[11] By the end of the Classical period, Christianity was the dominant religion in the western half of the Near East. This had an important impact on pigs. Christian leaders, especially Paul of Tarsus, advocated the elimination of dietary taboos in an effort to direct their followers' *orthodoxy* (believing purely) as opposed to *orthopraxy* (acting purely).

Pigs in Greek and Roman Cultures

Pigs were an important feature of Greek and Roman life. They were raised in urban and rural settings, and members of all social classes enjoyed pork. Swine also featured prominently in rituals and feasts, a situation that contrasted with the animals' more complicated roles in religious and celebratory activities in the Near East during the Iron Age (Figure 8.1). Northern Mediterranean cultures' decidedly more pork-friendly attitudes would set up

Figure 8.1. Marble funerary stela for a pig killed in an accident en route to a Dionysia festival. Pella, Greece, 2nd–3rd century AD. The inscription reads: "I, the Pig, beloved of all, a four footed youngster, am buried here. I left the land of Dalmatia, when I was given as a gift. I stormed Dyrrachion and yearned for Apollonia, and I crossed every land on foot, alone and invincible. But now I have departed the light on account of the violence of the wheel, longing to see Emathia and the wagon of the phallic procession. Now here I am buried in this spot, without having reached the time to pay my tribute to death." Translation by Onassis Cultural Center, New York, "A World of Emotions."

conflicts with some of the Near Eastern peoples conquered by the Hellenistic and Roman states.

Economic Roles

After their introduction in the Neolithic, pigs were major components of northern Mediterranean agricultural systems. In Greece and Italy, zooarchaeologists have shown that pig bones make up around 10–30 percent, and sometimes more, of the livestock in faunal assemblages from 6000 to 1200 BC.[12] But urbanism propelled swine management to new heights.[13] Cities first appeared in the northern Mediterranean with the Minoan and Mycenaean palatial states in the middle to late 2nd millennium BC. They became even more prominent after the 7th century BC and the emergence of the city-state (*polis*) in Greece, southern Italy, and Tuscany.[14] The Etruscan site of Poggio Colla provides a good example. Between the 8th and 3rd centuries BC, a period corresponding to the fluorescence of urbanism in the region, pigs increased at the site from 29 percent to 53 percent of the main livestock species.[15]

Pigs became even more important during the Roman period.[16] Swine provided an affordable source of meat for the growing urban masses, one that could be raised within cities.[17] To keep the poor fed, the Roman state also supported large-scale swineherding operations, which took advantage of the hardwood nut-bearing forests prevalent in Italy and other parts of Europe. As a result, in many urban centers throughout the 1st and 2nd centuries AD, pigs frequently made up 70–85 percent of livestock taxa.[18] Pork consumption also followed Romans and Greeks into the Near East. Their colonists and soldiers ate pork in frequencies not seen in the Near East since the Early Bronze Age.[19] Roman military commissaries, in particular, relied on pork, as excavations from dozens of forts installed throughout the Roman Empire have revealed.[20]

If pork was a staple for the poor, it was a delicacy for the wealthy. The Greek and Roman elite distinguished their haute cuisine from that of the lower social orders by elaborate and sometimes exotic preparation techniques, if the recipes that have survived to the present day are any indication. Many of them, in fact, test the border between animal cruelty and epicureanism. For example, "miscarried womb" (*vulva eiectitia*), a delicacy celebrated by the Greek philosopher Plutarch (ca. AD 46–120) and the Roman poet Martial (ca. AD 38–104), was prepared by beating a pregnant sow until she miscarried and then cooking her unborn litter.[21] Roman banquet-goers also prized sow's udder (*sumen*), which allegedly had a delightful milky taste.[22] Petronius's 1st century AD fiction *The Satyricon* describes another dish (or perhaps a fantasy of one), the "Trojan pig."[23] Conceptually similar to the modern turducken, the Trojan pig called for a hog to be slaughtered, gutted, and stuffed with sausages before being sewn back together, cooked, and served. When diners cut open the roast pig, its edible "intestines" spilled out.

Adventurous readers can try their hand at these and other recipes—if they dare. Archaeologist and amateur chef Eugenia Salza Prina Ricotti[24] has published a cookbook on Greek and Roman cuisine, featuring several enticing pork-based recipes such as "rose and brain pudding" and "stuffed suckling." Those unnerved by some of the delicacies described by Roman writers need not worry. The author does not include recipes that flagrantly violate modern animal cruelty standards.

Greek and Roman writers also celebrated swine husbandry. Early Greek and Roman natural historians and agricultural scientists wrote at length about pigs for an educated elite audience, the owners of manorial estates and large herds of livestock. Varro (116–27 BC), for example, advised his readers on the proper way to raise pigs. He suggested techniques for identifying good breeding stock (boars and sows should be in good physical condition, be born from litters with large numbers of piglets, and come from a region where fat swine are common; *On Agriculture* 2.4.4) and when to wean piglets (before two months, especially if one wants to sacrifice them; 2.1.32). Writing a century later, Columella (AD 4–70) advised his readers on when to castrate boars (six months or three to four years if used for breeding; 7.9.4–5) and how to keep sows in good health (provide them cooked barley and clean their sties regularly; 7.9.13–14).

Textual references and a bit of zooarchaeological sleuthing have also revealed that there were two distinct breeds in Roman Italy. For example, Columella (*On Agriculture* 7.9.1–3) described a small, black, and bristly breed that was herded in forests and fed on nuts, and a large, white, and hairless breed that was raised in sties. Zooarchaeologists have attempted to detect these different breeds from archaeological remains. Pig limb bones recovered from Roman period sites seem to fall into two groups, a larger group of animals that measured about 60–75 cm at the shoulder and a much smaller group that stood at around 80 cm. Hypothetically, the smaller pig bones might have belonged to the black bristly breed, which Roman writers identified as the main source of food for the commoners. These were the animals herded every autumn in hardwood forests to supply the poor with pork. The larger white hairless breed was less common in the archaeological record, but may have been more highly valued as a sacrificial animal.[25]

The Greek and especially Roman agricultural elite valued pigs in ways that their counterparts in the Near East did not. Large-scale pig husbandry and pork-curing operations helped create a market for pork, turning it into a commodity that could be transformed into wealth. The environmental conditions of the northern Mediterranean supported these endeavors. The nut-bearing forests allowed massive numbers of swine to be fattened before the winter, when colder conditions helped prevent spoilage during curing. Thus, from the outset, the role of pigs in Greek and Roman economies was different from that in much of the Near East.

Evidence that, in sharp contrast to peoples of the Near East, Greeks and Romans perceived pigs as animals translatable into wealth is pervasive. Palace texts from the Mycenaean period (Late Bronze Age) indicate that, from early on, pigs were incorporated into systems of agricultural wealth in the northern Mediterranean region.[26] As market economies expanded in the Roman Empire, pork products became even more valuable. Pigs and cured meats were produced on a large scale for profit,[27] were taxed,[28] and could even be used to pay off debts.[29] The value afforded to pork and its consistent demand in Roman markets made pig breeding a major source of income. Columella (*On Agriculture* 7.9) even advised readers living near towns to wean their piglets as early as possible so as to enable the sow to breed more quickly and thus increase their profits.

Roman authorities attempted to regulate this market in pork. In part, their goal was to keep the masses fed to prevent unrest. Tens of thousands of pigs, which were then fattened in hardwood forests and distributed to the masses, were doled out annually to the lower classes by the Roman state.[30] But controlling the pork market was also a part of other forms of economic regulation. For example, in an effort to create greater currency stability, the emperor Diocletian issued an edict in AD 301 setting the maximum prices for meat. It listed several types of pork but set its general price at 12 *denarii* per pound—higher than the price of mutton or beef at 8 *denarii* per pound.[31] Other cuts of pork were also regulated; the price of pigs' feet was set at a maximum of 4 *denarii* per pound, and fattened hog's liver at 16 *denarii* per pound. At the top of the list, sow's vulva and sow's udder were each set at 24 *denarii* per pound.[32]

Ritual and Cultural Significance

Beyond their economic value, pigs played important ritual roles in Greek and Roman cultures. In Greece, the archaeological recovery of burnt juvenile pig bones from the Mycenaean (Late Bronze Age) sanctuary of Ayios Konstantinos indicates that piglets long served as sacrificial victims in Greek religion.[33] By the Classical period, Greeks regularly sacrificed pigs along with sheep, goats, and cattle to their gods.[34] The Thesmophoria rituals, discussed in Chapter 5, offer another example. In this case, piglets helped symbolize and reconcile the dualism of life and death, negotiating humans' position between the power of the fertility goddess Demeter and that of the chthonic deity Hades.[35]

Piglets also served in rituals of purification. If a priestess of Demeter had been ritually profaned by entering the house of a dead person or walking into a place where a woman had recently given birth, she would slit a piglet's throat and drip the blood in a circle around her body to absorb the pollution.[36] These examples, offer remarkable parallels to the uses of pigs in purification rituals in the Bronze Age Near East, which were discussed in Chapter 5..

Roman rituals appear to have focused more on sacrificing pigs to ensure fertility and prosperity than on purification and substitution. Figure 8.2 shows Eros, god of love and sex, straddling a pig. Pigs were also a key component of the *suovetaurilia*—a portmanteau of *sus* (pig), *ovis* (sheep), and *taurus* (bull). These were sacrifices made specifically to Mars in order to purify the land.[37] Pig sacrifice was also central to the ancient Latin marriage ritual—so much so that Varro (*On Agriculture* 2.4.9–10) explained that the word *porcus* (pig), which was slang for a young woman's vagina, was simply a metonymic extension of the sacrifices of piglets intended to ensure fertility.

The ritual importance of pigs in Greek and Roman cultures paralleled their roles in mythology and literature, a situation that contrasts sharply with general lack of pigs in the myths and stories in the Near East.[38] For example, swine play prominent roles in *The Odyssey*, one of *the* core pieces of Greek literature. It is Eumaeus, Odysseus's faithful swineherd, who is the first to welcome the hero back to Ithaca. Eumaeus feeds and houses his master. He assists him in slaughtering the suitors who have invaded his home (*Odyssey* 14). In another example, in book 10, the deity Circe, having hosted Odysseus's crew, lays out a splendid feast. But her hospitality is a trap; upon eating the food, Odysseus's men metamorphose into pigs. Just as pigs can serve as substitutes

Figure 8.2. Terracotta statuette of Eros, god of love and sex, astride a pig. Southern Italy, 3rd century BC. Height 11.1 cm.

for humans in rituals, the gods can transform people into pigs. Finally, let us not forget that it is a wild boar that gores a young Odysseus (*Odyssey* 19; see Chapter 3, this volume). Though coming close to emasculating him, the boar provides the young hero his first battle scars and test of manhood.

Wild boar embodied the fierce, powerful, and fearless attitude that Greeks and Romans equated with masculinity on the battlefield. In *The Iliad* (17.323–326), Homer likens Ajax to a wild boar to describe the hero as he crashes through the Trojan lines. Similarly, Mycenaean warriors decorated their helmets with boars' tusks, perhaps to evoke the power and swiftness of swine. Several Roman legions also used the boar as their emblem, and boar hunts were a quintessential activity of the warrior elite.[39] But as much as the wild boar was a symbol of power, it was also one of chaos. For example, in two myths—the Caledonian Boar and the Erymanthian Boar (Ovid, *The Metamorphoses* 8 and 9)—desperate farmers call upon heroes to kill oversized swine that are ravaging their fields and terrorizing their families. Delivering the Erymanthian Boar alive was one of Herakles's twelve labors (Figure 8.3).

Perhaps the most unique role that pigs played in Greek and Roman cultures was on the battlefield. Swine were not merely symbols of aggression and chaos; they were actually deployed as weapons. Soldiers used these "war pigs" primarily as a way to deter elephant troopers—effectively making pigs the anti-tank weapon of the Classical period. The secret lay in elephants' alleged terror at the sound of pigs' squeals, something documented by numerous writers, including Pliny the Elder (*History of Nature* 8.9). There are several recorded examples of armies deploying pigs. In 266 BC, Antigonus II Gonata besieged Megara with soldiers and elephant-mounted troops. The Megarans poured pitch on pigs, lit them on fire, and sent them screaming toward Antigonus's lines. The burning swine so terrified the elephants that they panicked and trampled Antigonus's soldiers (Aelian, *On the Characteristics*

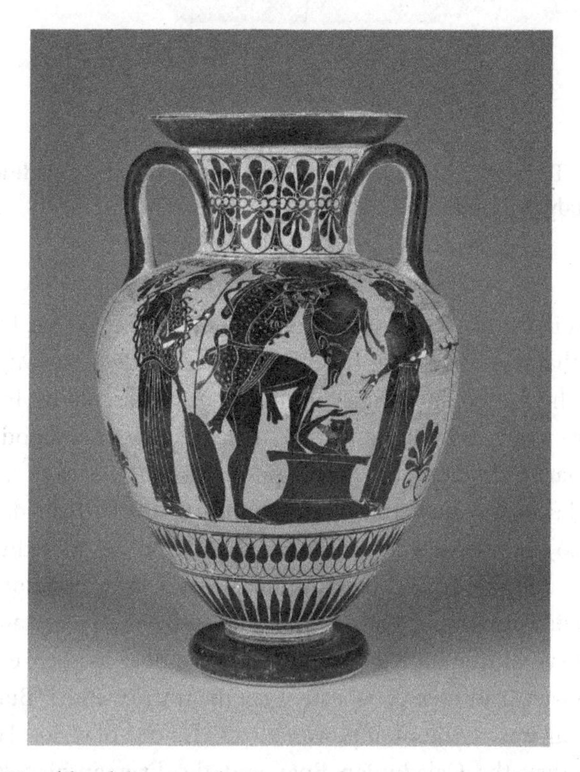

Figure 8.3. Herakles delivering the Erymanthean Boar alive to Eurystheus. Athens, ca. 510 BC. 43 × 28.2 cm.

of Animals 16.36). Similarly, in AD 544, the Byzantine defenders of Edessa staved off defeat by suspending a pig from the walls of the city, which squealed and so unnerved Sassanian elephants that the besiegers had to withdraw (Procopius, *History of the Wars* 8.14.30–43).

Zooarchaeology of the Greco-Roman Near East

The zooarchaeology of the Classical period Near East has yet to fully blossom. Only in the past decade or so have scholars begun to take serious interest in how food formed, maintained, and offered passage through cultural barriers at this time.[40] However, one feature that clearly stands out in the available zooarchaeological data from the Classical period is the renewed interest in swine husbandry, especially at military outposts and urban centers.

The impact on pig husbandry was most notable in those regions that had been under Greek and Roman hegemony for the longest period: Anatolia, Egypt, and parts of Syria and the Levant.[41] The shift was not uniform, however. Pigs remained unpopular (less than10 percent) at many sites, especially those located in the countryside. But in forts and cities, there was a spike in pork consumption that grew over the course of the Classical period. For example, the percentage of pigs recovered from the city of Pergamon in western Anatolia rose from 25 percent in the Hellenistic to up to 39 percent of the livestock species in the Roman period.[42] At other sites, especially those occupied in the Roman and Byzantine periods, pigs dominated the livestock assemblages. For example, at 4th–5th century AD Kom al Ahmar in Upper Egypt, pigs account for 69 percent of the livestock remains.[43]

The increase in pig husbandry was due to both the spread of Hellenistic culture and the settlement of Greek and Roman colonists, soldiers, and administrators in the Near East. For these newcomers, pork was a cherished food reminiscent of home fare. They ate pork and raised pigs for the same reason that Americans line up at Starbucks in Japan or McDonalds in Europe. But for native Near Easterners, eating pork, especially when it was cooked in Greek or Roman styles, was a way to connect with a cosmopolitan, pan-Mediterranean culture. For that reason, while people of all social classes raised and ate swine, pork was especially popular among the elite and upwardly mobile. For example, pig bones compose 60 percent of the bones of livestock mammals found in the kitchen waste of wealthy households at Ephesus in western Anatolia.[44] But pigs were also a cheap source of meat with which commanders provisioned their troops; many Hellenistic and Roman forts in the Near East contain higher percentages of pig bones than are found at nearby settlements, occasionally reaching as high as 80 percent.[45]

Outside the core regions of Greek and Roman influence, there were few changes in pig husbandry. In the Khabur Basin in northern Syria, swine continued to play a minor role in the agricultural economy, representing 10–15 percent of the remains of livestock.[46] However, even on the edges of the Greek and Roman world one finds potential evidence of its influences. For example, excavators at Tell Beydar uncovered a pit containing several pig fetuses.[47] The bones suggested sacrifices similar to those offered in the Thesmophoria rituals in Greece and two millennia earlier at Tell Mozan, 50 km to the northeast.

In Iran and Mesopotamia, areas that mostly fell under the control of the Parthian and Sassanaian Empires, the zooarchaeological data are extremely sparse. The role of pigs remains unclear. On the one hand, the almost complete lack of pig bones (less than 1 percent) from the faunal assemblage recovered from the city of Hatra[48] in central Iraq suggests that pork was eaten infrequently.[49] On the other hand, Parthian and Sassanian artists regularly depicted their royalty hunting wild boar. Additionally, at some military outposts near the Caspian Sea—forts associated with the 5th–6th century AD Gorgan Wall—excavations have revealed modest proportions of domestic pig remains (10–15 percent).[50]

Raising Pigs in the Classical Period

Zooarchaeological studies have shown that people changed how they raised pigs in the Classical period. One recent study investigated the dynamics of pig husbandry at Gordion Tepe in central Anatolia, a settlement occupied throughout the Hellenistic and Roman periods that would eventually support a Roman garrison.[51] At Gordion, the proportion of pig bones increased during this time from 12 percent to 26 percent, a change mirrored by a relative increase in bread wheat and cattle bones. This suggests that the town's economy reorganized itself in order to feed the garrison. Not only did the soldiers rely more heavily on pork, but pig husbandry also became more intensive. The age at which most pigs were slaughtered declined from about 18–30 months to 8–12 months, as shown in Figure 8.4. It seems that the soldiers fattened their pigs in sties—probably on cereal fodder—and slaughtered

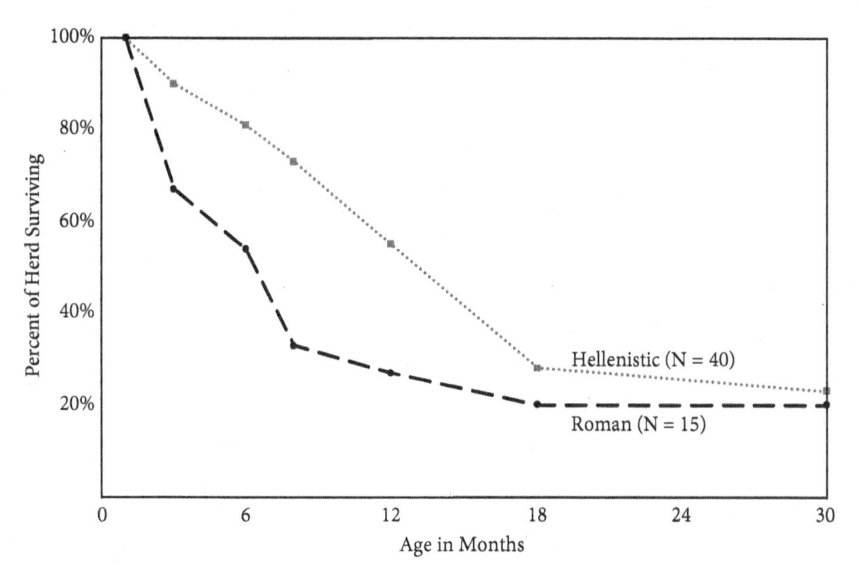

Figure 8.4. Intensification of pig husbandry in the Classical period at Gordion (central Anatolia).
Data reflect ages at death reconstructed from dental eruption and wear patterns in Roman and Hellenistic occupations. Lines represent the declining probabilities that a piglet born in either phase will reach successive age classes.

them shortly after weaning. This strategy may have been adopted as a way to increase the rate of pork production.

Zooarchaeological examinations of pig bones from Sagalassos, a prominent Hellenistic and Roman city in southwestern Anatolia, and its nearby satellite, Düzen Tepe, indicate changes in pig husbandry.[52] Similar to what occurred in Gordion, the relative abundance of pigs compared with other major domestic animals increased at these sites from 13 percent in the Hellenistic period to 32 percent in the Late Roman periods. Pigs were also slaughtered at younger ages.[53]

The shifts at Sagalassos and Düzen Tepe were accompanied by changes in the rates of dental hypoplasias. While the overall rate of hypoplastic defects decreased over time, suggesting husbandry strategies better able to minimize the dietary deficiencies that occurred during weaning, the rate increased on deciduous premolars, teeth formed during fetal development. This suggests that pigs in the Roman and Byzantine periods were under more stress in utero than those in the Hellenistic period. One could interpret the data as indicating a decline in sows' perinatal nutrition. Another interpretation is that during the Roman period farmers selectively bred their livestock for increased litter sizes, which would result in greater nutritional competition between littermate fetuses and thus higher rates of hypoplasias.[54]

Researchers working at Sagalassos and Düzen Tepe also measured ratios of stable isotopes of carbon, nitrogen, and oxygen from pig remains in order to examine shifting husbandry conditions. They found two main patterns. First, there was an increase over time in the ratio of the nitrogen-15 isotope to nitrogen-14 in pig bone collagen—the main skeletal protein.[55] This likely reflects one or a combination of two scenarios: the diets of livestock raised by Roman period pig breeders were richer in animal protein, and/or pigs were fed cereals grown in heavily fertilized fields. Both cases imply that food (and pork) production grew more intensive over the Classical period, with pigs regularly foddered with high-quality food.

Second, ratios of oxygen-18 to oxygen-16, which vary in drinking water between summer and winter, indicated intensive breeding strategies. Sequential sampling of the enamel along the length of pig incisors, when matched against the known ages in life at which those tissues form, enables researchers to reconstruct the seasons in which pigs were born. This technique was used to show that, at Sagalassos, sows in the Byzantine period gave birth throughout the year.[56] Getting one's sows to give birth multiple times per year is an important component of highly productive pig husbandry strategies. This evidence therefore adds further weight to the conclusion that Roman and Byzantine pig breeders practiced more intensive forms of management in order to meet the demands of a growing hunger for pork.[57]

These examples highlight the fact that in many corners of the Near East, pork became more popular and pig breeders ramped up swine production. But while the pork industry boomed in places, it increased the exposure of those who harbored pork taboos to an animal they abhorred.[58] Ground zero for the resulting clash of cultures was among the Jewish populations in the Levant.

Judaism and the Levant in the Classical Period

Before examining the clash between Greco-Roman and Jewish cultures over pigs, we need to take a slight detour into the historical context. The Classical period saw the development of a new type of Judaism, one that was increasingly nationalistic and identity-conscious. During this time, the Hebrew Bible was redacted into its final form, one that included new texts that debated the nature of Jewish identity—especially, the books of Ezra, Nehemiah, and Isaiah. These debates inspired the formation of different sects. The Jewish historian Josephus (ca. AD 37–100) described three of these sects: the Essenes, Sadducees, and Pharisees, the last of which were likely the traditional forebears of Rabbinic Judaism. Another sect would develop into its own religion—Christianity.[59] Each of these came into conflict with Greco-Roman culture. It is not difficult to see why. Greco-Roman and Jewish identities were both totalizing, both "supranational cultural systems that transcend[ed] birth."[60] And while Jewish and Greco-Roman philosophy borrowed heavily from each other, members of each group perceived themselves as diametrically opposed.[61]

But let's start at the beginning. Following on the heels of Cyrus's conquest of Babylon in 539 BC, contingents of exiled Jews returned to the Levant and rebuilt the Temple. Jews living in the newly reconstituted province of Judea thrived under Persian rule, whose rulers promoted the Torah and embraced it as an ethnospecific code of law for the Jews.[62] Respect for Jewish customs and *halakha* continued into the early Hellenistic period. In the aftermath of his victory over the Persians at Issus (333 BC), Alexander peacefully took over the Levant. A likely apocryphal story[63] told by Josephus describes Alexander visiting Jerusalem, paying homage to the Temple, and guaranteeing the right of Jews to follow their ancestral laws (*Antiquities of the Jews* 11.8). Indeed, while Greeks believed their own culture to be superior and delighted in spreading Hellenism, Greek imperialism exhibited a remarkable tolerance of local traditions. Alexander's successors founded colonies and created institutes of Greek learning, but they recognized and, perhaps begrudgingly, respected the cultural diversity of the lands they had conquered.[64]

In Judea, for reasons that remain unclear, the situation changed dramatically during the reign of the Seleucid king Antiochus IV (reigned 175–164 BC). Under his predecessor, Antiochus III (reigned 222–187 BC), the Greek-speaking population in the Levant had grown and the cultural tapestry of the region had become more Hellenistic. Many Jews adopted Greek names and customs. Continuing the tradition of Greek tolerance, Antiochus III asserted Jews' rights to their ancestral laws.[65] But Antiochus IV suddenly reversed this long-standing policy. According to 1 Maccabees (1:45–50), he forbade male circumcision, rededicated the Temple in Jerusalem to Zeus Olympios-Ba'al Shamim, and, just for good measure, sacrificed pigs at the Temple.[66]

These outrages sparked a holy war celebrated by Jews to this day as Hanukkah. As every Jewish child learns, the winter holiday commemorates the refusal of a priest named Mattathias to obey Antiochus IV and the revolt carried out by Mattathias's sons, especially Judah Maccabee. The historical reality is a bit more complex. Depending on one's perspective, the Maccabees may have been the quintessential religious freedom fighters extolled every November/December by rabbis across the globe. Or they may have been conservative reactionaries who found in Antiochus IV's desecrations a *casus belli* to purge Judea of Hellenistic influence.[67] Or they may have been power-hungry thugs who took advantage of a tense political situation to seize control of Judea.[68]

Whatever one's opinion of the Maccabees, they managed to wrest control of the southern Levant from the Seleucids—at least temporarily. Judah was killed in battle in 160 BC, despite an alliance with Rome (1 Maccabees 8), and the Seleucids were able to collect tribute and establish vassalage over Judea. But in 140 BC, Judah's brother, Simon, established the Hasmonean Dynasty.

The Hasmoneans ruled as Seleucid vassals until the empire collapsed in 110 BC, at which point Judea became independent. The Hasmonean Dynasty was expansionistic and oppressive, forcing the Jerusalem form of Judaism on peoples throughout their territory.[69] Beleaguered by revolts, Hasmonean independence did not last long. In 63 BC, during a war of succession between the brothers John Hyrcanus and Aristobulus (note the Greek names or epithets of the Jewish leaders), Pompey marched on Jerusalem and annexed Judea to Rome.[70]

The Mishnah, or the first part of the Talmud (redacted ca. AD 200), recounts the fateful moment when Judea, torn in civil war between Hyrcanus and Aristobulus, gave itself up for Roman annexation. In the parable, Aristobulus was besieging his brother in Jerusalem. The two sides, however, made arrangements so that the Temple sacrifices could continue as normal. The besieged would lower a basket filled with money to Aristobulus's men, who would replace the money with an animal. These animals, both sides knew, were supposed to be fit for sacrifice. That is, they should be sheep, goats, or cattle. But one of Aristobulus's soldiers, who knew "Greek wisdom," convinced his comrades that as long as the Temple continued to function, Jerusalem would not fall. So one day, instead of a sheep, the soldiers sent up a pig. As Hyrcanus's men lifted up the basket, the pig dug its feet into the walls of the city. This cosmic outrage caused the whole of Israel to shake. The priests who witnessed this scene declared, "Cursed be the man who raises pigs, and cursed be the man who teaches his son Greek wisdom" (Bava Kamma 82b).

The first few decades of Roman rule in the Levant were uncertain. Oppression of Jewish practice was not, at first, the order business. In the years after Pompey's conquest, Rome itself descended into two decades of civil war that ended with Octavian (Augustus) defeating Cleopatra and Mark Antony at Actium (31 BC) and becoming Rome's first emperor. Josephus described this as a time of intrigue and diplomacy for Jewish leaders, among them the still-active Hyrcanus and Aristobulus (*The Jewish War*, book 1). But power was ultimately consolidated under the kingship of Herod, son of an adviser to Hyrcanus.

Cunning, brutal, and paranoid, Herod (reigned 37–4 BC) knew how to balance his fealty to Rome with his own ambitions. Originally siding with Antony, he quickly pledged allegiance to Octavian after Actium. He happily accepted elements of Romanization and established Greek and Roman colonies such as Caesarea, a coastal city obsequiously named in honor of his Roman patron. He allowed Roman statues to be erected, built stadiums, and even, according to Josephus, sponsored the Olympic games (*The Jewish War* 1.21.12). Like other Near Eastern and Greek and Roman elites, he displayed his prowess through royal hunts—including ones for wild boar (*The Jewish War* 1.21.13). Under Herod's rule, the Jewish vassal state prospered and into the realm poured money that Herod spent lavishly on popular projects—for example, refurbishing the Temple.[71] But Herod was known as much for his cruelty as his competence. So infamous was he for executing members of his family that one story (likely apocryphal) recounts Octavian, taking a poke at Jewish dietary customs, quipping that he'd rather be Herod's pig than his son.[72]

Herod kept the peace by serving Roman imperial interests and cultivating Jewish popular sentiment. His policy promoted two processes—Romanization and Judaization—that were simultaneously entangled and paradoxical. These processes were *entangled* because they fed off one another.[73] The Romans viewed the local Jewish elite as a ready-made administrative apparatus. They were happy to encourage Jewish nationalism as long as it entailed ultimate loyalty to Rome. Meanwhile, Judea's position in the Roman Empire allowed it to profit from trade with Arabia and facilitated the collection of money from the sizable Jewish communities that had sprung up throughout the Mediterranean region. This money funded the refurbishment of the Temple in Jerusalem and the construction of synagogues.

Romanization and Judaization were also *paradoxical* because Jewish identity, which had become increasingly nationalistic since the Maccabean revolt, defined itself largely against the Greek/Roman "other." [74] The cultivated antagonism occasionally sparked hostilities (e.g., in Caesarea: *The Jewish War* 2.13.7) as more Greek- and Latin-speaking peoples populated the Levant.[75] Meanwhile, Roman authorities demanded allegiance to the imperial cult, which required sacrificing and bowing before its idols, in direct contradiction to the tenets of Judaism.

While Herod was able to manage these tensions during his reign, his successors were unable to do so. Things began to unravel. Tensions increased after a series of poorly handled crises, such as the desecration of a Torah scroll by a Roman soldier. In another episode, Pontius Pilate arrogantly paraded Roman military standards into Jerusalem, despite their condemnation by the priests as idolatrous effigies (*The Jewish War* 2.9.2–3). Jewish rebel groups formed and Roman authorities, rarely winners of hearts and minds, turned from tolerance to oppression.[76] Resentment simmered for decades but erupted in a major rebellion in AD 66–73. The "Jewish War," according to Josephus, left over one million dead (*The Jewish War* 6.9.3). This is probably an exaggeration, but it is nevertheless likely that the decimation of the Jewish population during the 1st and 2nd centuries AD was so vast that it would not recover until the 18th century.[77] However, the most painful episode for the Jews was the destruction of the Second Temple by Titus in AD 70. Judaism, as it had been, was "shattered."[78]

The Jewish War was succeeded by several other uprisings, each accompanied by a catastrophic loss of life and further oppression against Jews. They included the Diaspora Revolt (or Kitos War; AD 115–117), which was fought primarily in Cyrenaica, Cyprus, and Egypt, and the Bar Kochba Revolt in the Levant (AD 132–135). The Romans exacted revenge by several means: they razed towns and villages and executed civilians; Vespasian (reigned AD 69–79) introduced a special tax on Jews throughout the empire; and Hadrian (reigned AD 117–138) banned Jews from Jerusalem, renamed the province of Judea "Palaestina" after the Philistines, and even outlawed circumcision.[79]

While the results of these rebellions may have been predictable,[80] they were no less devastating. They left Judea stripped of the political autonomy it had enjoyed to varying degrees since the Persian period. On a cultural and religious level, the rebellions marked a turning point in Jewish history. The "shards" of Judaism[81] would be picked up and pieced back together by local rabbis in northern Palestine, Mesopotamia, Egypt, and other parts of the Old World. A new form of "Rabbinic" Judaism would emerge and persist over the next two millennia. Meanwhile, Jews were increasingly defined as a minority outcast group, especially after one of Judaism's sects, Christianity, was adopted as the Roman state religion in the early 4th century AD.[82]

Unholy of Unholies: Pigs and the Clash of Cultures

The conflicts between Jewish and Greco-Roman cultures often drew upon, and occasionally centered around, their divergent attitudes toward pigs and pork. In the context of imperialism and Jewish nationalism, pig hatred and Jewish identity became caught in a positive feedback loop driven, on the one hand, by the Greek and especially Roman fixation on Jews' aversion to swine and, on the other, by many Jews' distrust of Greek and Roman culture. Each perceived the other as an existential threat. In the process, the pork taboo was elevated to a position of prominence in the Jewish dietary laws.

The Greek and Roman love of pigs and pork expanded into the Near East in the Classical period. The Levant, including Judea, was no exception. Zooarchaeological data indicate an uptick in pig husbandry during the 4th–2nd centuries BC, with pigs comprising as much as 15–20 percent of the recovered bones of domestic livestock at some sites compared with the 1 percent or less typical of the Persian period (see Table A.5 in the appendix).[83] Under Roman and Byzantine rule, pig husbandry expanded even further, especially in regions like the Galilee and northern coastal region, where the temperate climate encouraged swine production.[84] For example, at 58 percent, pigs dominated the livestock bone assemblage at Caesarea.[85] By the Byzantine period in what is now northern Israel, 10–50 percent of the domestic mammals slaughtered in the majority of cities were pigs.[86]

This increase in pigs represented a radical break with tradition. In aggregate, people inhabiting the Levant raised pigs in quantities not seen since the 4th millennium BC. However, there was significant variation. Settlements where Jewish populations are historically attested had much lower proportions of pig remains (less than 5 percent and often less than 2 percent) throughout the Hellenistic and Roman period.[87] Similarly, relative abundances of pigs are much higher at sites with archaeological evidence of Hellenistic material culture than at sites with *mikvah* baths and other elements of Jewish material culture.[88]

It is simplistic to assume that the people who ate pork necessarily identified themselves—and were identified by others—as Greek, Roman, or otherwise non-Jewish. Nor can we assume that sites without pork were necessarily inhabited exclusively by Jews.[89] For one thing, this assumption ignores the so-called Hellenized Jews, those who had adopted elements of Greco-Roman culture and incorporated them into local Levantine customs. Most, perhaps all, Jews in the Classical period had adopted some elements of Hellenic culture.[90] Some Jews, while still firmly grasping their self-perception as Jews, must have succumbed to curiosity and tried pork, that meat held in such high esteem by the rest of the Hellenistic world. There is, in fact, scattered textual evidence for this rejection of tradition—namely Jews who raised pigs.[91] They include the Talmudic legal discussion of damages done by pigs (e.g., Bava Kama 17b) and the story of the swineherd and his pigs in the exorcism of the Gerasene demoniac (e.g., Matthew 8:28–34). But perhaps most convincing are the bones themselves. While sites associated with Jewish populations contain *low* relative numbers of pig remains, they rarely contain *no* pig remains.

Nevertheless, the politics of food surrounded conceptualizations of Jewish, Greek, and Roman identity in the Levant. Although debates on the social and political significance of food for Jewish identity took a variety of forms—for example, whether it was acceptable to share meals with non-Jews—nothing played a larger role than pork.[92] Pork became the metonym for relinquishing Jewish identity. Its power was fed by mutually reinforcing Greco-Roman and Jewish perceptions of each other.

One of the most significant developments in the Classical period was the weaponization of pork against Jews. The genesis of this type of behavior is not difficult to imagine; whenever one group has a unique taboo or custom, at least one person from a rival group will decide to provoke the former by breaking the taboo or—even better—tricking its members into breaking it. Such episodes occur more frequently when the two groups are at odds and when there is a greater temptation to vilify and mock the traditions of the "other."

Examples of the weaponization of pork abound. Many are likely apocryphal, but nevertheless their telling and retelling indicate how pork, or even the *idea* of pork, drove a wedge between Greeks/Romans and Jews. The earliest stories of soldiers or mobs forcing Jews to eat pork date to the Maccabean revolt. The books of Maccabees (2 Maccabees 6:18–7:42; 4 Maccabees 5–18) relate the story of a woman named Hannah and her seven sons, who, being forced at sword point to consume pork, chose to die instead of violating *halakha*. Antiochus IV also allegedly sacrificed swine at the altar of the Temple, perhaps in willful violation of Jewish custom (1 Maccabees 1: 45–50). That story bears a strong resemblance to that of Cleon of Gordiucome at Comana (Strabo, *Geography* 12.8.9). Finally, during an anti-Jewish riot in Alexandria in AD 38, the mob allegedly forced some Jewish captives to eat pork.[93]

Even when not directly violating Jews' taboo on pigs, Greeks and especially Romans took pains to mock Jewish custom. On his ambassadorial visit to the emperor Caligula in AD 40, Philo of Alexandria relates how the Roman leader mocked the Jewish delegation, asking, "Why do you refuse to eat pork?" (*Embassy to Gaius* 45.361). Deriding Jewish custom was a favorite pastime of Roman satirists and intellectuals, who especially harped upon Jews' avoidance of pork.[94] Petronius (AD 27–66), for example, mused that the Jews worshipped a "pig-god."[95] Writers as diverse as Juvenal, Seneca, Tacitus, Cicero, Plutarch, and Apion ridiculed Jews as descendants of lepers who, expelled from Egypt, had invented laws like pork avoidance so that they could self-righteously keep themselves separate from other peoples.[96] In deed and in writing, belittling the prohibition on pork became a cheap and easy way to terrorize Jews, a dress rehearsal for the long European tradition swine-focused anti-Semitism.[97]

The adoption of pig husbandry in the Levant, its contradiction to Jewish Law, and the use of pigs as a weapon against Jews all set in motion a process of increasing negative identification between Jews and pork. Historian Jordan Rosenblum[98] has articulated this process succinctly. His thesis hinges on the notion "you are what you eat," an admittedly glib aphorism that nevertheless captures an anthropological truism: food is an essential feature of social identity.[99] But food is much more than a badge of identity that people can affix to themselves or remove at other points. Through the act of ingestion, food, quite literally, becomes a part of the self. As Rosenblum[100] puts it, "[F]ood becomes *embodied* in each individual. It operates as a *metonym* for being part of the self" (emphasis in original).

For Rosenblum, the pig was a fundamental feature of Roman citizenship. Pigs figured prominently in Greco-Roman economies, mythology, and religion. Rosenblum argues that the Roman self-identification with pig husbandry was more pronounced even than that of the Greeks, observing that Latin texts spend more time discussing Jews' taboo on pig than Greek texts do.[101] Unlike the Greeks, Roman writers tended to ridicule the taboo as irrational and alien.[102] Swine were quintessential features of the Roman cosmopolitan identity; the taboo on pork therefore prevented Jews from being able to "ingest Romanness."[103]

For their part, Jews identified pigs not only with Romanness, but with its worst quality: something *appearing* to bestow wealth and happiness while in reality bringing oppression. Pork, in other words, epitomized the existential threat of Roman imperialism. Thus, the story of Aristobulus and Hyrcanus was retold again and again by the early rabbis. But they often switched the characters, casting the Romans ("the Evil Empire") as the besiegers instead of the Jewish Aristobulus.[104] According to the rabbis, just as the pig, which has cloven hooves but is not kosher, may appear to be good to eat, so too did Roman culture appear to offer benefits while in reality stripping Jews of their essential Jewishness.[105] Even today, the pig remains a potent symbol of the temptation to acquire material wealth at the expense of one's spiritual health. In a story from the Chassidic tradition (17th century to present), a rich man gives his poor yet pious brother a tour of his estate. Seemingly out of the blue, the pious man interrupts his wealthy sibling. A pig, he explains, wallows in the mud and can only dream of more mud. The parable's message is clear: the rich brother's worldly pleasures beget only a desire for more material wealth. They distract him from what is most important.[106]

In fact, the material wealth flowing into the Levant in the Classical period was significant, especially during the reign of Herod. While elevating Judaism to new glory, the marriage to Rome represented to some Jewish sects an existential threat, prosperity being purchased with the soul of Judaism. A classic scene in the 1979 film *Monty Python's Life of Brian* neatly captures this contradictory dynamic. In the film, the rebel leader (played by John Cleese) gathers a band of Jewish rebels to attack the Romans, condemning them for taking the Jewish homeland. "What have they ever given us in return?" he demands. Shyly, one of the rebels pokes his hand up and says, "The aqueduct?" The leader blinks and concedes the point. Then another rebel chimes in, "and the sanitation." That, too, is conceded. Another adds, "the roads." And on it goes . . .

The scene's historical insight is no less significant than its comedic value. It lays bare the contradictions of being Jewish in Roman Judea.[107] On the one hand, Roman occupation brought wealth (at least into the hands of the elites) that could be used to make the Temple more spectacular and support Jewish religious and social projects. But, on the other, it brought Roman soldiers, administrators, and culture (including pigs). That Romanization and Judaization went hand in hand did not sit well with many. The pig, an animal Romans loved to eat, became a useful metaphor for this contradiction and the insidiousness of Roman rule.

Roman administrators and soldiers were not blind to the evolving symbolism of pigs in Jewish thought. Rosenblum[108] argues that it was the recognition of the increasing anti-Roman quality of the pig taboo that inspired more frequent outbreaks of pork weaponization as well as the Roman intellectual fixation on mocking Jewish food laws. For both Jews and Romans, the coerced ingestion of pork came to be symbolic of the Roman demand for acquiescence to imperial authority and the surrender of one's Jewish identity.[109] Ingesting pork could unravel an individual's Jewishness, while on a deeper level threatening the orthopraxic foundation of Jewish identity.[110]

Forcing a Jew (or Muslim) to eat pork was, and remains today, a form of identity rape. It strips individuals of the personhood that they have cultivated over a lifetime and imposes another upon them. As such, it is an attack not only on the individual, but also on the constitution of the group to which they belong. When Donald Trump, as president of the United States, threatened to shoot jihadi terrorists with bullets dipped in pigs' blood,[111] he was not simply proposing a harsh treatment of criminals, but proposing a form of torture whose real target is Islam. By legitimizing the use of pork as a weapon, Trump assaults Muslims' right to practice their faith and derides Islam's central tenets. Whether or not they are acted upon, such threats recast religious taboo as personal weakness, something that can be exploited by those who do not subscribe to such prohibitions. This type of intolerance, which is as old as the Hellenistic period, is one of the most enduring legacies of the pork taboo.

But intolerance also breeds more deeply entrenched positions on the part of the oppressed. The increasing mobilization of pigs in the existential battle for identity in the Classical period created a feedback loop, with each party responding to developments in the other by ramping up their own attitudes toward swine.[112] Identity is defined not simply in terms of belonging, but also in contrast to an "other" and its practices. But this process rebounds on members of the "other" group. That is, when a group of people begin to focus on the specific norms (e.g., avoiding pork) of a rival group, they can inspire members of the rival group to more faithfully practice those norms (avoiding pork with greater intensity). That is why, when it becomes important to demonstrate their identity, members of ethnic groups often retreat into stereotypes of themselves. They are ready-made markers of difference. But in the process, they reinforce those stereotypes, imbuing them with greater energy and reproducing them as more salient symbols of self-identification.[113]

In the case of Jewish-Roman relations, the confrontation between pork-loving and pork-hating peoples entrenched these two ideologies further. Romans saw themselves as a pork-loving people, while Jews increasingly saw themselves, and were perceived, as pork haters. The outcome was an intensification of the pig taboo and its elevation to the type of food avoidance most important for reproducing Jewish identity. Along with observing the Sabbath and circumcising their sons, avoiding pork became one of *the* core features of daily Jewish practice throughout the world for centuries to come.[114]

The sanctions against pigs grew stronger in the Rabbinic period (AD 73–600). The writers of the Talmud (compiled ca. AD 200–600) elaborated on the Torah by adding additional restrictions on daily practices. In doing so, they laid the foundations for the body of laws that Jews now refer to as *kashrut*, which distinguish kosher from nonkosher foods.[115] Among other things, the rabbis augmented the taboo on pork, drawing on the intense emotions already surrounding pigs. By the middle centuries of the first millennium AD, the taboo had grown so prominent that it was extended to include pig husbandry and even the ownership of pigs. These were justified as part of a process to build a "fence" around *halakha* (Bava Kamma 82b). That is, they were measures taken to prevent the observant from even coming close to breaking Yahweh's commandments. Some rabbis even avoided uttering the Hebrew word for pig (*khazir*), using instead the phrase "another thing" (*davar acher*) when referring to the animal (e.g., Shabbat 129b).

Between Judaism and Christianity

Jesus's followers developed their own unique sect of Judaism. This new form of Jewish religious expression would enjoy success in proselytizing to people throughout the Roman world and beyond. But that success had an unforeseen consequence. As the sect grew, it diverged from Judaism, the religion of a specific ethnic group, to become a faith that could be readily adopted by "gentiles" (Hebrew *goyim*), or "(other) nations." This ultimately transformed it into a religion in its own right.[116] Although heavily persecuted for three centuries, Christianity would ultimately be adopted, and in the process modified further, by the emperor Constantine in the early 4th century AD. Several decades later, Christianity became the official state religion of the Roman and later Byzantine Empires by the Edict of Thessalonica in AD 380.[117]

This is not the place to discuss the evolution of Christianity and its impact on European history, but Christianity's relationship to Judaism and its appeal to people throughout the Roman Empire did have important consequences for pigs. Jesus and his early followers were Levantine Jews, and they almost certainly adhered to *halakha*.[118] But in the years following his death, Jesus's apostles began to debate the relevance of Jewish Law.[119] On one side, James, ostensibly Jesus's brother, and Peter advocated strict retention of *halakha*. On the other side was Paul, a tentmaker from Tarsus who converted from an anti-Christian Pharisee to a believer in Christ (Acts 9:1–4).

Paul justified his argument for abandoning much of *halakha* on principles adopted from Hellenistic philosophy. In this way he departed from other Jewish scholars who had been inspired to reconcile the practice-based features of Judaism with the spiritual-focused teachings of Plato and Aristotle. Philo of Alexandria, a contemporary of Paul, argued that the Torah and its laws contained within them hidden meanings and allegories. As we saw in Chapter 6, Philo's allegorical argument for the pig taboo was that its inability to ruminate reflected its disorderly life and intellectual limitations (*On Husbandry* 32). But Philo did not by any means suggest that one should abandon Jewish Law. In fact, his argument was that by following it one could learn of the inner truths contained within it.[120]

Paul and later Christians saw it differently. For them, Christianity embodied a community of people incorporated into the spirit of Christ (Ephesians 4:16). It therefore had nothing to do with *halakha*.[121] To be Christian was not to do, but to believe—and, especially, to believe properly. This philosophical position marked a subtle, but crucial divergence from Judaism. Christianity moved away from *orthopraxy* and toward *orthodoxy*.[122] While Jewish Talmudic scholars debated practices—what was and wasn't allowed by the Torah—Christian theologians would increasingly focus on the proper definition of Christ and how to reside in spiritual unity with him.[123]

But Paul must also have been thinking practically. He had set up Christian communities throughout Syria, Anatolia, and Mesopotamia and had converted a number of Jews and non-Jews to the faith.[124] While his followers were keen on adopting Christianity, many had reservations about *halakha*, especially two features: male circumcision and pork avoidance. Both of these had, historically, been barriers to the spread of Judaism.[125] Indeed, the decision to abandon much of Jewish Law was in part responsible for the spread Christianity, laying the foundations for its domination in Europe, North Africa, and the Near East. Between AD 200 and 250, the number of Christian adherents increased from around 200,000 to one million.[126]

The decision to put pork back on the table may also have been influenced by the specific populations targeted for proselytization. Christian missionaries promoted a universalist ("catholic") belief system intended to unite people across ethnic, linguistic, and political boundaries.[127] But the religion appealed especially to the marginalized elements of the Roman Empire. Christians glorified the poor, preached charity, and even elevated the status of women living under the decidedly masculinist Roman and Jewish cultures.[128] If attracting the lower classes of the Roman Empire was Christians' goal—as it appears to have been—then pork, a staple of the urban poor, would have to be tolerated. And if the Christian leaders were going to scratch the pork taboo, they might as well eliminate the vast majority of the food laws.

Eliminating the taboos was no easy task. Christian thinkers had to show that those parts of the Bible that enjoined people to follow the food laws were meaningless; in doing so, they not only upended centuries of tradition, but contradicted what was supposed to be the word of God filtered through Moses. For this reason, early Christians' transition away from the Pharisee sect of Judaism, which was led and largely shaped by Paul (Acts 23:6), may have been a halting, contradiction-riddled process marked by internal struggle.[129] But Paul's ultimate focus on orthodoxy and the limitations of orthopraxy resonated with Christians in his time and in the decades to come. The New Testament is filled, therefore, with admonishments against taking *halakha* too seriously and neglecting one's spiritual fitness:

> there is nothing from without a man, that entering into him can defile him: but the things which come out of him, those are they that defile the man. (Mark 7:15)

Going further, some Christian thinkers even argued that Jewish Law was harmful because it obscured the true message of Christ. For example:

> Now the Spirit speaketh expressly, that in the latter times some shall depart from the faith, giving heed to seducing spirits, and doctrines of devils; speaking lies in hypocrisy; having their conscience seared with a hot iron;

> forbidding to marry, and commanding to abstain from meats, which God hath created to be received with thanksgiving of them which believe and know the truth. For every creature of God is good, and nothing to be refused, if it be received with thanksgiving (1 Timothy 4:1–5).

The tone of these critiques of Jewish Law grew more radical as time went on. Within roughly a century after Jesus's execution, Christian leaders began forcefully endorsing the abandonment of *halakha*. The Epistle of Barnabas, written around the turn of the 1st–2nd centuries AD, condemned Jewish meat prohibitions. Barnabas argued that God had given them to Moses as metaphors for proper behavior, not actual restrictions on food.[130] Swine, for example, were emblematic of people who prayed only in times of trouble and neglected God when they were satiated and happy (Barnabas 10:3). Thus, like Philo, Barnabas emphasized the allegorical nature of Jewish Law and the Torah. Unlike Philo, he declared that obedience to the spirit of the law, as he defined it, was more important than submission to the letter of the law.

The elimination of the food laws also served political ends. Beginning in the 2nd century, Christian thinkers perceived a need to distance themselves from Judaism. At that time, many Christians were simply Jews who also followed the teachings of Jesus. They frequently attended synagogues and participated in Jewish festivals with other Jews.[131] But the prevailing opinion among church leaders was that Christianity was not simply a revitalization movement within Judaism, but a radically new ideology.[132] Recognizing that the sharing of food binds people together, Christian leaders advocated for the elimination of the taboo on pork and other foods as a means of driving a wedge between Christians and Jews.[133]

By the 4th century AD, Christian leaders were not only preaching against Jewish practices, but also admonishing their followers not to participate in Jewish festivals or eat Jewish food. Resorting to hyperbole, they demonized Jewish food as anti-Christian. For example, Ephrem the Syrian's "Hymn on Unleavened Bread" warned Christians not to eat Passover *matzo*, as it was made by hands "filthy with [Christ's] blood."[134] Over the next several centuries, Church Fathers repeatedly demanded that members of their flocks avoid eating with Jews. In effect, while Christianity had eliminated taboos on pork and other foods, it had placed one on any food prepared by or eaten in the company of a Jew. Later, the popes applied the same taboo on food consumed with "judaized" Muslims.[135]

This cultivated sense of disgust for traditional Jewish practices soon inspired more violent forms of anti-Jewish behavior. Church Fathers like John Chrysostom (ca. AD 349–407) actively preached against Jews and their "sinful" practices.[136] Pogroms and attacks on Jews soon followed, including one in AD 388 led by the Bishop of Callinicum (modern-day Raqqa, Syria) on a local synagogue.[137] In part because of these attacks, by the 6th century most Jews in Europe and the Near East had withdrawn from Christian-dominated civic society into their laws and local rabbinic authorities.[138] Thus began the long history of hostile Christian-Jewish relations, ghettoization, and persecution.

However, some Christians continued to avoid pork. For example, in his *Answer to Faustus, a Manichean* (ca. AD 410), Augustine of Hippo conceded diversity in Christian food practices, noting that while some followers ate any type of meat, some abstained from pork and others from the meat of quadrupeds (30.3).[139] Augustine defended his own consumption of pork by invoking the Christian interpretation that the Leviticus taboos were intended to prohibit only what swine *symbolized*—greed, gluttony, and thoughtlessness (the lack of spiritual rumination) (6.7).[140] While the Catholic Church and most other Christian denominations adopted Augustine's approach, various sects to this day avoid pork. They include members of the Ethiopian Orthodox Church, which dates back to the 4th century AD, and Seventh-Day Adventists, a 19th century American revitalization movement that, among other things, advocates a return to the dietary laws of the Old Testament.

The elimination of the pig taboo had important implications for Christian cultures and their relationship to Judaism. In Europe, communal pork consumption became an important feature of celebrations, including major holidays such as Easter and Christmas.[141] But pork was also used to force Christianity upon others. For example, the Spanish Inquisition demanded not only that Jews and Muslims eat pork to prove their conversion, but also that they enjoy it.[142] Indeed, the fact that pork was taboo to Jews and Muslims gave it a special power. Eating pork was an act of denouncing Judaism and proclaiming Christ. An 18th century song from Burgundy, reproduced by the historian Claudine Fabre-Vassas, declares:

> While the Jewish law
> Prohibits lard as heretical
> The same is not so in Christian lands.
> Let us eat fresh pork, Let us eat!

The more we enjoy the piglet
The better Catholics we become.[143]

Pigs in Christian Thought

With the exception of a few passages, the Hebrew Bible (or Old Testament) rarely features pigs.[144] Swine play a more prominent role in the New Testament. However, their significance is decidedly mixed.

In the parable of the Prodigal Son (Luke 15:11–32), a young man demands and squanders his inheritance on worldly pleasures. Destitute, he ends up a swineherd. Raising pigs epitomizes the son's wantonness and the misery it has brought him. However, it is in his poverty that the son regains piety. Starved and desirous even of the food his pigs are eating, he comes to his senses and returns to his father, who welcomes his wayward child back with open arms.

Pigs also appear as unclean things alongside dogs. As we saw in Chapter 5, the connection between swine and canines dates back to the Bronze Age, but it remained a theme in early Christian writings.[145] In the book of Matthew, for example, Jesus demands:

> Give not that which is holy unto the dogs, neither cast ye your pearls before swine, lest they trample them under their feet, and turn again and rend you. (Matthew 7:6)[146]

Metaphorically, pigs and dogs represent that which can corrupt the wisdom of Christ's teachings; the passage is a warning against hypocrisy. But in mobilizing these images, Jesus is referencing the polluting nature of pigs and dogs, something his audience would have understood.

The flip side of pigs' polluting nature is their use as substitutes, something common to many Near Eastern traditions as well as Roman and Greek religions. The New Testament also plays on the imagery of swine soaking up sin. In Matthew (8:28–34), two individuals (or one in Mark 5:1–21 and Luke 8:26–40) become possessed by demons. They confront Jesus in the Galilean countryside, where a herd of swine is feeding. The demons, knowing Jesus is about to exorcise the men, request, "[I]f thou cast us out, suffer us to go away into the herd of swine" (Matthew 8:31). Jesus does just this, although perhaps not as the demons had intended, for he sends the herd charging off a cliff into the Sea of Galilee.

Pigs appear in perhaps their most positive light as companions of the Egyptian ascetic monk Antony (or St. Anthony, ca. AD 251–356; Figure 8.5). The patron saint of skin diseases and livestock, he is often viewed as the father of Christian monasticism.[147] His association with pigs is somewhat unclear, but seems to relate to his battling demons in the desert. If so, then swine may represent the vessels that demons inhabited in order to torment Antony.[148] Whatever its genesis, the pig was later depicted as a faithful companion of

Figure 8.5. *St. Anthony*, 1564. Engraving by Hieronymus (Jerome) Wierix (ca. 1553–1619).

Antony and served as a symbol of his ministry. His missions, which doubled as hospitals, raised pigs,[149] and dedications of parts of pigs (ears, heads, feet, and even sausages) on his behalf are still common in southern Europe as offerings for souls stuck in purgatory.[150]

The Classical period represents an incredibly dynamic era in the history of pigs in the Near East. On an economic level, pig husbandry expanded in many parts of the Near East. Zooarchaeological data show that, in some places (e.g., the Levant), people ate pork at rates not seen for almost 3,000 years. This increase in the number of pigs raised in the Near East can be directly tied to the animals' popularity in Greek and Roman cultures.

In the context of expanded swine production and consumption, the Jewish taboo evolved significantly. While the origins of the taboo must be sought in earlier times, it was the antagonistic encounters between Jews and their Greek and Roman overlords that refashioned pork avoidance into one of *the* core features of Judaism. The Torah applied no special status to the abstention of pork compared with, for example, that of rabbits or fish without scales and fins. But by the end of the 6th century AD, swine represented the taboo animal *par excellence* for the Jewish peoples, redolent of the existential threat posed by living under foreign domination and, increasingly, their Christian neighbors.

For Christians, pigs came to signify their faith's divergence from Judaism. Theologically speaking, the meat taboos in Leviticus and Deuteronomy represented the orthopraxic doctrine that early Church Fathers rejected in favor of orthodoxy. For them, then, the existential threat was the pork taboo and other elements of *halakha*, which they feared could distract their followers from Christ's truth. But rejecting the pork taboo presented other benefits to Christian leaders. It helped win converts and spread Christianity. On a political level, the rejection of the pig taboo helped Christians distinguish themselves from Jews as members of a separate (and in their eyes, superior) religion.

9
Islam and the Modern Period

Between AD 602 and 628, a final and largely inconclusive war between the Byzantine and Sassanian Empires left both sides weakened. The resulting power vacuum set the stage for the Arab Conquest. Along with it came the spread of a new religion—Islam—that would include a unique take on the pig taboo developed in Judaism and rejected by Christianity.

Arab peoples had long featured in pan–Near Eastern politics. In 853 BC at Qarqur in northern Syria, camel-mounted Arab troops fought alongside their Levantine allies, including the kingdom of Israel, against a Neo-Assyrian army. In the Classical period, the Nabataean kingdom and other Arab polities grew rich from camel caravan trade and their positions as buffer states between the Roman/Byzantine and Parthian/Sassanian Empires.[1] By the 6th century AD, the trading hub at Mecca began attracting large amounts of wealth to the Hejaz region. The pouring in of riches and the newfound internationalism brought to the fore spiritual questions, especially among those who felt left behind by the economic boom.

Muhammad (AD ca. 570–632) was the right person at the right time in the right place. Tying together Jewish, Christian, and traditional Arab religious and social concepts, he forged Islam ("submission [to Allah]"). But Muhammad's philosophical insight was matched or even exceeded by his political tact and military prowess. He married Islam to the state that he founded and, in a series of campaigns, conquered the Arabian Peninsula. Following Muhammad's death, Islam and its state exploded into the power vacuum of the Near East, taking Damascus in 635 and decisively defeating the Byzantines at the Battle of Yarmouk in 636. Soon after, Arab armies dispatched the Sassanian state by capturing its capital, Ctesiphon (637). Arab armies captured Jerusalem (638), Egypt (639–642), Mosul (641), Persepolis (650), and Armenia (652). By the 8th century, the Umayyad Caliphate (661–749) stretched from the Indus River to the Pyrenees Mountains.[2]

The Islamic empires that followed developed new forms of statecraft that sought unity in a common Muslim theology, which imperial authorities heavily promoted even while they tolerated Christianity and Judaism.

Evolution of a Taboo. Max D. Price, Oxford University Press (2020). © Oxford University Press.
DOI: 10.1093/oso/9780197543276.001.0001.

But political and religious factionalism continually tugged at the seams of the empires. The medieval Near East saw numerous civil wars, dynastic changes, and religious struggles, including the split between Sunni and Shia. Foreign invasions also threatened Islamic hegemony. The Crusades (1095–1291) set up a Christian stronghold in the Levant at roughly the same time that Turkic peoples were migrating in large numbers into the Middle East. Meanwhile, Mongol invasions ravaged the armies and cities of the Near East, especially during the sack of Baghdad in 1258 and the murder of perhaps a million civilians in the process.[3] In the centuries that followed, a Turkic people, the Ottomans, set up a state in western Anatolia that would conquer Constantinople (1453) and later control most of the Near East, with the exception of Persia, until the 19th century.

Islam: Orthopraxy, Food Laws, and the Pig Taboo

The Quran, along with its commentary in the *sunna* and *hadith*, contains the bulk of Muhammad's theological legacy. As with Judaism and Christianity, at the heart of Islam is a concern for egalitarianism, piety, social justice, and pursuing divine truth. Like Judaism, Islam is essentially orthopraxic in orientation;[4] it stresses adherence to laws of moral behavior as a means to achieve spiritual purity.

If Christianity represents a shift in Abrahamic tradition from orthopraxy to orthodoxy, the Quran and *hadith* recast orthopraxy by incorporating orthodoxic elements while ultimately remaining a religion based on practices.[5] The most important practices of Islam are its so-called five pillars: the profession of faith (*shahada*), praying five times a day (*salah*), charity (*zakat*), fasting during the holy month of Ramadan (*sawm*), and going on pilgrimage to Mecca (*hajj*). In addition to these fundamental rules, the Quran, *sunna*, and *hadith* spell out several other core practices. They include male circumcision, female modesty (however interpreted), avoiding alcohol and gambling, not charging interest on loans (or at least at usurious rates), and not eating pork.[6]

Many of these proscriptions are reminiscent of *halakha*, especially circumcision and the pork taboo, and in fact betray the strong influence of Judaism on Muhammad's thinking. Muhammad was attempting to forge a new religion out of existing theological threads. Judaism and Christianity were important features of the general religious milieu of the Near East in the

6th–7th centuries AD. But the influence of Judaism on Islam was also born out of the pervasiveness of Judaism in Arabia and East Africa. Jewish communities existed as far away as Ethiopia and India by the early centuries AD In Muhammad's day the Hejaz and Yemen contained several Jewish towns. The early history of Islam is replete with interactions between members of the new faith and their Jewish neighbors. At times, these relations were peaceful. For example, Jews welcomed Muhammad and his community (*umma*) into Yathrib (later called Medina) after his escape from Mecca, an event referred to as the Hijra (migration). At other times, early Jewish-Muslim relations were marked by hostility. For example, Jews were expelled from Badr and other conquered towns and cities in the Hejaz.[7]

On a theological level, Muslim leaders, like early Christians, were careful to balance, on the one hand, their acknowledgment of the spiritual legacy of Judaism, which gave them legitimacy, and, on the other hand, their desire to distinguish Islam as a superior faith. The food taboos outlined in the Quran are prime examples of Islam's revitalization of *halakha* as well as its differences from Jewish traditions. [8] In particular, the rules strip *halakha* down to what Muhammad thought were its essential elements—bans on pork, carrion, blood, and animal sacrificed to other gods:

> Prohibited to you are dead animals, blood, the flesh of swine, and that which has been dedicated to other than Allah, and [those animals] killed by strangling or by a violent blow or by a head-long fall or by the goring of horns, and those from which a wild animal has eaten, except what you [are able to] slaughter [before its death], and those which are sacrificed on stone altars, and [prohibited is] that you seek decision through divining arrows. That is grave disobedience. This day those who disbelieve have despaired of [defeating] your religion; so fear them not, but fear Me. This day I have perfected for you your religion and completed My favor upon you and have approved for you Islam as religion. But whoever is forced by severe hunger with no inclination to sin—then indeed, Allah is Forgiving and Merciful. (Quran 5:3)[9]

These rules are repeated a number of times (Quran 2:173, 5:3, 5:60, 6:145, 16:115). All of these taboos have precedents in the Hebrew Bible—for example, those on pork (e.g., Leviticus 11:7), blood (e.g., Leviticus 17:13), carrion (e.g., Deuteronomy 14:21), and meat sacrificed to other gods (Exodus 34:15).[10] Note also that the Quran makes an exception for cases of life-threatening emergencies or accidental ingestion (Quran 6:145, 16:115), as does the rabbinic tradition.

The food taboos reflect Islam's unique approach to the Abrahamic tradition. By relaxing the food laws but keeping the taboo on pork and several other key features of Jewish Law, the writers of the Quran sought to establish the "golden mean between two undesirable extremes."[11] On the one hand, Islam rejected Christianity's suspension of the taboos and its shift to orthodoxy. On the other hand, it eschewed the excessiveness of Jewish Law as unnecessary and distracting. While the early Christian theologians tried to balance their inheritance of Jewish thought with the desire to present a new and radical truth, the first Muslim thinkers worked toward the same end by threading the needle between Christian and Jewish tradition.

In general, the Muslim leaders were more explicitly tolerant than the Church Fathers. Thus, in contrast to the early Christians, Muhammad encouraged Muslims to eat with and respect Christians and Jews, the "people of the Book" (*Ahl al-Kitab*; Quran 5:5). While Barnabas and Ephrem the Syrian viewed Jewish customs as an existential threat to their faith, the Quran espoused a more nuanced perspective. It placed Jews and Christians on a spectrum situated between the ideal of Islam and the alleged backwardness of pagans.[12] In other words, if you had to choose a religion besides Islam, Judaism and Christianity were better than the alternatives.

Islam's viewpoint on the other Abrahamic faiths reflects Muhammad's desire to position Islam as a more purified form of the religion "revealed" to the ancient Israelites. He painted the other Abrahamic religions as essentially true, but bastardized in form by contemporary Jews and Christians. These adulterations, however, should be tolerated. For instance, the Quran explained that the strict laws of the Torah were not false per se, but merely exclusive to Jews owing to the Israelites' alleged transgressions against Yahweh/Allah (Quran 6:146–147).[13]

Nevertheless, the writers of the Quran were careful to keep Jews and Christians in the category of disbelievers. For example, the Quran states that "those who do not believe in the verses of Allah—Allah will not guide them, and for them there is a painful punishment" (Quran 16:104). While this passage suggests divine justice rather than legal persecution, medieval Islamic states did not shy away from applying oppressive rules to non-Muslims under their hegemony. They levied a special tax, the *jizya*, on Christians and Jews, who were sometimes referred to as *dhimmi*, or "protected people." Part tithe, part protection racket, the *jizya* emphasized the ambiguity with which early Muslims held Jews and Christians. Their religion should be respected—but not too much!

The Quran's retention of only a handful of the Torah's food taboos, including that on pork, reflects the struggle to both legitimize Islam within Abrahamic tradition and differentiate it from competing faiths. But why choose to keep the ban on pigs as opposed to, for example, that on camels? The main reason was that, by Muhammad's time, Jews had already focused their orthopraxic lens on three key elements: observing the Sabbath, avoiding pork, and circumcising male children.[14] Anyone attempting in the 6th–7th century AD to revitalize Abrahamic orthopraxy and strip it down to its most essential features would have found these elements at the core. Only by reading the Torah would one realize that, as it is written, the pig taboo held no special place. Accessing a copy of the Hebrew Bible was no mean feat in the era before the printing press and, anyway, Muslim tradition has long held that Muhammad was illiterate (*ummi*).[15] It is therefore likely that the prophet had to rely on his observations of Jewish practice rather than careful readings of the scriptures.

The Muslim pig taboo was also practical. Pigs were extremely uncommon in the Arabian Peninsula. Zooarchaeological data indicate a near or total absence of pigs at almost every site in the Arabian Peninsula since the Neolithic.[16] While there is no evidence for a taboo on pork among Arab peoples, Western writers during the Classical period took note the absence of swine. For example, the 5th century Christian historian Sozomen noted that "Saracens" (i.e., Arabs) did not eat pork.[17] Zooarchaeological evidence clearly shows, however, that other animals proscribed by *halakha*, such as camels, were eaten on the Arabian Peninsula.[18] Therefore, while Muhammad might have had trouble convincing local Arabs to adopt all the food laws of Leviticus, he would have met little resistance in declaring pork *haram* (forbidden). Muslim authorities would have had more trouble after Islam spread to other parts of the Near East where pigs were more popular. In fact, the zooarchaeological data attest to the vicissitudes of pig production and consumption in the wake of the Islamic Conquest.

Zooarchaeological and Historical Data on Pigs

Zooarchaeological data deriving from Near Eastern sites occupied over the past 1,500 years are extremely scarce. This is a result of both a lack of archaeological research on the Islamic and Ottoman periods and a bias among many researchers who tend to ignore what they perceive as uninteresting features of everyday life. As one expert remarked, "[E]xcavated material has contributed to our understanding of the formation and development of Islamic art, but we know very little about what was eaten."[19] The unfortunate result is that there are large gaps in our understanding of pig husbandry and other forms of livestock production in the Islamic period.

The available zooarchaeological data show a general decline of pig exploitation during the Islamic period, but not its total elimination. Data from Mesopotamia and Iran largely show an abandonment of pig consumption.[20] However, in central and western Anatolia, which remained under Byzantine control until the 14th century, swine husbandry remained a prominent feature of livestock production.[21] For example, pig remains represent 48 percent of the livestock remains at Gritille Höyük in the 11th–13th centuries.[22] In Egypt, too, pig husbandry remained common, especially at trading emporia and Christian settlements. Excavators have recovered large numbers of pig bones, as well as their coprolites and tethering posts, from the 11th–13th century Islamic fortress of Hisn al-Bab south of Aswan.[23] These data corroborate textual records, which indicate that in Alexandria and Cairo, pig husbandry operations run by Greek and Venetian merchants thrived throughout the Islamic period, despite the often hefty taxes imposed by Muslim authorities.[24]

In the Levant, pig husbandry declined significantly after the 7th century. But it persisted in those areas with sizable Christian communities. For example, pig bones amounted to 17–52 percent of the livestock remains in various levels of the Islamic period occupation of Caesarea.[25] But even some Christians in the Muslim-dominated Levant began to give up pork.[26] At Pella, a town in the northern Jordan River Valley that hosted a sizable Christian community,[27] the relative abundance of pig remains declined from 10 percent of the livestock remains in Late Byzantine times to less than 1 percent by the time of the Abbasid Caliphate.[28]

The Crusades temporarily altered this trajectory in the Levant. Most Crusader sites contain modest proportions of pig remains (5–20 percent of livestock), but the contrast with sites outside the Crusaders' dominion is as stark as that between Philistine and Israelite sites two millennia earlier.[29] Northern Europeans' taste for pork no doubt contributed to this uptick in swine production. But pigs' unique ability to thrive exclusively in urban environments also played a role during the many sieges that took place throughout the Crusader period. A good example comes from the excavations of Arsur, a city captured in 1101 by the Crusader Baldwin I and later destroyed by the Mamluks in 1265 following a lengthy siege. Zooarchaeologists working at Arsur compared animal bones from the early 13th century with those deposited during the final siege. While the proportion of domestic pigs was initially only 3 percent, it increased to 61 percent of the livestock remains during the siege.[30]

While most Muslims and their neighbors avoided raising domestic pigs, they continued to hunt wild boar, at least occasionally. For example, in North Africa, the Amazigh were reported to have hunted wild boar as late as the 20th century.[31] Zooarchaeological indications of boar hunting can also be seen at Nazareth, where the assemblage of mammal bones dating to the Mamluk and Ottoman periods comprised 8 percent to 4 percent wild-sized *Sus scrofa* remains.[32] Pig or wild boar remains (it is not clear which) made up 6 percent of the entire assemblage of bones found in 19th and 20th century deposits from the Palestinian Arab village of Majdal Yaba, a settlement with no historical records of a non-Muslim population.[33]

Although it is reasonable to assume that pig bones found in archaeological deposits usually reflect the consumption of pork, there were other uses of pigs that may explain their presence. One of these was magic. Consistent with traditions rooted in the 3rd millennium BC (Chapter 5), pigs were thought to be powerful medico-magical elements in extracting disease or warding off evil, especially in veterinary contexts. For example, 19th century traveler accounts allege that horse breeders in Egypt and Morocco raised wild or feral pigs with their horses in order to ward off malevolent spirits, or *jinn*.[34] Apparently, pork also held magical powers. British archaeologist George Murray[35] described encountering a Bedouin sheikh begging for pork outside the Monastery of St. Catherine in the Sinai Peninsula. Finding this behavior peculiar for a high-status Muslim, Murray questioned the man and discovered that he did not intend to eat the pork but to use it as a cure for his camel's blindness.

Today, official statistics collected by the Food and Agriculture Organization of the United Nations (FAO) reveal a small but vibrant pig husbandry economy in the Near East.[36] Some countries, like Saudi Arabia, have officially banned the importation and sale of pork, barring any possibility of pig husbandry or consumption.[37] But many countries, like Turkey (1,642 head) and Lebanon (7,000) report small numbers of pigs, while Israel (200,000 head) is experiencing a rapid expansion in swine husbandry.

The official records often leave out pigs raised informally and/or in secrecy, open or otherwise. The FAO, for example, does not record any pigs in Jordan even though archaeologists working in that country have indicated that some Christians raise small numbers of pigs.[38] But perhaps the most glaring example of unreliable pig records comes from Egypt, where the FAO reported 37,000 head of pigs in the country in 2008, just one year before the Egyptian government slaughtered around 300,000 pigs in its attempt to combat the H1N1 flu.[39] This order-of-magnitude difference between what was reported and what existed on the ground calls attention to the informal economies that currently surround pigs.

Raising Domestic Pigs in the Near East Today

The number of pigs being raised in the Near East today pales in comparison with, for example, the number of sheep, which number in the millions in many countries in the region.[40] Although it is difficult to measure, per capita pork consumption across the entire Near East is probably at its lowest point since the Neolithic. Nevertheless, pig husbandry remains a vibrant activity in several communities. Among them are the Coptic Christians in Egypt, who tend to raise pigs in an off-the-books "informal" manner, and large-scale pig factory farms in Israel, which are part of the formal sector.

Informal Economics: Pig Husbandry in Egypt

In Chapter 5, I introduced the term "informal economy" to refer to any type of off-the-books or under-the-table transaction. Such informal transactions rely on face-to-face interactions among people who develop personal relationships with one another, in contrast to the more discrete and anonymous exchanges typical of the formal economy. While often disregarded, the informal sector of the economy accounts for well over half of nonagricultural labor in many parts of the developing world. It is a significant component of the overall global economy.[41] While I have speculated that pigs may have formed a part of the informal economy in the Bronze Age, there is definitive evidence that they do so in modern contexts as well. The Zabaleen in Cairo provide a prime example of this.

As we briefly saw in Chapter 1, the Zabaleen (plural of *zabal*, "garbage") are Coptic Christians who run an informal—yet incredibly efficient and sustainable—waste management system in and around Cairo. While operating in the informal sector, the Zabaleen are highly organized and specialized, with parents passing down techniques to their children. They collect garbage in donkey-drawn carts or pickup trucks from homes around Cairo and bring it back to the settlement of Manshiet Nasser, with its "Garbage City," on the outskirts of the metropolis.[42] There, they sort the waste into over a dozen categories—for example, glass, aluminum, plastic, paper, organic waste—and process it. Much of it they sell to dealers in raw materials, but a good portion of the organic material goes to feed their pigs. The pigs are then sold for their pork, often via informal transactions.[43]

The history of the Zabaleen dates back to the 1930s, when Christians migrated to Cairo from southern Egypt and began purchasing the rights from another group (the Muslim Wahiya) to collect garbage, much of which they used for pig feed.[44] The Zabaleen kept pigs in enclosed yards (*zeriba*) adjacent to families' houses, a management practice still common today.[45] Soon, and with the Wahiya serving as their middlemen, the Zabaleen began to take over the waste management trade. Over the next few generations, the Zabaleen became a regular and critical feature of Cairo as the city grew in population and areal extent.

The informal nature of the Zabaleen waste management collection has caused conflict with authorities in Cairo. Mai Iskander's film *Garbage Dreams* (2009) documents how the Zabaleen, who live below the poverty line and contend with occupational risks ranging from high infant mortality to tetanus, have struggled in the wake of the Egyptian government's 2003 decision to sell waste collection contracts to international corporations. This decision pushed the traditional Zabaleen way of life to the margins. But the hired workers of the waste management companies have been far less efficient. On average, they have achieved a 20 percent recycling rate compared with the Zabaleen's 80 percent rate. The numbers testify to the potential of family-organized ingenuity to pick up the slack in the margins of capitalism and its unfulfilled promises of progress and efficiency.

One of the biggest and most publicized blows to the Zabaleen way of life came in April 2009, when the H1N1 swine flu scare prompted the Egyptian Agricultural Ministry to slaughter around 300,000 pigs,[46] the vast majority of the Zabaleen's herds. The move was chastised by international organizations, including the World Health Organization, which pointed out that, despite the name, swine were not a major vector of H1N1.[47] Disregarding these recommendations, the government went ahead with the cull and ripped away a major source of income for Zabaleen families.

The Zabaleen's objections to the cull were twofold. First, they perceived the move as part of a government attempt to Islamize the country. Second, they were incensed by the loss of valuable livestock. Prior to the pig cull, families had been able to sell 5–15 pigs from their litters every six months at a rate of about 450 Egyptian pounds (at the time around US$80) per pig. And while the government compensated the Zabaleen for their losses, it offered them only up to 250 Egyptian pounds per pig. Moreover, the elimination of the breeding stock had effectively removed a reliable and predictable source of income.[48] In reaction, the Zabaleen staged a large-scale protest in which they refused to pick up the garbage in Cairo. The protest brought the city to a standstill, garnered international media coverage, and proved humiliating for the Mubarak government.

The ousting of Hosni Mubarak in 2011 and the rise of the Muslim Brotherhood in 2012 proved both hopeful and challenging to the Zabaleen. While recent Egyptian governments have promised to work with the Zabaleen,[49] the strengthening of Islamist ideology represents a direct threat to raising swine. For the moment, however, pig husbandry has regained its momentum, even though raising pigs remains (technically) illegal in Egypt. Still, the informal economy can thrive outside the law, especially if officials are unwilling to enforce it. In 2014, fully 50,000 pigs were reported in the Zabaleen district of Manshiet Nasser.[50] Tentatively, it seems, Egypt's pig farmers are rebuilding their herds.

Formal Economics: "White Steak" and Pig Husbandry in Israel

Informal pig husbandry in Egypt contrasts with the formal pork economy in Israel. While formal, the production of pigs in Israel is not without controversy. Soon after the country achieved independence in 1948, a debate erupted over whether Israeli citizens could import pork or raise pigs. The argument, which pitted left-wing socialist Zionists against right-wing religious leaders, caused considerable strife and even occasional street fights in the new Jewish democracy.[51] In the end, despite vigorous opposition from the Left, Israel passed laws banning pork and pig production, although provisions were made for medical uses of pigs (e.g., in research and as a source of insulin in injections for diabetics).[52] In 1956, the Local Authorities (Special Enablement) Law gave municipalities the right to outlaw possession of and trade in pork. In 1962, Israel adopted the Pig-Raising Prohibition Law, making it illegal to raise pigs except in areas heavily populated by Christians.[53]

In her book, *Outlawed Pigs*, the Israeli Supreme Court judge Daphne Barak-Erez notes that these laws provide a unique example of a "basically secular legal system adopting a specific religious norm."[54] This contradiction has led to considerable tolerance on the part of authorities for the activities of clever pig farmers in Israel. In particular, a legal loophole allows the sale of meat from pigs raised for research. As long as the pigs are classified as research animals, they can be sent to the butcher and slaughtered. The "researchers" can then sell their meat. This has enabled the production of pork in large quantities on kibbutzim such as Mizra and, more recently, Lahav, where roughly 20,000 "research animals" are slaughtered every year.[55]

A recent surge in demand for pork has made pig husbandry in Israel a more lucrative but no less contentious form of agriculture. Since the 1970s, some restaurants have offered "white steak" on their menus, a speakeasy term for pork. For decades, the market was small and decidedly niche. But in the 1980s and 1990s, the influx of immigrants from the former USSR brought a new demand for pork.[56] As a result, the number of pigs raised by commercial farms in Israel has almost doubled from 120,000 head in the mid-1990s to almost 200,000 today.[57] Many of these farms are located in Christian villages, where pig production is legal. The sale of pork to restaurants and distributors is more complicated. While the market for "white steak" has grown, its legal status has remained opaque. However, in 2004, the Israeli Supreme Court sidestepped the issue by recognizing the legitimacy of the pig laws while nevertheless deferring to local authorities, especially in areas heavily populated by Christians, on matters of pork production and sale.[58]

As pig husbandry takes off, so too do concerns for animal welfare. In 2005, Tamara Traubman reported in the newspaper *Haaretz* on pig farms in A'abalin, a Christian village in the north of the country.[59] "It felt like we had arrived in hell," she wrote. She described how farmers castrated male piglets without the use of anesthesia and kept their sows in "isolation cages" (also called "gestation crates"), which confine peripartum sows to just over one square meter of space. In no way unique to Israel, such practices, which are difficult to describe as anything but inhumane yet nevertheless enhance the efficiency of industrial-scale meat production, are common among commercial pig farms the world over. But formal economies are subject to regulation. Indeed, unlike the situation in informal economies, abuses in formal economies can often be dealt with only in a top-down manner. Like Americans, Chinese, and Europeans, Israelis must decide whether an animal—even one that should never be eaten or even touched—deserves ethical treatment.

Wild Boar in the Near East Today

In her 1881 book, *Pilgrimage to Nejd*, the British traveler Anne Blunt described her encounter with a wild boar in the marshes of southern Iraq. Deciding to hunt the animals on horseback, she, her fellow travelers, and their local guides were soon disabused of any illusions of an easy chase:

> The island was half under water, and droves of pigs, boars, sows, and little ones, turned out of the bushes, where they generally lie in the day-time, were grunting and trotting and splashing about. We singled out a great red boar, and all gave chase [. . .] At last, he charged, and was hit, but not enough to stop though it turned him, and then we had another gallop, and another shot rolled him over. The people on foot, who were following, rushed in, but just as they got near him up he jumped, and bolted towards some deep water [. . . but the boar] suddenly changed his mind, [. . .] and before we were aware, had charged right in among us. Wilfrid turned his mare, but alas, not fast enough. To my horror, I saw the hideous beast catch Ariel [the mare] and give her a toss, such as I have seen in the bull-ring by a bull. He seemed to lift horse and rider clean off the ground.[60]

Blunt's depiction of wild boar teeming in parts of the Near East is just as applicable today.[61] Wild boar populations are thriving across the region. In Turkey, the animal's notorious ferocity has inspired a small industry of hunting tour companies targeted primarily at wealthy male Europeans and Americans. Not unlike Odysseus's hunt, killing wild boar provides a powerful test of manhood among the elite. But wild boar are dangerous quarry whether or not one is hunting them. A story published in several media outlets in 2017 recounted how three ISIS fighters were killed by wild boar near the town of Hawija in northern Iraq. Apparently, the jihadists had been hiding in the reeds, waiting to attack the town, when they were surprised by a sounder of wild boar.[62]

Wild boar remain crop pests in the region. Since at least the 19th century, governments have attempted to control wild boar populations. In 1846, the Egyptian government enlisted 832 soldiers to march through Lower Egypt,[63] shooting any and all wild boar they encountered. The Egyptians were largely successful in their campaign; the last reported wild boar in Egypt was killed at the beginning of the 20th century.[64] Other governments, however, have been less successful. Wild boar and feral pig populations can be contained in places like Egypt, where their only habitat is a narrow strip of green around the Nile. It is far more difficult to manage them in less circumscribed regions, like the marshy or hilly regions of Iraq, Turkey, Syria, Lebanon, Israel, and Palestine.[65]

Reporting for *Vice News* in 2014, journalist Ben Hattem talked to Palestinian farmers in the West Bank village of Salfit about an Israeli settlement dumping sewage on their land.[66] But Hattem soon learned that a far greater menace to the Palestinians was swine. The number of wild boar, according to local villagers, had increased around 20-fold in recent years. The damage to their crops was causing many to abandon agriculture altogether.

But Hattem's investigation uncovered another story, one that places pigs squarely at the intersection of nationalist and ethnic politics. The villagers of Salfit complained not only about the wild boar, but also about how authorities prevented them from taking care of the problem. Both the Israeli government and Palestinian Authority prohibit the use of firearms as well as the poison strychnine. Other than chasing them off their land, the Salfit villagers had no way to prevent the wild boar from ravaging their fields. In frustration, the villagers began to voice their suspicions of a conspiracy. Hattem reported their allegations, unverified and officially dismissed as rumors, that the Israeli government or Jewish settlers had unleashed the wild boar. The president of the Palestinian Authority, Mahmoud Abbas, had mentioned similar allegations in a speech in 2012.[67] Although there is no evidence to support these claims, many Palestinians living in the West Bank find it believable enough that Jewish settlers would weaponize swine in order to drive them from their homes.

Swine, Bigotry, and Intolerance

The accusation that the Israeli government or settlers unleashed a horde of swine on Palestinian communities recalls the long-standing role of pigs in the politics of Near Eastern peoples. In previous chapters, we have seen how pigs (or their absence) have negotiated class differences, demarcated religious spaces/people, and defined ethnic groups. It is precisely because of the heightened degree of cultural sentiment afforded to them that pigs embody enough power to have become weaponized. Pigs, for all the good they have heaped upon humanity, have also been pawns in all too human acts of intolerance. These episodes, occurring within the Near East and without, have colored Jewish-Christian-Muslim relations for the past 1,500 years.

In Chapter 8, we saw the origins of swine-related acts of intolerance in the Classical period. By the medieval period, the use of pigs in interfaith politics became even more pronounced. While medieval Islamic states were generally more tolerant than their European Christian counterparts, the reality was that their "tolerance" often amounted to little more than abstention from direct persecution. In fact, Muslim authorities placed strong limitations on the activities of Christians and Jews. The Pact of Umar, ostensibly dating to the reign of the Caliph Umar (634–644), laid out a series of regulations for Jews and Christians, including the *jizya* tax, prohibitions on building or repairing places of worship without permission, the public display of religious paraphernalia such as crosses, and public ritual activity.[68] Christians and Jews had to wear special clothes so that they could be easily identified, and they were banned from raising or selling pigs.[69]

Many authorities chose to relax the enforcement of legislation such as the Pact of Umar. Maintaining peaceful relations with Christian and Jewish communities and/or more gently persuading them to convert to Islam were typical tactics of everyday Muslims in the medieval Near East. In this spirit, the prohibitions on Christians raising pigs were frequently overlooked. Nevertheless, the proximity of pigs to their homes and places of worship was often perceived by Muslims as offensive. Christians who ate pork were considered polluted and could be barred from social or political advancement in the Islamic empires.[70] Within and around the Near East, Muslim soldiers or authorities occasionally killed pigs, either as acts of casual violence or as part of organized programs to eliminate the tabooed animal, as the 2009 Cairo incident exemplifies.[71] These episodes stripped pig owners of valuable livestock and sometimes their livelihoods.

Christian bigotry, too, has leaned heavily on pigs and pork in its oppression of Jews and Muslims. During the Spanish Inquisition (1478–1834), when Jews and Muslims were told to convert or die, eating pork was taken as proof of conversion.[72] In response, many Muslims and Jews chose to play

Christian by eating pork publicly and attending church, while carrying on their traditions in secret.[73] The so-called crypto-Jews and crypto-Muslims continued to practice their faiths, even settling in the Spanish colonies of the New World to escape the Inquisition. From afar, Muslim and Jewish authorities advocated resistance to the Inquisition, even if it meant violating the pig taboo. For instance, in 1504, the mufti in Oran handed down a *fatwa* (legal opinion) that provided tactics for combating forced conversion: "If they force pork on you, eat it, but in your heart reject it" (Harvey 2005:62).

If pigs were weaponized in the Inquisition as a litmus test of conversion, more casual episodes of pork terrorism have pervaded Europe in the medieval period up to the present day. A children's game allegedly once common on the island of Majorca involved surrounding a *xueta* (descendant of a converted Jew or crypto-Jew) and slipping pork into his or her pocket.[74] Even though the individual was Christian, the prank was a declaration by the community that it remembered the *xueta*'s outcast status. Other episodes of pelting Jews or Muslims with pork or tricking them into touching it were common throughout Europe and remain so up to the present day. In May 2016, a closed-circuit television camera caught a man throwing rotten pork at a mosque in London.[75] The incident is a depressing reminder of how far Western society has yet to go in terms of embracing a policy of tolerance.

Some of the most glaring episodes of intolerance between the three religions involve stories, jokes, and folktales. A somewhat counterintuitive, yet commonly held folk belief in Europe was that Jews were, or at least were somehow essentially linked to, pigs. Jews were said to have ears like pigs, to smell bad like pigs, and, like pigs, to have shorter pregnancies than Christians.[76] They were ostensibly greedy and lusty like swine.[77] This explained Jews' abstention from pork—it was a form of self-preservation. The connection between Jews and pigs was enshrined in artistic representation. For example, starting in the 14th century, German artists began depicting *Judensau* (Jews' sow) in sculpture and woodcuts, which typically show Jews suckling from a sow and eating from its anus/genitals.[78] Figure 9.1 is one such woodcut. More controversially, the *Judensau* images decorate the outsides of over a dozen churches and cathedrals throughout Germany, Switzerland, and France to this day.

Jewish bigots also used stories or metaphors of pigs to attack Christians. On a basic level, many Ashkenazi Jews—those who lived in central and eastern Europe and who spoke the German-derived language Yiddish—equated their Christian neighbors with "impurity." For example, the Yiddish

Figure 9.1. Woodcut entitled *Das grosse Judenschwein* (The Jews' Big Pig). Germany. 15th century.

word *shikse* refers to a young non-Jewish woman, usually an attractive one who might tempt a Jewish man. The word is the feminized form of *sheygets*, in Hebrew *sheketz*, a term used in Leviticus to refer to a class of impure animals, such as slithering reptiles, fish without scales or fins, and vultures (although not pigs) (Leviticus 11:9–20).[79]

Even more directly related to swine are the folk traditions and jokes in Ashkenazi culture that depict non-Jews (*goyim*) in ways all too reminiscent of swine. Goyim are joked about as being hopelessly wayward, slovenly, fat, dull, and hedonistic. A Yiddish folksong sums up this image: "He is drunk / He has no choice but to drink / Because he is a goy."[80] A Yiddish children's finger rhyme (ironically similar to the "this little piggy" rhyme common in America), starts with the pinky, the *yidele* ("little Jew"), and ends with the thumb, the *grober goy* (big fat Gentile).[81] Gentiles are not explicitly called pigs; they don't have to be. The images used to denigrate them are the exact ones that have been used by Jewish scholars to depict swine for centuries (e.g., Philo, *On Husbandry* 32.143–145).

The descriptions of Gentiles play on the symbolic significance of pigs in Ashkenazi culture. In Yiddish, a *khazer* (pig) is someone who is greedy and unpleasant; *khazerai* is unhealthy food or junk. All of these qualities describe the Gentile in traditional Ashkenazi thought. Despite frequent attempts to dismiss these and other examples of Jewish prejudice as simple jokes, a small bit of humor to lighten the load on the shoulders of a historically oppressed people, they ultimately reflect the cyclical perpetuation of bigotry. Cut from the same cloth as the *Judensau* imagery, Yiddish stereotypes of swinish Gentiles are exemplary of the all too human tendency to indoctrinate children from a young age into an ideology that views the members of other groups as foreign, bizarre, and, to varying degrees, less than human.

Pig-featured bigotry also found a place in medieval Jewish scholarship. A number of polemics, written by Jews from the 12th century and later, mock the hypocrisy of, on the one hand, Christians' acceptance of the Hebrew Bible and, on the other, their neglect of its laws.[82] The anonymously written *Sefer Nizzahon Yashan* (Ancient Book of Victory), dating to around 1300, contains some prime examples of the anti-Christian arguments common among Jews in France and Germany. In one passage (103), the author identifies Christians as the blasphemers whom the prophet Isaiah had condemned. Specifically, Isaiah 65:1–4 referenced "obstinate people" who turn away from God, offer sacrifices in gardens (monasteries, according to the *Nizzahon*), burn incense on altars, and eat pork.[83] Another passage in the *Nizzahon* (227) contains an indecorous rumor about the Muslim prophet:

> Their god Muhammad got drunk from wine and was thrown into the garbage, and when the pigs came and passed through the dump, they found him, dragged him, surrounded him, killed him, and ate him; how, then, could he be divine?[84]

Islamic bigotry has also referenced pigs in its attacks on Christians and, less frequently, Jews. For example, the Arab writer al-Jahiz (777–869) in his book *Al Radd 'ala l-Nasara* (Refutation of the Christians) concluded that "the Christian is at heart a dirty and foul creature. Why? Because he is uncircumcised, does not wash after intercourse, and eats pig meat."[85] Pigs were also used as metaphors for non-Muslims and their spiritual impoverishment. Some of the most unusual passages in the Quran (36:63–67 and 5:60) discuss divine punishment in which sinners, often interpreted as Jews and Christians, are transformed into apes and pigs.[86]

> Say, "Shall I inform you of [what is] worse than that as penalty from Allah? [It is that of] those whom Allah has cursed and with whom He became angry and made of them apes and pigs and slaves of Taghut [Satan]. Those are worse in position and further astray from the sound way." (Quran 5:60)

When viewed against the backdrop of the symbolic significance of swine in Near Eastern cultures, this passage reveals more than just an equation between non-*halal* meat and those who might eat it. It also seems to play on an uncanny resemblance between pigs, apes, and people. Just as pigs have served as substitutes for humans in rituals dating back to the Bronze Age, sins magically transferred from human to suid, so too can sinners become trapped in the bodies of swine. To refuse to convert to Islam, according to some interpretations of this passage, is to be a pig. Or, perhaps, to turn away from spiritual enlightenment is to condemn oneself to one's inner swinishness.

Transgression

When Christians, Jews, and Muslims employ pigs as vehicles for hurling abuses at one another, they unwittingly increase the power of pigs, reinforcing the equation of pork consumption or avoidance with religious identity. This can make taboos stronger, as we saw in the case of the evolution of the pig taboo in the Classical period. But it also marginalizes those seeking cultural hybridity. Those who reject the taboo are branded transgressors. Many find life in the interstices between cultural extremes a difficult one.

The perceived threat posed by transgression to one's sense of identity develops at an early age. For example, in a study of Turkish German children growing up in Berlin, a team of early childhood education researchers interviewed a preschooler named Ayla, whose family, Alevi Muslims, ate pork. Ayla identified her self-perceived chubbiness as related to her pork consumption and its pollution of her Muslim identity. According to the researchers, for her, "eating pork equals being a pig, equals being unlikeable, equals non-Muslim/non-Turkish, equals being a sausage-eating German, equals getting fat, equals Ayla's self-perceived body image."[87] Even though Ayla's parents belonged to a Muslim minority sect that does not subscribe to the pig taboo, the prevalence of the equations Muslim = no pork and German = pork pervaded her sense of personhood. Even to a young child, the uneasiness of transgression can translate into a fear of loss of identity, which can be manifested as self-disgust.

These strongly internalized linkages between one's sense of self and pork avoidance act as strong barriers against breaking the pig taboo. Nevertheless, as they did historically, some Jews and Muslims today eat pork, willfully transgressing the ancient texts' unequivocal commandments. Some of the best examples are drawn from places outside the Near East where Muslims and Jews are in frequent contact with Christian or other pork-eating communities. A landmark zooarchaeological study of 16th–18th century Amsterdam, for example, found small but variable numbers of pig bones (0–5 percent) and other nonkosher food remains such as mollusk shells in cesspits associated with Jewish households.[88] Similarly, ethnographic accounts describe how Hui Muslims in China eat pork as a way to blend in with their non-Muslim neighbors. To avoid openly violating the Quran, they refer to the meat as "mutton."[89] Historical documents also attest to swine husbandry and pork consumption among Muslims in North and sub-Saharan Africa.[90] In fact, pig husbandry was encouraged in Sudan under the Sennar Sultanate (1504–1821) as a way to settle mobile pastoralists into villages.[91]

Many cases of transgression are based on personal choices, not top-down directives. One factor may be simple curiosity; if Christians and other groups around the world relish pork and speak so fondly of it, it must be something worth trying. One could even argue that pork, *because* it is so taboo, can be irresistibly titillating in the same way that tabooed sexual acts can be the focus of fetishes. Such "transgression fetishes," the subject of many jokes involving otherwise pious rabbis secretly eating swine flesh, may explain why many Jews seem to enjoy pork so much.[92]

Another factor involved in the decision to eat pork may be a desire to fit in with a dominant culture by accepting a food whose avoidance has typically been used to brand Muslims and Jews as "backward," "traditional," or simply "other." Eating pork may seem, to some, to signify an acceptance of modernity and a rejection of the ways of the "Old Country." Similarly, eating pork may appeal to some as a way of constructing a hybrid identity, mixing elements of their Jewishness/Muslimness with those of another culture with which they feel a sense of belonging (e.g., the case of the Hui Muslims). This sentiment might be particularly prevalent among immigrants, children of so-called mixed marriages, or simply Jews or Muslims who grow up outside Israel or Muslim-majority countries.

There may also be a political angle to transgression. Some Jews or Muslims may eat pork to demonstrate a more cosmopolitan identity and politically liberal philosophy by embracing the traditions of other cultures, even those that run counter to their ancestors' beliefs. In other words, some may transgress a taboo in order to make a statement about tolerance and diversity. Eating pork may also be a way of breaking with traditions that seem irrelevant or even oppressive. For example, the biographies of several Jewish atheists, socialist, and communists in the early 20th century include a story (whether apocryphal or true) of them eating pork, sometimes on Yom Kippur.[93] The example in Chapter 6 of how LGBT activism successfully challenged a taboo on homosexuality in the West shows how such scenarios can play out. Indeed, many Jews and Muslims strongly reject the taboo on homosexuality, despite its clear proscription and punishment by death (Leviticus 18:22 and 20:13). And while refusing to eat pork is certainly not comparable to murdering someone for being gay, the symbolic act of transgression in both cases might be intended as a rallying cry for a radical reconsideration of biblical and Quranic rules in light of recent social progress.

The political implications of transgression can be read in multiple ways. While some may see transgression as an act of liberation, others may view it as a betrayal of their people and an alliance with an oppressive dominant culture. The Jewish studies scholar Jonathan Schorsch[94] argues that those Jewish "foodies," especially Michael Pollan, who advocate enjoying pork are adopting a neocolonial, hyperrational approach to food that is inherently rooted in anti-Semitism. In other words, Pollan and other foodies, despite their self-proclaimed desire to be open and accepting, are participating in the harassment of their own people. But beneath the thin veneer of decolonialist rhetoric, Schorsch's argument smacks of a reactionary attack intended to brand those who eat and enjoy pork—and, worse, show no shame in the process—as transgressors. Dolled up in academic garb and disguised as an appeal to the liberal-minded, it is simply another rehashing of the argument that Jews who eat pork are sellouts to the majority, traitors ingesting the identity of the Hellenistic/Roman/Christian/Western "other." Those who make such arguments defy the anthropological reality of cultural hybridization. They arrogantly see themselves as the gatekeepers of identity, judges of who is and who is not a "real" Jew or Muslim.

The fact of the matter is that pork consumption has long been a matter of debate, even if the words of the Torah or Quran are unambiguous. The zooarchaeological evidence shows that episodes of transgression, however infrequent on a day-to-day basis, were pervasive in Israelite, Judahite, Jewish, and Muslim worlds. The historical data corroborate this picture of low-level but persistent transgression within many communities of Jews and Muslims.

Today, Muslims and Jews who eat pork argue that dietary laws written centuries ago are largely irrelevant. The essences of Islam and Judaism are their moral lessons. In fact, this argument is not too dissimilar from those found in the Gospels and the writings of Paul. But unlike the avowed purpose of early Christians, the goal of Jews and Muslims who reject the taboos is not to form a new religion, but to reclaim what they perceive as the most fundamental elements of Judaism and Islam. The most striking example of this is the abolition of food laws in Reform Judaism.[95] Reform Judaism began in Germany in the 19th century, but it has become one of the main forms of Jewish practice and worship in the United States. Reform rabbis explicitly reject the food laws in favor of more modern interpretations of the Hebrew Bible, the Talmud, and later rabbinic teachings. In fact, today, the majority (57 percent) of self-identified American Jews eat pork, according to a Pew Research poll.[96]

Even though pork is allowed in the Reform Jewish movement and many American Jews eat pork, there remains a degree of ambivalence. A cousin of mine, a long-practicing American Reform Jew, admitted to eating pork, but only in small amounts and only when he didn't have to think about it. It tastes great, he confessed, but "it gives me the willies." When interrogated further, he denied following any taboo or being influenced by the Torah, but added that consuming bacon was like "eating a pig's skin." It reminded him too much of the animal. Even though he did not have such qualms about roasted chicken, fish, or other forms of meat in which the animals' skin is visible, pork was somehow different.

Similar ambivalence can be seen in the types of dishes prepared. Oftentimes those dishes that look less obviously like pork are considered more acceptable. Another personal example shows this quite clearly. In Detroit in the 1950s and 1960s, my father's mother would occasionally serve bacon and ham, but not at special events and never as pork roast or pork chops, which were considered too obviously piglike—or, perhaps more important, too obviously *Gentile*. Indeed, many may tolerate transgression, but the outright flaunting of kosher laws is often perceived as sacrilege, or at least going too far. In fact, such flaunting of the laws is often cited as the reason for the break between Reform and Conservative Jews in the United States. At a celebration of the first graduating class of the Hebrew Union College in 1883, waiters set down shrimp and clams (an abomination according to Leviticus

11:10) before the rabbis. The outrage, which became known as the "Trefa Banquet" (after *tref*, "nonkosher"), allegedly inspired the development of the more tradition-focused Conservative movement in the United States.[97]

A mainstream reform movement eschewing the food taboos has not yet caught on in Islam, although some minority Muslim sects, like the Alevi, allow pork to be eaten. According to a Pew poll, 9 percent of self-identified Muslims in the US claimed to eat pork.[98] Dietary laws and other issues confronting Muslims in the 21st century are frequent topics of discussion on the podcast *Good Muslim, Bad Muslim*.[99] One of the hosts, Zahra Noorbakhsh, discusses her own consumption of alcohol and pork, a personal choice that she elegantly frames within the discussion of centuries-old questions: What are these dietary laws and why do we follow them? What is the *essence* of being Muslim or Jewish?

In summary, the emergence and development of Islam, as well as the changes in global cosmopolitan culture, have added nuances to the pig taboo. Islam's essentially orthopraxic orientation placed emphasis on certain rules of behavior, but Muhammad and his followers recast Jewish Law, boiling it down to what they perceived as its most essential elements. The pig taboo, which by the end of the Classical period had risen to the top of the hierarchy of Jewish food laws, was one of these elements. On the other hand, by rejecting the other elements of *halakha*, Muslims were able to differentiate themselves from Jews. By tacking back and forth between Christian and Jewish theologies, those who practiced Islam could claim theirs as the truest form of Allah's revelations to humanity.

Muhammad's political and military revolutions were as important as his religious ones. The success of Islam rested on the religion being made part and parcel of the Arab armies and empires that conquered the majority of the Near East. For pig farmers in the Near East, this spelled the end of the boost that they had received during the Classical period. Pigs no longer lent political elites a kind of cachet. Pork was once again the food of marginal people in much of the region. Yet the persistence of Christian communities in the Near East, and especially in the Levant, Anatolia, and Egypt, meant that pigs never left.

As communities of Jews, Muslims, and Christians mingled in the Near East, Europe, and around the globe, swine were increasingly marshaled to guard their borders. Of course, pigs and pork gained their greatest significance on the frontier between pork-eating Christians and pork-avoiding Jews and Muslims, but swine have also rooted their way into Jewish-Muslim relations. Acts of intolerance involving pigs and bigotry mobilizing the imagery of pigs were and remain to this day unfortunate by-products. Each episode of pork terrorism, or each threat by a US president to use pig blood against Islamist criminals, reproduces and magnifies the sentiments surrounding swine. As a result, the cultural and religious significance of pigs and pork has, in some cases, grown stronger, becoming further embedded in the individual's sense of self. But secularization is also gaining strength in many circles. As a result, members of all three faiths are increasingly questioning what it means to be Jewish, Muslim, or Christian and the differences that have separated them for centuries. What this means for pigs is anyone's guess; they have a history of following unexpected routes.

10
The Complexity of Swine

Swine in Retrospect: A Window onto Complexity

As we approach the end of the first quarter of the 21st century, a number of interconnected problems present themselves to the people of the Near East: the devastating effects of climate change, hypernationalism, the international oil trade, the withering effects of global capitalism, radical Islam, military intervention by foreign powers, and civil war. Human populations in the region have never been so large, the strain on agricultural and other resources never so great. The number of factions, and the power dynamics among them, seem to change each day. To many living within and outside the Near East, it is difficult to comprehend the situation from moment to moment, especially amid ongoing crises such as the Syrian civil war, the struggles for Kurdish and Palestinian national identities, and the tangle of alliances forged between atypical bedfellows, including the awkward Israeli-Saudi-American triangle.

As difficult as it is to wrap one's head around the present, it feels almost impossible to predict with any clarity what the future may hold for the Near East. Many, scholars and lay readers alike, seek shelter in what they believe to be a less complex past, one in which the lines were more clearly drawn and the players more consistent in their actions. Grand narratives, those that reduce history to, for example, the movements and migrations of people[1] or the geographic settings of different societies,[2] may assuage the anxieties of the present. Such stories, however, are myths. Admittedly, they are myths extrapolated from hard data, but their proponents frame them in such a way as to make those data far less complex than they really are. These stories are simplistic because they excise inconvenient truths.

This book, if it has one goal, is intended to add a dose of humility to the examination of the past. For even when a single variable is isolated—the interactions between people and pigs in the Near East—complexities emerge that quickly reveal how limited the human mind is in its attempts to understand not only why people do what they do, but also how over time the effects of people's actions can accumulate, in concert and in conflict with those of

Evolution of a Taboo. Max D. Price, Oxford University Press (2020). © Oxford University Press.
DOI: 10.1093/oso/9780197543276.001.0001.

others, to produce large-scale cultural processes that themselves mutate and evolve.

The story of swine, from wild boar to domestic pig to forbidden flesh, encapsulates complexity. Central to the drama are the "pig principles" laid out in Chapter 2. These are abstractions of ethnographic observations and historical patterns. But these principles, which include, for example, pigs' ability to thrive in urban environments and their intolerance of arid environments, are not so much laws as guidelines. Historical contingencies—what the Greeks called *tyche* (chance)—also played critical roles. Also relevant are the broader social and environmental processes that extended well beyond the spheres of pig production and pork consumption—urbanism, climate change, ethnogenesis, state formation, deforestation, class antagonism, gender dynamics, and many others. These features (the pig principles, chance, and large-scale processes), operating at different temporal scales and interacting with one another in different ways, are difficult to account for in any grand narrative of history.

Unpredictability lurks behind every turn in the evolution of the pig in Near Eastern cultures. Even the most analytically minded *Homo erectus* hunters, those who occasionally bagged, or perhaps scavenged, a ferocious boar, could never have imagined that their younger, larger-brained cousins, *Homo sapiens*, would one day develop rituals around these animals, enshrining them as symbols of masculinity. Nor could the hunter-gatherers who introduced wild boar to Cyprus almost 12,000 years ago have predicted that within a few thousand years, their descendants would be raising a peculiar, *domestic* form of *Sus scrofa*—one that was more variably colored, shorter-snouted, significantly less aggressive, and better suited to living among humans. Even those people who were, from our retrospective viewpoint, domesticating pigs from wild boar would be surprised to learn that by managing animals, they were selecting for novel phenotypes. And neither the Neolithic villagers, nor the emerging political elites of the Chalcolithic, nor the first city dwellers of the Early Bronze Age, nor even the imperial subjects of the Middle and Late Bronze Age could have predicted that a taboo on pork would develop among a people in the Levant, who would enshrine that taboo as law in a holy text. They would be shocked to learn that the commandments written in this text would come to represent a people's commitment to its god and that, though they had been conquered and oppressed, the tenacity with which they stuck to their laws would only intensify over time and influence other peoples, one of which would conquer almost the entire Near East. They could never have dreamed of a world in which over a billion people eschewed pork as one might feces and that the flesh of swine could be used to terrorize and persecute.

Swine offer a window onto complexity. No book can completely capture the entirety of their history—indeed, each subsection of this work could itself probably be turned into a shelf of weighty tomes. There is no single story about pigs in the Near East. There is no single answer to why they came to hold the power they did. But in summing up, we can draw on a few themes to help us organize our thoughts.

Domestication

Domestic animals are those that have evolved to inhabit human-modified ecosystems and to reproduce under conditions of exploitation by humans, the latter of whom consider those animals the property of either individuals or the community at large. Domestication is also a partnership. The domestication of pigs was a coevolutionary process, not a willful act of domination over nature by Pre-Pottery Neolithic peoples in northern Mesopotamia. Although zooarchaeologists have yet to document the exact mechanisms by which wild boar evolved into domestic pigs in the PPNB, three features helped place swine on the road to domestication.

(1) People began to intensify their exploitation of wild boar by the beginning of the Holocene period, a time when human populations were expanding and likely sought to enhance the reliability of their food resources. For hunting wild boar, the addition of domestic dogs would have been a game-changer, allowing people to target this prey more frequently and safely. However, our most concrete evidence of evolving animal exploitation is twofold. First, when hunter-gatherers colonized new locations like Cyprus, they brought with them animal populations to ensure a dependable source of meat. Second, hunters began targeting animals of specific ages and sexes to enhance the reproductive capability of their herds—as at Hallan Çemi and Çayönü Tepesi by the 10th millennium BC. This was the beginning of animal management, something that hunter-gatherers applied to populations of wild boar as well as wild sheep, goats, and cattle.

(2) At the same time, wild boar saw human settlements, which had become more permanent on the landscape, as reliable sources of food. Deposits of garbage and cultivated stands of cereals and legumes attracted wild boar to

the early villages in northern Mesopotamia. Those wild boar that were more tolerant of being around people were able to exploit this new resource patch most effectively. These acclimatizing wild boar may well have interbred with those animals already under management by humans, leading to a hybrid pathway to domestication consisting of both Melinda Zeder's[3] commensal and prey pathways.

(3) However, the transformation of wild boar into domestic pigs took a long time—measurements of bones and teeth from Çayönü show a gradual shift to smaller animals over the course of three millennia.[4] The process almost certainly included continual interbreeding between managed and nonmanaged stocks. It probably included many failures along the way, times when the mutations that had built up in a subset of a population of wild boar hit an evolutionary dead end. Any number of factors could have abruptly ended the long-term process of domestication: the abandonment of settlements, selection against the "domestication syndrome" mutations, epidemics or harsh winters wiping out villages' herds, and/or enough people simply losing interest (for a generation or two) in managing swine.

When domestic pigs finally emerged on the scene—something we can detect via measurements on pig bones and teeth in deposits dating to the late 9th–early 8th millennium BC—people across the Near East were slow to adopt them. Domestic plants, sheep, goats, and cattle, in general, spread across the region centuries before pigs.[5] One can explain the slow spread of domestic swine in a number of ways. Perhaps the herding of sheep, goats, and cattle was an inherently more mobile process, making it easier for people to pick up and move to a new village. Perhaps the environmental conditions of the Near East, especially its vast areas of open and arid grasslands, placed limits on the spread of pig husbandry. Or perhaps ruminant pastoralism more closely matched hunter-gatherers' aesthetic sense of how humans should procure meat.

Pig husbandry ultimately did spread throughout the Near East and beyond. By 5000 BC, pigs were being raised by village societies from Egypt to Iran. Pigs were also increasingly raised in pens and fed household waste as opposed to being allowed to roam freely through forests and meadows. Pigs adapted well to this new ecosystem, as they would to urban environments by the 3rd millennium BC. But in doing so they changed people's perception of them, creating the opportunity for new ritual uses and meanings of swine.

The Ritual Significance of Swine

One of the most fascinating aspects of swine in the Near East is how these animals figured in rituals and symbolically important images. Pervasive throughout the entire sweep of history and across cultures is the symbolic connection between boar hunting, masculinity, and political power. While the earliest definitive representations of this connection might be the 4th millennium cylinder seals from Uruk in southern Mesopotamia,[6] it likely has a much earlier history. The large-scale feasts centering on wild boar at PPNA Hallan Çemi and Asiab and the depiction of wild boar on stone pillars at Göbekli Tepe are indications of boar's symbolic significance in the deep past. Boar hunting remained symbolic of kingliness and masculinity, with notable representations of wild boar in Sassanian art. Even today, the hunting of wild boar for sport in Turkey by wealthy Western men continues this long tradition.

Domestic pigs possessed a somewhat more eclectic symbolic repertoire than their wild cousins. Because of their rapid rates of reproduction, they served as potent images of fertility. From at least the Early Bronze Age through the Classical period, people drew on the image of sows to represent female fecundity. Similarly, the sacrifice of piglets to underworld deities played on the symbolic significance of fertility's dialectical opposite, death. At the same time, and perhaps connected to their fertility/chthonic symbolism, pigs and especially piglets acted as substitutes in rituals. Similarly, they were thought to be receptacles for otherworldly forces and to soak up pollution like a sponge. If 2nd millennium BC Hittite texts provide some of the earliest concrete examples of pigs being ritually used as substitutes,[7] travelers' accounts of pigs or pork being used to treat animals in veterinary magic are evidence of its continuation in at least some circles of the Near East up to the modern era.[8]

The powerful ritual significance of pigs had been turned on its head by the Late Bronze Age. If the pig could absorb pollution, then it could itself become polluted. If the pig could be used in magic rites relating to fertility/death, perhaps its power had no place in temples. Moreover, if the animal did not represent wealth, perhaps its sacrifice would be offensive to the gods. These threads evolved into the idea that swine were, by nature, polluted. This reconceptualization, while probably rooted in the logic of magic and supernatural forces, was no doubt aided by the animals' unflinching willingness to eat garbage and feces. The idea developed by the Late Bronze Age into a number of proscriptions on pigs entering temples or being eaten by priests. These taboos, restricted to the religious sphere, may have contributed to the development of the taboo on pigs and pork among the Israelites in the Levant in the Iron Age.

The Unpredictable Evolution of the Pig Taboo

Given the messy, torturous history of pigs in the Near East, it is tempting to make sense of it by isolating a key variable, a single factor that can explain human-suid interaction. The various theories that have been put forward to explain the origins of the taboo on pork best exemplify this reductionist line of thinking. Early on, the authors of Leviticus drew upon pigs' unique physiology to explain the taboo. More recently, pigs' ecological sensitivities and water requirements[9] or the fact that swine are ideal for small-scale household-level production[10] have been marshaled as explanations for the emergence and persistence of the pig taboo. Meanwhile, the discovery of foodborne illnesses like trichinosis has spawned popular theories that the pig taboo was intended to prevent disease, despite disease never having been mentioned in relation to the pig taboo in Leviticus, Deuteronomy, or the Quran.

In Chapter 5, I pointed out that while all of these and other claims have been exaggerated, many of the factors they endorse did play a role in the evolution of the taboo. However, by excluding from the discussion the other factors involved in pigs' long trajectory from domestication to taboo, and indeed the contingencies that brought them all together at various points in time, these arguments have the effect of making history seem inevitable. The pig, they imply, *had* to become taboo. The evolutionary perspective I adopt in this book rejects this fatalism.

In reality, the pig taboo was not a foregone conclusion. It is only by examining the long history of swine in the Near East, from the Paleolithic to the present, that we can see how wrong these reductionist arguments are. Multiple factors came together to build a foundation on which a taboo could emerge. In addition to this "perfect storm"[11] of underlying factors, chance played a crucial role in the initial appearance of the taboo. And the taboo would not have persisted without this element of chance or the coming together of different long-term cultural processes. Rather than isolating a single moment or process that shaped the pig taboo, we can suggest a more realistic, more honest answer to the question How did the taboo come into being? Namely, it *evolved*. The taboo formed slowly over time and was subject to competing forces, both external and internal to the cultures in which developed. These forces themselves changed over time. One can trace this history in broad strokes, as we have done, but in no way does it provide a path to a simple explanation. There is no one reason that the taboo came into existence because it did emerge fully formed.

One could probably begin the story of the pig taboo in the Chalcolithic and Early Bronze Age, when pigs' association with urbanism and their symbolic/ritual power developed in new ways at the same time that the secondary products revolution effectively excluded pigs as animals conveying wealth. (But even here we have to pause and recognize that the secondary products revolution could never have happened without the innovations in livestock husbandry developed by Late Neolithic farmers, who themselves built on the knowledge and traditions of their Pre-Pottery Neolithic forerunners.) By the 3rd millennium BC, these factors were working against pigs being a major part of Bronze Age agriculture in the Levant and setting up a tradition of passive exclusion of pork from the diet. Meanwhile, in certain religious contexts, people began to perceive pigs and pork as polluting elements. By the Late Bronze Age, a taboo on pigs was being observed in some temples and by some priests. But this was not the all-encompassing pork taboo that later defined pigs' place in Near Eastern cultures. It is not entirely clear how much influence these religious-specific restrictions had on the taboo codified in the Torah.

The ethnogenesis of the Israelites in the wake of the Late Bronze Age collapse was a critical moment for the taboo, as scholars have long recognized. On one level, Israelites inherited a tradition of pork nonconsumption. Their recourse to pastoral imagery and the glorification of life spent herding sheep and goats in the desert provided an additional reason to exclude pigs from the diet. So too did confrontations with Philistines, who ate pork more frequently, on average, than other Levantines. This sparked the first stirrings of an anti-pork sentiment and a consciously recognized taboo among the Israelites.

The critical moment for the pig taboo came later, when the leaders and priests of the kingdom of Judah initiated a conservative revitalization movement focused on the Temple of Yahweh in Jerusalem and a specific set of practices that bound people to it. The practices they laid out for their people revitalized a tribal ideal, complete with its pastoral imagery, as well as the practices and beliefs that they imagined defined the lives of their ancestors, whom they believed lived in a more glorious era. In doing so, they found existing food traditions convenient for connecting contemporary Hebrew-speaking peoples to their past without contradicting existing social inequalities and political power structures. These traditions included the taboo on pork that was developed during the Israelite-Philistine conflicts and that was probably waning by the 8th century BC. They may have bolstered this taboo by borrowing the concept that the pig could pollute sacred spaces—an idea common throughout much of the Near East by the Late Bronze and Iron Ages. In this way, the biblical authors gave new meaning to the pig taboo. They made the additional revolutionary move of inscribing the taboo and other food laws into holy texts.

Historical circumstances continued to drive the evolution of the pig taboo. The conquest by the Greeks, the militant (if abortive) Hellenization project of Antiochus IV, skirmishes with Greek colonists, and the conquest of Judea by the Romans set in motion a new set of "selection pressures" on pigs and pork. The counterplay between the processes of Judaization and Romanization in the southern Levant and the outbreak of several bloody revolts led Romans and Jews to view each other as implacable enemies. Invariably, cultural elements would be drawn into these battles. Pork consumption became one of the focal points in this culture war; it became weaponized. As a result of these decidedly negative encounters with pigs and pork, especially after the conquest of the Levant and the bloody suppression of several Jewish revolts, many Jews began to perceive pigs not just as the objects of a taboo, but as oppressive tools of the enemies who had conquered them.

One sect of Jews, the Christians, took a different tact and embraced pork consumption and other elements of Greco-Roman culture. Their *orthodoxic* outlook (focused on proper beliefs) had no real place for most of the elements of *halakha*. As Christianity emerged as a separate religion and took hold in the Near East and Europe, the pig taboo became a symbolic wall separating Jews and Christians. Christians could point to the taboo as an example of Jews' obsession with specific practices over spiritual well-being; Jews could highlight Christians' failure to abide by what had become one of the most salient markers of Jewish identity as evidence of the new group's waywardness. Islam added a new force to this debate. When it emerged in the 7th century, its leaders used Christianity and Judaism as foils that proved the correctness of Muhammad's vision.[12] They chose to keep a small number of what they saw as critically important food laws, one of the main ones being the taboo on pork. This would be a critical development for the pig. Islam's rapid expansion on the backs of empires spread the pig taboo from Spain to Central Asia.

The Uniqueness of the Pig Taboo

It is the complexity surrounding the pig taboo as well as the intensity with which it is observed that have made it one of the most unique forms of food prohibition. For that reason, scholars since the Classical period have speculated on its origins. But the pig taboo distinguishes itself in other ways. For example, in contrast to most ethnographic examples of food taboos, the proscription on pigs is written in texts. The codification of the taboo in the Hebrew Bible and the Quran distinguishes it even more from the taboos of "text-heavy" societies, such as modern Western ones. Although there are occasional legal restrictions against obtaining/consuming tabooed foods, such as swans in the UK,[13] many foods are simply avoided by custom. Readers may be surprised to learn, for example, that there are no laws in the US against cannibalism, per se (although laws against murder and corpse desecration make it practically difficult),[14] nor are there legal restrictions on eating dogs or cats. But laws can be challenged in terms of their legality and customs can be broken by adventuresome people willing to risk social stigma. Over time, many fade and disappear. In contrast, the recording of a taboo as a law handed down by God (or one of his prophets) helps make it more permanent.

Yet another thing that distinguishes the pig taboo from other proscriptions is its role as a major barrier between three world religions. Taboos often serve to separate people from one another, both within societies (e.g., pregnant women, ceremonial initiates, holy people) and between them (entire ethnic groups or groups of ethnic groups). But rarely are taboos so deeply entrenched that they can create divisions between peoples for centuries across several continents. The power of the pig taboo to divide Jews, Christians, and Muslims ultimately derives from the ways that pigs and pork were integrated into the social projects of these religions. The avoidance of pigs came to represent one's commitment to the tenets of one's faith and, therefore, one's obligations to its other members. By the same token, for those seeking to convert others, explore new traditions, or reject the customs of their communities, pork became a form of ingestible identity.

The power of pigs to separate Jews, Christians, and Muslims is continually refueled when these groups interact. As such, it has become a key weapon for acts of intolerance between Christians, Jews, and Muslims. While Christian bigots may throw pork at Jews or Muslims, Jewish or Muslim bigots mock Christians or other pork-eating peoples as unclean. Or they massacre Christians' livestock and thereby threaten their livelihoods. Meanwhile, Jews or Muslims who choose to eat pork are castigated as wayward, self-hating, or fake by other members of their faith. Few are willing to accept what the zooarchaeological data make eminently clear: small numbers of Israelites, Jews, and Muslims have, in all periods of time, eaten pork. There was *always* some element of negotiation. If the pig taboo is a historic tradition of Jews and Muslims, so too is its transgression.

One need not look solely to pigs to see the ugliness of human intolerance. But pork's unique role has been to separate religions that are otherwise quite similar in message and tone. Social justice, protection for those who suffer, egalitarianism, piety, and righteous conduct in the world are core principles held in common by Judaism, Christianity, and Islam. Yet prejudice breeds contradiction. Swine have been a major focal point of difference among these faiths, and people have defined themselves in part by their attitude to the pork taboo. But by lending themselves to acts of bigotry, pigs have also exposed the difficulties of adhering to the core principles of the Abrahamic religious tradition.

Tradition and Fate

> The Puritan wanted to work in a calling; we are forced to do so [. . .] In Baxter's view, the care for external goods should only lie on the shoulders of the "saint like a light cloak, which can be thrown aside at any moment." But fate decreed that the cloak should become an iron cage.
>
> —Max Weber, *The Protestant Ethic and the Spirit of Capitalism*[15]

We have investigated the pig taboo on the level of societies, collections of individuals that operate through and in many ways above their constituent parts. Our understanding of it and explanations for its origins therefore deal with social structures. But if we bring our analysis to the scale of the individual, *the* most important explanation for the perpetuation of the pig taboo becomes obvious: tradition. At any moment in the taboo's long history, the single most common reason Jews or Muslims would cite for avoiding pork was that they were following food traditions passed on to them by their parents and communities.

The statement that tradition is the reason individuals observe the pork taboo is no mere platitude; rather it articulates a fundamental, if somewhat obvious social fact with relevance to the persistence and morphosis of the taboo. The tradition of avoiding pork is tied to one's place in a community and a perceived relationship with God. The tradition is reproduced daily—each time a meal is served without pork or, even more emphatically, when a dish is advertised as being certified pork-free by an authority. Each expression of disgust, each insult hurled in frustration that equates others with pigs or pork consumption, reproduces this tradition.

Tradition endures because inertia is as powerful as any factor in motivating human behavior. Left unchallenged, some traditions persist for millennia; others are forgotten and die of natural causes. When challenged, especially by members of an external group, a tradition can be reinforced rather than eliminated. In the case of the pig taboo, the persistence of a traditional diet lacking in pork, which can be traced zooarchaeologically in the Levant to the Bronze Age, faced external confrontations at several key moments. Paradoxically, these threats to tradition strengthened it. They ultimately served to link pork avoidance and swine revulsion ever more closely with Jewish and later Muslim identity.

Like capitalism's emergence from Weber's Protestant ethic, the pig taboo's evolution over millennia of encounters between humans and pigs in the Near East eventually came to dominate those interactions, trapping peoples with differing attitudes toward pork in relationships defined by disgust and mutual distrust. What was once a cultural trait weakly—or perhaps not at all—perceived by the people who practiced it transformed over a long span of time into a definitive marker of identity and being. As a result, what was, and remains, ultimately a form of food avoidance daily practiced by individual people developed a life of its own. That long and unimaginably complex arm of history formed by the interactions of millions of people spread across dozens of generations—what, for lack of a better word, we can call "fate"—decreed that the pig taboo would solidify into an iron cage. The taboo drew around itself ever more layers of meaning that symbolized not only social identity, but also resistance to the perceived threats posed by other cultures and religions. In so doing, this particular tradition transformed into something that neither Bronze Age urbanites nor even the writers of Leviticus could have ever imagined.

Appendix

Table A.1 Daily Drinking Water Requirements for Pigs and Other Domestic Animals

Animal	Body Mass (kg)	Water Intake (Liters/Day)
Pig		
Weaning Pig	7–22	1–3.2
Growing Pig	23–36	3.2–4.5
Growing Pig	36–70	4.5–7.3
Growing Pig	70–110	7.3–10
Boar	>100	13.6–17.2
Pregnant Sow	>100	13.6–17.2
Lactating Sow	>100	18.1–22.7
Sheep		
Feeder Lamb	27–50	3.6–5.2
Lactating Dairy Ewe	90	9.4–11.4
Cattle		
Feedlot Beef Cattle	350–650	27–55
Dairy Cow[a]	500–600	68–83
Horse		
Medium-Sized	450	26–39

[a] Numbers are for cows producing 13.6 kg of milk per day.

Source: Data from the Ontario Ministry of Agricultural, Food, and Rural Affairs (Ward and McKague 2007).

Table A.2 Pigs as a Percentage of Major Livestock (Sheep, Goats, Cattle, and Pigs) in Mesopotamian and Syrian Cities, Organized by Time Period and Subregion

Archaeological Site or City	Pigs (%)	Region
Early Bronze Age, 3000–2000 BC		
Eshnunna	37	Southern Mesopotamia
Uruk	33	Southern Mesopotamia
Lagash	22	Southern Mesopotamia
Tell Hamoukar	53	Northern Mesopotamia (Khabur)
Tell Leilan, Lower Town (Nonelite)	50	Northern Mesopotamia (Khabur)
Tell Leilan, Upper Town (Elite)	35	Northern Mesopotamia (Khabur)
Tell Arbid	44	Northern Mesopotamia (Khabur)
Tell Brak	25	Northern Mesopotamia (Khabur)
Tell Mozan	24	Northern Mesopotamia (Khabur)
Tell Beydar	2	Northern Mesopotamia (Khabur)
Tell Taya	28	Northern Mesopotamia (Northern Iraq)
Ebla	5	Northern Mesopotamian (W. Syria/Upper Euphrates)
Umm el Marra	3	Northern Mesopotamian (W. Syria/Upper Euphrates)
Titris	1	Northern Mesopotamian (W. Syria/Upper Euphrates)
Tell es-Sweyhat	<1	Northern Mesopotamian (W. Syria/Upper Euphrates)
Tell Chuera	<1	Northern Mesopotamian (W. Syria/Upper Euphrates)
Emar	None	Northern Mesopotamian (W. Syria/Upper Euphrates)
Middle Bronze Age, 2000–1600 BC		
Mashkan-Shapir	40	Southern Mesopotamia
Sippar-Amnanum	40	Southern Mesopotamia
Uruk	32	Southern Mesopotamia
Nippur	23	Southern Mesopotamia
Isin	20	Southern Mesopotamia
Ur	18	Southern Mesopotamia
Tell Brak	40–45	Northern Mesopotamia (Khabur)
Kurd Qaburstan	29	Northern Mesopotamia (Northern Iraq)

Table A.2 *Continued*

Archaeological Site or City	Pigs (%)	Region
Tell Mozan	17–28	Northern Mesopotamia (Khabur)
Umm el Marra (E&M)	< 5	Northern Mesopotamian (W. Syria/ Upper Euphrates)
Tell es-Sweyhat	None	Northern Mesopotamian (W. Syria/ Upper Euphrates)
Emar	None	Northern Mesopotamian (W. Syria/ Upper Euphrates)

Source: Data from Mashkan-Shapir, Old Babylonian (Brellas 2016; Redding 2015:330); Sippar-Amnanum, Old Babylonian (Bökönyi 1978a); Uruk, Post-Akkadian to Old Babylonian (Böck et al. 1993); Eshnunna, ED I-(post-)Akkadian (Hilzheimer 1941); Nippur, Old Babylonian (Boessneck 1978; Twiss 2017); Lagash, ED III (Mudar 1982); Ur, Isin-Larsa and Old Babylonian (Twiss, personal communication); Isin, Old Babylonian (Boessneck 1977; Twiss 2017); Tell Leilan, EJ II–V (fauna have not been fully published, but Zeder [2003] provides estimates of Upper and Lower Town assemblages; see also Rufolo 2011:525–530; Weiss et al. 1993:fn. 30; Zeder 1998a); Tell Ḥamoukar, EJ III–V (Grossman 2013); Tell Arbid, EJ II–V (Piątkowska-Małecka and Smogorzewska 2010; Piątkowska-Małecka and Smogorzewska 2013); Tell Brak, EJ III–MBA (Weber in Dobney et al. 2003; Schwartz et al. 2017; Weber 2001); Tell Mozan, EJ III–OJ III (Doll 2010); Tell Beydar, EJ III–IV (Van Neer and De Cupere 2000); Tell Taya, Levels IX–VI, EJ III–V (Bökönyi in Reade 1973:184–185); Kurd Qaburstan, Old Babylonian (Weber in Schwartz et al. 2017); Ebla, EB III–MB II (Minniti 2013; Minniti and Peyronel 2005); Umm el Marra, EB IV–MB II (Weber 2006:260); Titriş Höyük, EB III–IV (Greenfield 2002; Trella 2010); Tell Chuera, EJ II–V (Vila 1995, 2010); Tell es-Sweyhat, Period VI–IV, EB III–IV (Buitenhuis 1985); Emar EB IV–MBA (EB IV is a temple area and MB deposits from Upper Town only; Gündem 2010).

Table A.3 Pigs at Iron I Sites in the Levant

Archaeological Site	Pigs (%)[a]	Site Type	Cultural Affiliation
Ashdod (E)	11	Urban	Philistine
Tel es-Safi (E&L)	13	Urban	Philistine
Ashkelon (E&L)	2–14	Urban	Philistine
Miqne-Ekron (E&L)	7–20	Urban	Philistine
Qabur el-Waleyide (E)	None	Nonurban	Philistine[b]
Qasile (E or L)	1	Nonurban	Philistine
Aphek (L)	<1	Nonurban	Philistine
Khirbet Qeyafa (L)	None	Nonurban	Israelite?[c]
Tel Massos (E)	None	Nonurban	Israelite
Mount Ebal (E)	None	Nonurban /Ritual	Israelite
Beersheba (L)	None	Nonurban	Israelite
Shiloh (E)	<1	Nonurban	Israelite
Khirbet Raddana (L)	<1	Nonurban	Israelite
Izbet Sartah (E)	1	Nonurban	Israelite
Tel Dan (E&L)	None	Urban	Canaanite
Bet Shemesh (E&L)	<1	Urban	Canaanite?
Tel Rehov (L)	1	Urban	Canaanite
Megiddo (E&L)	1–2	Urban	Canaanite
Tel Dor (E&L)	1–2	Urban	*Sikil* (Sea Peoples)

Note: "E" indicates bones dating to Early Iron I (1200–1050 BC); "L" indicates bones dating to late Iron I (1050–950 BC). Values represent the proportion of pigs among the total number of identified livestock specimens (NISP).

[a] Sapir-Hen et al. (2013) include equids in their tally of livestock NISPs, which departs from my general calculation of pigs as a percentage of the combined total of sheep, goats, cattle, and pigs. However, the relative numbers of equid remains are very small; they do not alter the percentage of pigs by more than a few tenths of a percent. [b] Pig remains from the Iron I Philistine village at Qabur el-Waleyide are unpublished but reported by Sapir-Hen et al (2013:fn. 32) as a personal communication from the site's excavator, G. Lehmann, in 2012.

[c] Kh. Qeyafa is generally understood to be Israelite, but Garfinkel (2017) has questioned this designation.

Source: Data from Sapir-Hen et al. (2013), with additional data from Hesse and Fulton (forthcoming) and Lev-Tov (2012).

Table A.4 Pigs at Iron II Sites Within the Kingdoms of Judah and Israel

Archaeological Site	Pigs (%)[a]	Political Affiliation
Iron IIA (950–780 BC)		
Hazor	3	Kingdom of Israel
Tel Yoqneam	1–2	Kingdom of Israel
Megiddo	1	Kingdom of Israel
Lachish	<1	Kingdom of Judah
Beersheba	0–1	Kingdom of Judah
Iron IIB (780–680 BC)[b]		
Bet Shean	8	Kingdom of Israel
Megiddo	8	Kingdom of Israel
Tel Yoqneam	5	Kingdom of Israel
Hazor	3	Kingdom of Israel
Lachish	1	Kingdom of Judah
Beersheba	<1	Kingdom of Judah
Jerusalem	<1	Kingdom of Judah
Mosa	<1	Kingdom of Judah
Tel Halif	None	Kingdom of Judah

[a] Sapir-Hen et al. (2013) include equids, although they make up a small proportion of the faunal remains and do not significantly impact the percentage of pigs.

[b] Kingdom of Israel invaded by Assyria in 732 BC; fully conquered in 722 BC.

Source: Data from Sapir-Hen et al. 2013.

Table A.5 Pigs at Classical Period Sites in the Southern Levant Organized by Period and Settlement Type

Archaeological Site	Pigs (%)	Site Type
Hellenistic (4th–2nd Century BC)		
Tel Dor	18	Urban
Maresha	11	Urban
Tell Jemmeh	1	Urban
Tel Michal (Strata XIV–XII)	None	Urban
Tel Anafa	13	Military
Shaar Haamakim	6	Military
Roman (1st Century BC–3rd Century AD)		
Umm Qais (Gadara)	70	Urban
Caesarea	58	Urban
Tel Hesban	6	Urban
Sepphoris	5	Urban
Petra	3	Urban
Jerusalem	None	Urban
Tel Anafa	22	Rural
Horvat Rimmon	1	Rural
Qumran	None	Rural
Lejjun	3	Military
Byzantine (4th–7th Century AD)		
Caesarea	51	Urban
Pella	11–39	Urban
Petra	28	Urban
Sepphoris	28	Urban
Tell Hesban	10	Urban
Bab el Hawa	7	Rural
Horvat Rimmon	<1	Rural
Dajaniya	17	Military
Upper Zohar	13	Military
Lejjun	4	Military

Source: Data from summaries published by Horwitz and Studer (2005) and Perry-Gal et al. (2015a).

Notes

Chapter 1

1. Kelley et al. 2015.
2. Fahmi and Sutton 2010; Hessler 2014.
3. Slackman 2009a, 2009b.
4. Slackman 2009a.
5. Hessler 2014.
6. Russell 2012; Sykes 2015.
7. Albarella et al. 2017.
8. Barak-Erez 2007.
9. Landau 2010.
10. Yoskowitz 2010.
11. Several global histories of pigs have been written for popular audiences (e.g., Essig 2015; Malcolmson and Mastoris 1998; Mizelle 2011; Watson 2004).
12. Food and Agriculture Organization of the United Nations 2017.
13. Izadi 2015.

Chapter 2

1. TAVO Map A IV 4.
2. Data for US climate from NOAA 2019.
3. TAVO Map A IV 4; TAVO Map A VI 2.
4. TAVO Map A IV 4; TAVO Map A VI 2; Wilkinson 2003:18.
5. TAVO Map A VI 2; van Zeist and Bottema 1991.
6. Food and Agriculture Organization of the United Nations 2017.
7. Food and Agriculture Organization of the United Nations 2017.
8. Suid evolution can be traced with genetic and morphological data (Frantz et al. 2016; Gongora et al. 2011; Groves 1981, 2007; Orliac et al. 2010).
9. Frantz et al. 2016; Groves 1981.
10. Frantz et al. 2016:65.
11. Frantz et al. 2016:69.
12. Frantz et al. 2016:75.
13. E.g., Romanes 1883:339.
14. E.g., Kornum and Knudsen 2011.
15. Spinka 2009; Taylor et al. 1998.
16. Bieber and Ruf 2005.

17. This includes three to four weeks of nursing and one to two weeks post-weaning (Pond and Mersmann 2001; Taylor et al. 1998).
18. Bazer et al. 2001; Bywater et al. 2010; Ramos-Onsins et al. 2014.
19. Gimenez-Anaya et al. 2008; Herrero et al. 2006; Schley and Roper 2003; Wilcox and Van Vuren 2009.
20. E.g., Nemeth 1998.
21. Schley et al. 2008; Schley and Roper 2003.
22. Pimental 2007:4.
23. Nannoni et al. 2013. Pigs drink about 50 percent more water when it's 35°C than when it's 12°C (Almond 1995).
24. Choquenot and Ruscoe 2003; Mount 1968.
25. E.g., Bleed 2006; Clutton-Brock 1992; Ducos 1978; Hemmer 1990; Vigne 2011b; Zeder 2015.
26. Darwin 1868:6.
27. Price and Hongo, in press; Zeder 2018.
28. Bleed 2006.
29. Zeder 2018.
30. Kruska 2005.
31. Hemmer 1990.
32. Albert et al. 2012.
33. Belyaev 1969; Trut et al. 2009.
34. Wilkins et al. 2014.
35. Meadow 1984.
36. Zeder 2012b:249.
37. Vigne et al. 2009b.
38. Hongo et al. 2007.
39. Rowley-Conwy and Dobney 2007.
40. Albarella et al. 2006b.
41. Price and Hongo, in press.
42. Larson et al. 2007a; Ottoni et al. 2012.
43. Larson et al. 2007b, 2010.
44. Frantz et al. 2015:1146; Larson et al. 2005.
45. White 2011.
46. Albarella et al. 2006a; Cucchi et al. 2011; Ervynck et al. 2001; Evin et al. 2013; Flad et al. 2007; Flannery 1983; Payne and Bull 1988; Price and Evin 2019; Rowley-Conwy et al. 2012.
47. Fang et al. 2009; Krause-Kyora et al. 2013; Meiri et al. 2013.
48. Dobney and Ervynck 2000; Dobney et al. 2007.
49. Frémondeau et al. 2012; Hamilton et al. 2009.
50. Ervynck et al. 2001; Lemoine et al. 2014.
51. E.g., Burrin 2001; McGlone and Curtis 1985; Studnitz et al. 2007.
52. Intensively raised pigs gain weight faster than extensively raised ones. Hadjikoumis (2012) studied pigs raised under semi-free-range and free-range conditions in Iberia

and found that those under fully free-range conditions were typically slaughtered at around 16–18 months, while those under semi-free-range conditions (involving some degree of penning) achieved slaughter weights around 12 months. Industrial-raised pigs can achieve slaughter weight (100 kg) at around 9 months or less. But slaughter timing is variable and culturally determined. Albarella and colleagues' (2011) study of Sardinian swineherds showed that the desire for suckling pigs at Christmas resulted in a heavy cull of young animals. Pugliese et al. (2003) compared the growth performance of Nero Siciliano breed pigs under intensive and extensive conditions and found that intensively managed pigs gained weight faster and produced less lean meat than extensively managed ones. Extensively managed ones, on average, were slaughtered at 486 days old at 88 kg, while intensively raised ones were slaughtered at 448 days at 102 kg.

53. For example, Columella (*On Agriculture* 7.9) in the 1st century AD discussed the construction of large stalls consisting of several parallel sties.
54. Malcolmson and Mastoris 1998:39; White 2011.
55. Blackshaw et al. 1994.
56. Bittman 2014.
57. Hemmer 1990:147.
58. For summaries of ethnographic work on New Guinea pig-keeping, see Blanton and Taylor 1995; Boyd 1985; Dwyer 1996; Dwyer and Minnegal 2005; Hide 2003; Kelly 1988; Sillitoe 2007.
59. Albarella et al. 2007, 2011; Halstead and Isaakidou 2011b.
60. Lemonnier 2012:23–30.
61. E.g., Albarella et al. 2011:149.
62. Neonatal pigs are especially vulnerable to hypothermia (Mount 1968:19–20).
63. E.g., Rappaport 1968:160–162.
64. Albarella et al. 2011; Diener and Robkin 1978:498.
65. Grigson 1982; Malcolmson and Mastoris 1998; Wealleans 2013; White 2011.
66. Hadjikoumis 2012; Parsons 1962.
67. Albarella et al. 2007, 2011.
68. Halstead and Isaakidou 2011b.
69. Kapoor Sharma 2002 (translated and quoted by Masseti 2007:166).
70. New York Times 1859.
71. Baldwin 1978; Kagira et al. 2010; Masseti 2007:169.
72. Hesse 1990; Hesse and Wapnish 1997, 1998. To their list I have added a few more "principles" in light of more recent research.
73. E.g., Harris 1974.
74. The deforestation hypothesis was first articulated by Carleton Coon (1951:346).
75. Historically, pork was cheaper and readily available to the poor. To my knowledge, the first scholars to suggest the connection between pigs and class in the Near East were Paul Diener and Eugene Robkin (1978).
76. This idea was proposed by Richard Redding (1991) in his work on ancient Egypt. Goats are able to subsist on chaff, stubble, and lower-quality grasses and are thus more

ideal complements to cereal production than sheep. Cattle, meanwhile, provide traction power for plowing and hauling, which compensates for their need for higher-quality graze and fodder.

77. This idea derives from the work of Mary Douglas (1966), who used it to explain the pig taboo in Judaism.
78. The inability of authorities to tax pig production is part of Diener and Robkin's (1978) explanation for the Islamic pig taboo.
79. This idea was particularly favored by Melinda Zeder (1996, 1998b), although she has adopted a more nuanced perspective in other publications (e.g., Zeder 2003).

Chapter 3

1. The earliest fossil evidence for *Sus scrofa* dates to around 800,000 years ago (Horwitz and Monchot 2007:93; Rabinovich and Biton 2011; Tchernov 1979).
2. Gabunia et al. 2000.
3. Smith 2007.
4. Tchernov and Valla 1997.
5. Bar-Yosef 2002a; Hartman et al. 2016; Rosen 2010. Climate affected wild boar physiology as well; after the Late Glacial Maximum, wild boar developed smaller body sizes (Davis 1981).
6. For example, 1 percent of medium and large mammal remains recovered from Gesher Benot Ya'aqov (Rabinovich and Biton 2011); <1 percent at Holon (Horwitz and Monchot 2002); 2 percent at Qesem Cave (Stiner et al. 2011).
7. Speth 2012; Speth and Tchernov 1998.
8. For example, Shanidar Cave (<1 percent); Qafzeh (6 percent), Misliya (5 percent), Kebara (3 percent), and Hayonim (4 percent). For data, see Evins 1982; Perkins 1964; Rabinovich and Tchernov 1995; Speth 2012; Speth and Tchernov 1998; Stiner and Tchernov 1998; Tchernov 1998; Yeshurun 2013; Yeshurun et al. 2007.
9. Chase and Dibble 1987:275.
10. Stiner 2009.
11. Bates 2013:1–2; Lemonnier 2002:135.
12. Robert Fagles's translation of Homer's *Odyssey* (1996:404–405).
13. Klein 2009:562.
14. Bar-Yosef 2002b; Gladfelter 1997; Shea 2006.
15. Evidence for the use of nets in hunting in the Upper Paleolithic: increase in the number of small fast game (e.g., rabbits) and use-wear on stone tools indicating rope production (Lupo and Schmitt 2002; Soffer 2004).
16. Wolf 1988.
17. These dates cover the Aurignacian and Ahmarian periods. Ksar Akil (XX–VI), Kebara (E–D), Hayonim (D), and Boker Tachtit (A–C), and other sites contain 0–5 percent *Sus* remains (Bar-Yosef and Belfer-Cohen 1988; Rabinovich 2003). At Üçağızlı I Cave, *Sus scrofa* remains are somewhat higher (7 percent), similar to the Middle Paleolithic levels at Üçağızlı II (Kuhn et al. 2009; Stiner 2009).

18. Rabinovich 2003:43.
19. Snir et al. 2015; Weiss et al. 2004.
20. Wild boar was <1 percent of the medium and large mammal assemblage (Rabinovich and Nadel 2005).
21. Bakken 2000; Bar-Yosef et al. 1992; Davis et al. 1988; Turnbull and Reed 1974.
22. Bar-Yosef 1998; Makarewicz 2012; Munro 2004.
23. Bar-Yosef 1998; Munro 2004; Starkovich and Stiner 2009.
24. E.g., Munro and Grosman 2010.
25. In the Levant, El-Wad Terrace, Hilazon Tachtit, Hayonim Cave, Ein Gev II, and Neve-David all contain very low numbers of *Sus* bones—often 1 percent or less (Bar-Oz 2004:37; Grosman et al. 2016; Munro 2004). Wild boar remains are uncommon at Karain B and Öküzini in the Taurus region (Atici 2009), Abu Hureyra on the Euphrates (Legge and Rowley-Conwy 2000), and Shubayqa 1 in Jordan (Yeomans et al. 2017).
26. The percentage of wild boar varies over time at 'Ain Mallaha and the counts of mammal bones differ among authors. Ducos (1968:73) gives a figure of 14 percent *Sus* as a total proportion of the identified faunal remains for all phases of the site (NISP = 1,425). This equals 23 percent of the medium and large mammals. Bouchud (1987:17) indicates that 6 percent of medium and large mammals were wild boar in the Early Natufian (Niveaux II–IV; NISP = 1,039) and 14 percent in the Late Natufian (Niv. I; NISP = 553). Bridault et al.'s (2008) work on material recovered from the most recent excavations indicates 23 percent of the medium and large mammal remains were wild boar in the Final Natufian (Niv. Ib; NISP = 524). Ducos (1967:387) notes a high proportion of young wild boar in the assemblages; Bridault et al. (2008:113) also note perinatal bones.
27. Valla 1988.
28. Davis and Valla 1978.
29. There is a growing consensus that domestication processes took place over a wide area and not in a core region (Arranz-Otaegui et al. 2016; Fuller et al. 2011).
30. Arranz-Otaegui et al. 2016; Kozlowski and Aurenche 2005; Willcox et al. 2008; Willcox and Stordeur 2012.
31. Zeder 2009, 2012a, 2015.
32. O'Brien and Laland 2012; see also Smith 2011.
33. Smith 2007.
34. Bar-Yosef 2011; Binford 1968; Flannery 1969; Munro 2004; Munro et al. 2018; Starkovich and Stiner 2009.
35. Munro et al. 2018; Stiner and Kuhn 2016.
36. For the "domino effect" of sedentism, see Kelly 1995:152.
37. Bleed 2006; Bleed and Matsui 2010; Fuller and Stevens 2017.
38. Zeder 2018.
39. Budiansky 1992; Rindos 1984.
40. Zeder 2018.
41. Shipman 2010.

42. These include "Round House" phases at Çayönü Tepesi (49 percent of the medium and large mammals; Hongo et al. 2009), Jericho (13 percent; Clutton-Brock 1979), Hasankeyf (15 percent; Hitomi Hongo, personal communication), and Hallan Çemi (25 percent; Peasnall et al. 1998; Redding and Rosenberg 1998; Rosenberg 1994; Rosenberg and Redding 2000; Starkovich and Stiner 2009; Zeder and Spitzer 2016). For summaries of NISP data, see Marom and Bar-Oz 2009 and Arbuckle 2013.
43. For dating, see Starkovich and Stiner 2009; for evidence of year-round occupation at Hallan Çemi, see Zeder and Spitzer 2016.
44. Rosenberg and Redding 1998.
45. Peasnall et al. 1998; Redding and Rosenberg 1998; Rosenberg 1994; Rosenberg and Redding 2000; Starkovich and Stiner 2009; Zeder and Spitzer 2016.
46. Redding and Rosenberg 1998.
47. Albarella et al. 2011; Hide 2003; Sillitoe 2007.
48. Wilford 1994.
49. Redding and Rosenberg 1998:70.
50. About half the wild boar were killed before three years of age (Redding and Rosenberg 1998:69).
51. Redding and Rosenberg 1998.
52. Archaeological examples of hunters targeting younger wild boar have been suggested for sites in the Crimean peninsula (Benecke 1993) and France (Leduc et al. 2015).
53. Ximena Lemoine (2012) makes a strong case for the hunting of farrowing sows at Hallan Çemi.
54. Wild piglet capture strategies are known among the Kubo in New Guinea (Dwyer and Minnegal 2005).
55. The bones could also be from jackals (Rosenberg 1994:130).
56. Rosenberg and Davis 1992:Figure 8.2.
57. Knapp 2010; Simmons 1988; Vigne 2015.
58. Vigne 2015.
59. Vigne et al. 2009b, 2011.
60. Vigne et al. 2009b:16135. Vigne and his team also recovered six distal limb bones from a layer dating to 10,800–10,500 BC, but the bones were not directly dated (Vigne 2015; Vigne et al. 2009a:S256).
61. Vigne et al. 2009b:16135.
62. *Sus scrofa* remains are uncommon at Akrotiri Aetokremnos but constitute 46 percent at Agia Varvara Asprokremnos (8800–8600 BC; Manning et al. 2010; McCartney et al. 2007), over 90 percent at Klimonas (9100–8600 BC; Vigne et al. 2012), and over 90 percent in the earliest levels at Shillourokambos (8300 BC; Vigne 2011a). Measurements of *Sus scrofa* bones and teeth from Klimonas and Shillourokambos indicate that these animals were quite small (Vigne et al. 2012). However, the older age at death has led Vigne et al. (2012) to conclude that they belonged to a population of hunted wild boar that had undergone the "island effect."
63. Bar-Yosef 2001; Goring-Morris and Belfer-Cohen 2002; Makarewicz and Finlayson 2018; Twiss 2008.

64. Dietler 2001; Dietler and Hayden 2001.
65. Dietler 2001; Jaffe et al. 2018; Mintz and Du Bois 2002; Twiss 2008.
66. Russell 2012:358–394.
67. Russell 2012:157–170.
68. Rosenberg and Redding 1998.
69. Bangsgaard et al. 2019; Darabi et al. 2018.
70. Bangsgaard et al. 2019.
71. Dietrich et al. 2012; Peters and Schmidt 2004; Schmidt 2000.
72. Peters and Schmidt 2004.
73. Zeder 2012b.
74. Zeder's (2012b) third pathway involves domesticating animals with the goal of domestication in mind. The evidence for a drawn-out process of pig and other animal domestication in the PPNB strongly argues against such intentional domestication (although for an alternative perspective, see Müller 2005).
75. Price and Hongo, in press.
76. Geiger et al. 2018; Trut 1999.
77. Conolly et al. 2011; Peters et al. 2005; Vigne 2015; Zeder 2011.
78. E.g., Nevalı Çori (ca. 10 percent in Early PPNB [EPPNB] to 20 percent in Middle PPNB [MPPNB]; Peters et al. 2005) and Cafer Höyük (10–26 percent; Helmer 2008).
79. Dental and postcranial measurements falling below the accepted lower limits of early Holocene wild boar occur at MPPNB Tell Aswad, Mezraa-Teleilat, Cafer Höyük, and Çayönü Tepesi. For data see Ervynck et al. 2001; Helmer 2008; Helmer and Gourichon 2008, 2017; Hongo and Meadow 1998b; Ilgezdi 2008.
80. Ervynck et al. 2001; Hongo and Meadow 1998a, 1998b.
81. Ervynck et al. 2001.
82. Ervynck et al. 2001:54.
83. Ervynck et al. 2001:63; Price and Hongo, in press.
84. Dobney et al. 2007; Ervynck and Dobney 1999.
85. Wood et al. 1992.
86. The Channeled Phase (late EPPNB).
87. There is evidence for domestic pigs at the end of the 8th millennium BC at, among others, Tell Halula, Gürcütepe, Gritille Höyük, Mezraa-Telielat, Hayaz Höyük, Çayönü Tepesi, and, perhaps, Jarmo. For data, see Ervynck et al. 2001; Kuşatman 1991; Monahan 2000; Peters et al. 2005; Price and Arbuckle 2015.
88. Redding and Rosenberg 1998.
89. Bleed 2006.
90. Bocquet-Appel 2009.

Chapter 4

1. For simplicity, I am excluding the PPNC (or Final PPNB), which dates to the first half of the 7th millennium BC.
2. E.g., Kuijt and Goring-Morris 2002; Rollefson 1989.

3. Some "megasites," such as Çatalhöyük, persisted beyond the Pre-Pottery Neolithic. But on a regional level, the number declined dramatically. The causes of this "collapse" might include climate change (Berger and Guilaine 2009), landscape/resource depletion (Rollefson and Kohler-Rollefson 1992), or general social breakdown (Banning 1998; Bogaard and Isaakidou 2010:198).
4. Arbuckle 2013; Düring 2013; Zeder 2008, 2017.
5. Bernbeck 2017; Nieuwenhuyse et al. 2010.
6. Bernbeck 1995; Frangipane 2007.
7. McMahon et al. 2011; Rowan and Golden 2009:71.
8. Stein 2012a; Ur 2010.
9. Algaze 1993; Stein 1999.
10. Galili et al. 1997; Miller 2008; Rossel et al. 2008.
11. Graham 2011.
12. Styring et al. 2017.
13. Sherratt 1981, 1983.
14. Greenfield 2010.
15. Evershed et al. 2008.
16. Greenfield 2010.
17. Arbuckle and Hammer 2019; Arbuckle et al. 2014; Zeder 2017.
18. McCorriston and Martin 2009; Uerpmann et al. 2000
19. Linseele et al. 2014.
20. The percentage of pigs compared with that of other medium and large mammals: Çayönü Large Room-PN, 24–37 percent (Hongo et al. 2009); Gritille Höyük, 15 percent (Monahan 2000); Mezraa-Teleilat LPPNB-PN, 12–16 percent (Ilgezdi 2008:85); Hayaz Höyük, 20 percent (Peters et al. 1999); Gürcütepe, 19 percent (Peters et al. 1999); and PPN-PN Jarmo, 2–7 percent (Flannery 1983; Price and Arbuckle 2015; Price and Evin 2019; Stampfli 1983). Pigs were uncommon at some sites, such as Bouqras and Tell es-Sinn (<1 percent; Clason 1979–1980), Umm Dabaghiya (1 percent; Bökönyi 1973), and the PN levels (B10 and E8) at Abu Hureyra (0 percent; Legge and Rowley-Conwy 2000). Percentage of pigs in the 6th millennium: Umm Qseir, 13 percent (Zeder 1994): Domuztepe, 28 percent (Kansa et al. 2009b); Tell Kurdu, 17 percent (Yener et al. 2000); Höyücek, 13 percent (De Cupere and Duru 2003); Gird Banahilk, 16 percent (Laffer 1983); PN-Halaf Tell Sabi Abyad I, 4–9 percent; (Russell 2010); and Hajji Firuz, 30 percent (Meadow 1983).
21. Some have argued for local domestication of pigs in the Levant (Haber and Dayan 2004; Makarewicz 2016; Marom and Bar-Oz 2013; Munro, et al. 2018).
22. Arbuckle 2013; Arbuckle et al. 2014; Düring 2013.
23. Arbuckle et al. 2014.
24. Bökönyi 1978b.
25. Mashkour 2006; Price and Arbuckle 2015.
26. Linseele et al. 2014.
27. Pigs were common in Neolithic and Pre-Dynastic Egypt (Bertini 2016; Redding 2015). There are not many data from southern Iraq, but pigs represent

48 percent of the medium and large mammals from Ubaid 0-3 Tell el-Oueili (Desse 1983).

28. Grigson 2007; Price et al. 2013; Raban-Gerstel and Bar-Oz 2010.
29. Pigs usually make up 1–5 percent of the medium and large mammals (Bökönyi 1973, 1977; Martin 1999; Russell 2010).
30. Price and Arbuckle 2015:450.
31. Rowley-Conwy 2011; Tresset and Vigne 2007; Zeder 2008.
32. Haak et al. 2010.
33. Caliebe et al. 2017; Evin et al. 2013; Frantz et al. 2019; Girdland-Fink and Larson 2011; Larson et al. 2007a; Ottoni et al. 2012.
34. Manunza et al. 2013.
35. Balasse et al. 2016; Evin et al. 2015; Frantz et al. 2015, 2019.
36. Caliebe et al. 2017.
37. Some scholars working in Anatolia and the southern Levant refer to the latter part of this time frame as the "Early Chalcolithic." I avoid the potentially confusing terminology here. Note that I am including the Halaf tradition as part of the Late Neolithic.
38. Biehl and Nieuwenhuyse 2016.
39. Çakırlar 2012; Evershed et al. 2008; Nieuwenhuyse et al. 2015; Rooijakkers 2012.
40. The sites are Umm Qseir, Banahilk, and Domuztepe (Price 2016).
41. Price and Evin 2019.
42. Price, in press.
43. Price 2016; Weber and Price 2016.
44. Ilgezdi 2008; Özdoğan et al. 2011.
45. Ilgezdi 2008:161.
46. Bogaard 2005.
47. Campbell et al. 2014:46. Pig bones were also found in the "Death Pit" at Domuztepe, but curiously at only 11 percent of the main livestock taxa compared with 28 percent for the whole site (Campbell et al. 2014:46; Kansa et al. 2009a).
48. Ben-Shlomo et al. 2009. These values represent the proportion of pigs compared with all specimens found in these contexts, including those not identified to taxon. Hill's dissertation provides NISPs for Tel Tsaf, which indicates that pigs represent 36 percent of medium and large mammals at the site (Hill 2011:107).
49. Hadjikoumis 2012:357.
50. Blanton and Taylor 1995; Rappaport 1968; Sillitoe 2007; Watson 1977.
51. Dietler 2001.
52. Frangipane 2007.
53. Rollefson and Kohler-Rollefson 1993:38.
54. Campbell 2007–2008:131.
55. Twiss 2006.
56. Oates 1969:130. Other examples include a vessel in the shape of a pig in a ritual hearth at Yarim Tepe II (Oates 1978:120) and a possible terracotta pig (or hedgehog) in the Burnt House at Arpachiyah (Campbell 2000).
57. Hendrickx 2011.

58. Englund 1995.
59. Sealing technology provides another example of something that was once used to ensure egalitarianism, but was in the Chalcolithic transformed into a tool of the emerging elite (Frangipane 2000).
60. Cf. "patron-role" feasts (Dietler 2001).
61. Helwing 2003; Oates et al. 2007.
62. Paulette 2016.
63. Cf. "diacritical feasts" (Dietler 2001).
64. Arbuckle 2012a; D'Anna 2012; Helwing 2003; Hill et al. 2016; McMahon et al. 2011.
65. Bartosiewicz 2010.
66. D'Anna 2012.
67. Halstead and Isaakidou 2011a.
68. Halstead (2014:42) states that manual cultivation allows the preparation of 0.01–0.03 hectare/day, while cattle-drawn plowing allows 0.1–0.3 hectare/day.
69. Greenfield 2010, 2014; McCorriston 1997; Payne 1988; Sherratt 1983; Vila 1998.
70. For a description of how cattle may have contributed to inequality, see Bogucki 1993.
71. Foster 2014; Sallaberger 2014.
72. Payne 1973.
73. Arbuckle 2012a; McCorriston 1997; Zeder 1988.
74. Bartosiewicz 2010; Bigelow 1999; Dobney et al. 2003; von den Driesch 1993.
75. Grigson 2007.
76. Price et al. 2013.
77. Bartosiewicz et al. 2013.
78. Bartosiewicz 2005; Frangipane et al. 2002.
79. Hecker 1982; Redding 1991.

Chapter 5

1. Adams 1981; Bachhuber 2015; Lawrence and Wilkinson 2015; Ur 2010; Wilkinson 1999.
2. Postgate 1992:51–58; Postgate et al. 1995.
3. E.g., Ur 2010.
4. Diamond 2005; McAnany and Yoffee 2010.
5. Kennedy 2016; Peltenburg 2000; Weiss 2012; Wossink 2009.
6. Van de Mieroop 2007:122–125.
7. Cline 2014.
8. E.g., Adams 1981; Algaze 2008; Arbuckle 2014; Pollock 1999; Waetzoldt 1972; Zeder 1991.
9. Arbuckle 2015:290.
10. Roth 1980.
11. Appadurai 1981.
12. Paulette 2016.
13. E.g., Van Lerberghe 1996.

14. Arbuckle 2015; McInerney 2010.
15. Goats were also plucked (Colonna d'Istria 2014; Foster 2014), and I use the term "wool" for sheep or goat fiber.
16. Evidence for the expansion of wool production includes an increase in the relative abundance of sheep/goats, beginning in the Uruk period (Vila 1998:90; Vila and Helmer 2014); sheep/goat slaughter at later ages (Arbuckle 2014; Payne 1988; Vila 1998); increase in sheep size, suggesting a greater proportion of males (Vila and Helmer 2014); and iconographic evidence beginning in late 4th millennium BC (McCorriston 1997:520; Vila and Helmer 2014).
17. McCorriston 1997.
18. Adams 1981:11; Larsen 2015; Sallaberger 2014.
19. Adams 1978; Breniquet and Michel 2014; Zagarell 1986.
20. Rossel et al. 2008; Weber 2008.
21. Moorey 1986.
22. E.g., Ehituv 1978.
23. Richard Redding (1991), for example, has suggested that intensive grain production would lead to a reduction in pigs. However, this was not always the case (e.g., Early Bronze Age northern Mesopotamia).
24. Scott 1998, 2017.
25. Some states may have attempted to tax pigs. In the Old Babylonian period, certain high-ranking officials were obligated to deliver a pig annually or even monthly to the king. However, this was probably a token payment rather than a serious attempt at taxation (Van Koppen 2006:190–191).
26. Wallace 1938:145.
27. Harper 1904:18.
28. Laws 80–86 in Meek 1969.
29. Van Koppen 2006:187–189.
30. E.g., Van Koppen 2006:188.
31. E.g., Van Koppen 2006:185.
32. Hecker 1982; Price et al. 2017.
33. That is, through "wealth finance," to use the terminology of D'Altroy and Earle 1985.
34. Payne 1988; Vila 2006:140.
35. McCorriston 1997.
36. Stein 1999:132–145.
37. Note that I have switched from representing the percentage of pigs as a proportion of medium and large mammals to a proportion of sheep, goats, cattle, and pigs.
38. For published faunal data from the Khabur in the Early Bronze Age, see Kolinski 2012; Kolinski and Piątkowska-Małecka 2008; Price et al. 2017. For the Middle Bronze Age, see Berthon 2011; Doll 2010:215; Weber in Schwartz et al. 2017:244.
39. For pigs congregating at sewers, see George 2015. The presence of shed deciduous teeth in streets at, e.g., Early Bronze Age Abu Salabikh provides evidence of pigs roaming free in cities (Matthews et al. 1994). Texts suggest penning in institutional settings (Lion and Michel 2006). Potential evidence of penning also includes healed

cranial injuries on pig skulls, possibly reflecting fighting in crowded pens, at Tell Shiukh Fawqani, Abu Salabikh, and other sites (Vila 2005, 2006:142).

40. Lobban 1994; Price et al. 2017; Redding 2015.
41. Zooarchaeological data quantify the relative, *not absolute*, abundance of meat in the diet. This is important to consider when we have reason to suspect that one group may have eaten more meat overall than another, as is the case with social classes. For example, suppose in a given year (1) an upper-class individual ate 30 kg mutton, 15 kg beef, and 5 kg pork while (2) a lower-class individual consumed 4 kg mutton, 1 kg beef, and 5 kg pork. In terms of absolute abundance, the two individuals ate the same amount of pork (5 kg). But in terms of relative abundance, the upper-class person consumed less pork (10 percent vs. 50 percent).
42. Mudar 1982; Redding 2015; Zeder 2003. One exception is Tell Arbid, where pigs make up 52 percent of the livestock species from the Building of the Plastered Platform dating to the Ninevite V period (Piątkowska-Małecka and Smogorzewska 2010).
43. E.g., Faust 2005:117; Ristvet 2012:153.
44. Price et al. 2017.
45. Hart 1973; Portes and Haller 2005.
46. Çevik 2007; Frangipane 2010.
47. At Troy, the relative abundance of pigs was 24–25 percent in the 3rd millennium BC (Çakırlar 2016; Uerpmann 2003).
48. E.g., Çakırlar 2012.
49. Egypt was not devoid of cities as some early scholars argued, but at ca. 10–20 hectares in size, they were modest in comparison with those in Mesopotamia (Cowgill 2004:530; Wilkinson 1999:323).
50. E.g., Kom el-Hisn, 56 percent (Redding 1992), and Buto, 58 percent (von den Driesch 1997).
51. Nadine Moeller, personal communication
52. See syntheses by Hecker (1982), Redding (2015), and Bertini (2016).
53. Hecker 1982; Lobban 1994.
54. Mashkour 2006; Zeder 1988, 1991.
55. Mashkour 2006.
56. Mashkour 2006.
57. Price et al. 2013.
58. Horwitz 1997, 2003.
59. Horwitz 2003.
60. Berger 2018.
61. Bigelow 1999; von den Driesch 1993.
62. Price et al. 2017.
63. Allentuck and Greenfield 2010; Hesse and Wapnish 2001; Horwitz and Tchernov 1989.
64. Matthiae and Marchetti 2003.
65. Lawrence et al. 2016; Ur 2010.
66. When large cities finally appeared in the Levant in the Middle Bronze Age, pig husbandry received a slight bump: e.g., 12 percent at Tell Jemmeh (Wapnish and Hesse 1988). For more data, see Vila and Dalix 2004.

67. Fleming 2004; Schloen 2001; Stein 2004.
68. Fall et al. 1998; Faust 2005; Wattenmaker 1987.
69. Rosen 2007.
70. E.g., Coon 1951; Grigson 2007; Harris 1974, 1985.
71. Allentuck 2013; Price et al. 2017.
72. Deckers and Pessin 2010; Styring et al. 2017.
73. E.g., Tell Arbid, Level VIIA, 42 percent vs. Level VI, 40 percent (Piątkowska-Małecka and Smogorzewska 2010); Tell Mozan, EJ IIIb, 34 percent vs. EJ IV, 30 percent vs. EJ V, 19 percent (Doll 2010); Tell Brak, Areas FS, SS Akkadian, 28 percent vs. Post-Akkadian, 28 percent (Weber 2001).
74. Bar-Matthews and Ayalon 2011; Rosen 2007:82–103.
75. Price et al. 2017.
76. Allentuck 2013:167; Price et al. 2017.
77. Riehl 2008.
78. For data from pollen cores, see, e.g., Deckers and Pessin 2010; Rosen 2007:100; Yasuda et al. 2000.
79. Postgate 1992:166–167; Sallaberger 2014; Stepien 1996; Waetzoldt 1972.
80. Hecker 1982:59; Price et al. 2017.
81. Uruk Text W 23948 in Englund 1995.
82. Dahl 2006.
83. Dahl 2006:33–35.
84. Price et al. 2017:55.
85. Ikram 1995; Lobban 1994.
86. Lobban 1994:63–64.
87. Renni of El Kab, 17th Dynasty.
88. Kees 1961:87.
89. Ikram 1995:31.
90. Fisher 2013.
91. Ikram 1995:31; Van Koppen 2006:188.
92. Foster 2006:284.
93. For Mesopotamia, see Van Koppen 2006:188. In Anatolia, texts from the Assyrian merchant colony at Kültepe-Kanesh mention the purchase and shipment of several kilograms of lard (Michel 2006:174). In Egypt, excavators at Amarna recovered a jar bearing the label "pig fat of the herds of the estate of the Aten in/from the Western (?) River . . ." (Leahy 1985:67).
94. Dahl 2006:37; Foster 2006:284; Stepien 1996:31; Van Koppen 2001.
95. MS Doc. 829 from Zabala located at the Cambridge University Library, https://cudl.lib.cam.ac.uk/view/MS-DOC-00829/1.
96. Parayre 2000:168.
97. Moran 1992.
98. Pigs represent 49 percent of the livestock taxa found in the Workmen's Village (Hecker 1984).
99. Kemp 1984.
100. Shaw 1984.

101. Panagiotakopulu 1999.
102. Bertini 2016; Bertini and Cruz-Rivera 2014.
103. For the limited relative abundance data, see Parayre 2000:197; Weber in Schwartz et al. 2017:244. Becker (2005, 2008) also presents faunal data for Late Bronze Sheikh Hamad (9 percent pigs) and Tell Bderi (<1 percent). Pigs are well represented (37 percent) in Late Bronze levels at recently excavated Kurd Qaburstan (Weber in Schwartz et al. 2017).
104. See Vila 2006:143; Vila and Dalix 2004. At Tell Hazor, pigs represent 2 percent of the main livestock taxa (Marom and Zuckerman 2012). One exception is 14th century BC Alalakh (23 percent pigs; Çakırlar et al. 2014).
105. For data on Late Bronze Age Anatolia, see Arbuckle 2012b; Boessneck and von den Driesch 1974; Çakırlar et al. 2014; Hongo 1998b; Slim et al. 2020; von den Driesch and Boessneck 1981.
106. Lawrence and colleagues' (2016) aggregate settlement density data for the Jezireh, the Orontes Valley, and the Euphrates River Valley all decline beginning around 2000 BC and reach their local minima at around 1200 BC. Survey data in the southern Levant also suggest population decline in the Late Bronze Age (Wilkinson 2003:131–135).
107. Van Koppen 2006:191–192.
108. Foster 2016:1, 36.
109. Foster 2006:284–285.
110. See Bottero 2004; Ristvet 2014:40–42.
111. Lobban 1994:63–64.
112. Hecker 1982:60.
113. De Cupere and Van Neer 2014; Ikram 1995; Ikram 2008; Polcaro et al. 2014:11; Porter 2002; Ristvet 2014; Schwartz et al. 2000.
114. Ikram 1995:291–292.
115. E.g., Tell es-Sweyhat (Weber 2006:224) and Umm el-Marra (Schwartz 2013:501) in Syria.
116. E.g., Hassek Hoyuk (Boessneck 1992:65), Tell Abqa' (Amberger 1987), and Hirbemerdon Tepe (Laneri 2011:84).
117. Parayre 2000.
118. Quoted by Collins 2006:162–163.
119. Cooper 1996:51; Foster 2006:289.
120. Wasserman 2016: 27–34, lines 13–22. Original translation by Claus Wilcke 1985.
121. Buccellati and Kelly-Buccellati 2005.
122. Collins 2004; Kelly-Buccellati 2005.
123. The excavators of the site interpreted the pit as an *abi* associated with the Hurrian god Kumarbi (Buccellati and Kelly-Buccellati 2005:29).
124. Collins 2002, 2004, 2006.
125. Bones of young pigs, interpreted as sacrifices, were also found in an underground ritual space at Middle Bronze Age Tell el-Farah (ancient Tirzah) in the southern Levant (De Vaux 1971:252).

126. E.g., Old Babylonian and Hittite spells using pigs to expel disease (Collins 2006:173–176; Geller 1991:111).
127. Collins 2004:55, 2006:174–176.
128. Schwemer 2007:31.
129. Parayre 2000:171.
130. van der Toorn 1999:141.
131. Dolansky 2013:62; Wiggerman 2010.
132. Wiggerman 2010.
133. Dalix Meier 2006.
134. Dalix and Vila 2006; Vila and Dalix 2004.
135. Lions, bulls, and goats are more common animal motifs in the Levant and Mesopotamia. Pigs and wild boar were more commonly employed as symbols of power in Anatolia and the Aegean (Dalix and Vila 2006:364–370).
136. Ikram 1995:32; Lobban 1994.
137. Allen's translation is from an 18th Dynasty text (Allen 1974:91, fn 195).
138. Lobban 1994.
139. That is, the 21st–25th Dynasties (Lobban 1994:68).
140. Bertini 2016; Redding 2015.
141. Mortuary and foundation deposits all seem to date to the Middle Bronze Age or earlier.
142. Falconer 1995; Lev-Tov and McGeough 2007; Salleberger and Pruß 2015; Van Koppen 2006:190; Zuckerman 2007.
143. Parayre 2000:168–171.
144. Collins 2006:157; Parayre 2000; Scurlock 2002:393.
145. Collins 2006:156–157.
146. Richardson 2007:197.

Chapter 6

1. Radcliffe-Brown (1939:8–9) suggested that the term "taboo," due to its origins in Polynesian cultures, was not amenable to generalization; he preferred the term "ritual prohibition."
2. Shirres 1982.
3. Frazer 1911.
4. Radcliffe-Brown 1939.
5. Durkheim 1995 [1912].
6. Freud 1918.
7. I am building on Rappaport's (1968:208) definition of taboos as "supernaturally sanctioned proscriptions of physically feasible behavior" as well as Allan and Burridge's (2006:27) concept of "radioactivity." Discussions by Valeri (2000) and Fowles (2008) have also influenced my definition. Earlier anthropological definitions tended to focus on the vacillation between sacred and profane. Frazer (1911:224–225) argued that the

sacred is dangerous and therefore must be tabooed; Durkheim (1995 [1912]:306–313) argued that taboos prevent cross-contamination between the sacred and profane.

8. Miller 2004.
9. Rozin et al. 1997a:79.
10. Allan and Burridge 2006:27.
11. Akimichi 1998.
12. Hastorf (2016:184–186) provides a discussion of food taboos applied to women.
13. Henrich and Henrich 2010; Meyer-Rochow 2009; Schieffelin 2005 [1976]:63–65; Turner 1967:59–92; Van Gennep 1960.
14. DeBoer 1987; Valeri 2000:412; Whitehead 2000.
15. Anderson 2014:229–249.
16. DeBoer 1987; Valeri 2000.
17. Howell 2012.
18. Sometimes taboos are applied only to certain portions of meat. Kosher families traditionally avoid the hindlimbs of animals because of their association with the sciatic nerve (*gid hanasheh*; see Genesis 32:33), although the nerve can be removed, rendering this part of the body kosher. Gustavo Politis (2007:299) observed that Nukak hunters would not eat the heads of monkeys or peccaries out of a fear of illness or future hunting impotence. *Mixtures*, especially of meat or animal parts, are also frequently tabooed. Valeri (2000:378) noted these types of taboos among the Huaulu, who would not eat certain meats together because doing so would entail combining contrasting colors (e.g., red and black). Another example is the injunction against eating meat and dairy in *Kashrut*, which derives from the Mishnaic interpretation of "do not cook the kid in its mother's milk" (Exodus 23:19, 34:26 and Deuteronomy 14:21).
19. Fessler and Navarrete 2003.
20. On the social power of animals in human cultures, see Bulliet 2005; Fagan 2015; Ingold 1988; Serpell 1996 [1986].
21. Fagan 2015:13.
22. Hunting, slaughtering, and sacrificing are often connected to the reproduction of political authority, the display masculinity, and the maintenance of social cohesion (Fiddes 1991; Russell 2012:155–164; Speth 2009; Whitehead 2000).
23. E.g., Politis 2007:299.
24. E.g., Henrich and Henrich 2010; Malinowski 1932:336; Reichel-Dolmatoff 1985:116–117.
25. Kathryn O'Neil Weber articulated this thought to me in June 2018. I am paraphrasing her.
26. Freud 1918.
27. E.g., Schorsch 2018:2.
28. Goossaert 2005.
29. E.g., Scarborough 1982.
30. See, e.g., Clark 2004.
31. I am using the English translations of the original Hebrew available on the Sefaria website, https://www.sefaria.org/texts. I reject their translation of *tame* as "unclean" for reasons I explain in the text.

32. E.g., Cohen 2006:125.
33. Some Classical writers reported speculations on the origin of the pig taboo in Egypt. According to Plutarch (AD 46–120), the taboo was thought by some to have arisen because swine mated during the waning of the moon and people who drank pig milk developed leprosy (*Moralia, Isis and Osiris* 8.353–354). Aelian also referenced the claim that leprosy could be acquired from pig milk (*On Animals* 10.16), citing the now-lost works of the Egyptian Manetho.
34. Schäfer 1997:77–78.
35. See Fabre-Vassas 1997.
36. "Leprosy" probably referred to skin diseases in general. Some of the most common infections of the skin contracted from pigs are ringworm and erysipeloid, but they have many vectors besides pigs (Neumann et al. 2010).
37. Seeskin 2017.
38. Neghina et al. 2012; Simoons 1994:65–69.
39. E.g., Heisen 1891; Wallace 1844.
40. E.g., Neghina et al. 2012:503.
41. Simoons (1994) has written the most extensive rebuttal to the trichinosis hypothesis.
42. Whitehead 2000:96.
43. Rozin et al. 1997b.
44. Foer 2009.
45. The list includes Simoons (1994:65–71), Douglas (2002 [1966]), and Harris (1974).
46. Frazer 1912.
47. Frazer 1912:17–18.
48. Frazer 1912:22.
49. Frazer 1912:24.
50. Frazer 1912:25–30.
51. In the case of the Egyptians, Frazer (1912:33–34) suggested that they transferred the association between pigs and the god Osiris to his enemy Typhon.
52. Collins 2004.
53. Milgrom 1998.
54. Milgrom 1998:649–653.
55. Milgrom 1998:766–768.
56. Collins 2004.
57. Douglas 2002 [1966].
58. Douglas 2002 [1966]:56.
59. E.g., Milgrom 1998:721.
60. Bulmer 1967:21; Tambiah 1969.
61. For dating of the Torah see, e.g., Dever 2003:7–8; Emerton 2004; Finkelstein and Silberman 2001:12.
62. Douglas 1975:276–318.
63. Douglas 1999:149.
64. Douglas 2002 [1966]:viii.
65. E.g., White 1959:8.
66. Coon 1951.

67. Harris 1974, 1985.
68. The timing of deforestation in the Near East varied from region to region, but in general began in the Bronze Age and accelerated in the Iron Age and Classical periods (Deckers and Pessin 2010; Rosen 2007; Wilkinson 2003).
69. Harris 1985:78.
70. Diener and Robkin 1978:496.
71. Diener and Robkin 1978:498; Miller 1990; Nemeth 1998.
72. Diener and Robkin 1978.
73. Diener and Robkin 1978:501.
74. Diener and Robkin 1978:501.
75. Diener and Robkin 1978:502.
76. Falconer 1995; Price 2016; Price et al. 2017; Redding 1992, 2010, 2015; Wattenmaker 1987; Zeder 1991, 1996, 1998b.
77. E.g., Hesse 1990; Zeder 1996.
78. Unless Diener and Robkin (1978) are subtly arguing that the pig taboo was added to the Quran *after* the conquests of Arabia, the Levant, and Mesopotamia.
79. See Aslan 2011.
80. Simoons 1994.
81. Simoons (1994:65–71) also refuted the health/trichinosis hypothesis, the ecological hypotheses of Carleton Coon and Marvin Harris (Simoons 1994:71–85), and the "symbolic and cultic hypotheses" of Mary Douglas (Simoons 1994:85–92).
82. Douglas 1975:307–308.
83. Soler 1979.
84. Simoons 1994:87–89.
85. Crabtree 1990; Goody 1982; Jaffe et al. 2018; Smith 2003; Stein 2012b; Twiss 2012.
86. Anderson 2014:225–249.
87. Barth (1969) often receives the lion's share of credit for this theory, but a number of other scholars have employed and developed the instrumentalist approach to ethnicity (e.g., Comaroff and Comaroff 1992:235–263; Emberling 1997; Faust 2016a; Smith 2003:1–10).
88. E.g., Smith 2003.
89. Liebmann 2015; Silliman 2013.
90. Simoons 1994:99–101.
91. Simoons 1994:93–94.
92. Simoons 1994:93. The mixed group model for ancient Israelites has support among other scholars (e.g., Killebrew 2014).
93. E.g., Valeri 2000.
94. Fowles 2008:21; Valeri 2000.
95. Even when pastoralists view pork with suspicion, they may not have an outright taboo. For example, Simoons (1994:45) admits that the Mongols, though they harbor negative sentiments against pigs, eat pork and even raise pigs on occasion.
96. More mobile forms of extensive pig husbandry, including transhumance between mountain and lowland feeding grounds, have been documented in, for example, the Mediterranean (Albarella et al. 2997, 2011).

97. E.g. Dyson-Hudson and Dyson-Hudson 1980.
98. The dichotomy lies at the heart of Ibn Khaldun's historiography in *The Muqaddimah*.
99. Redding 2015.
100. Redding 2015:351. In earlier publications, Redding argued that pigs were associated with lower-status households in Egypt and Mesopotamia and were of no interest to institutions (Redding 1991, 1992, 2010).
101. Redding 2015:356.
102. Chickens may have been raised in Iran in the 4th millennium BC, and they have been identified at a number of 3rd millennium sites in Syria, southern Turkey, and northern Iraq (Piątkowska-Małecka 2015; Redding 2015:337). The situation is less clear in Egypt: the oldest chicken bones, according to Redding, date to the 7th century BC (Redding 2015:337), while the oldest textual and artistic references date to the 2nd millennium BC (Redding 2015:338–339).
103. Redding (2015:38–40) rightly points out that the number of chickens may be underrepresented due to the lack of sieving at many Bronze Age sites.
104. Perry-Gal et al. 2015b.
105. Redding 2015:356.
106. Rappaport 1968:208.
107. Goody 1982.
108. De Beauvoir 2011 [1949]:754.
109. Sykes 2015:165–167.

Chapter 7

1. Cline 2014; Kaniewski et al. 2013; Langgut et al. 2013; Liverani 1987; Oren 2000; Wiener 2017.
2. Dothan 1998; Maeir and Hitchcock 2017.
3. Akkermans and Schwartz 2003:367–377; Dever 2003; Faust 2016b; Finkelstein 1988; Finkelstein and Silberman 2001; Killebrew 2005; Stager 1985.
4. Finkelstein and Silberman 2001:17–19.
5. With some exceptions (Friedman 2017), almost all scholars agree that the archaeological data argue in favor of a local Levantine ethnogenesis for the Israelites (Dever 2003; Faust 2016a; Finkelstein 1988; Killebrew 2005).
6. E.g., Liverani 2017.
7. Akkermans and Schwartz 2003:360–398; Briant 2002.
8. Cohen 2006:175–176.
9. Finkelstein and Silberman 2001:12.
10. See, e.g., Coogan 2017; Dever 2003; Emerton 2004.
11. Most date D as preceding P; many assign P to an Exilic date (e.g., Coogan 2017).
12. The dating of Leviticus (the P and H source) and other works of the Hebrew Bible is a contentious issue. Milgrom (1998:3–30) favors a date for P and H around the time of Hezekiah's reforms. Many others also argue that the main parts of the text are pre-Exilic, even if redaction continued into Exilic and even post-Exilic times

(Finkelstein and Silberman 2001:5–14; Haran 1981; Hurvitz 1988). Other scholars favor an Exilic or even Persian period date for P and H (e.g., Cohen 1999:124; Coogan 2017; Rhyder 2019).

13. The Hebrew Bible was written primarily from the perspective of the southern kingdom, but probably incorporated elements of the texts of the kingdom of Israel (Fleming 2012).
14. Finkelstein and Silberman 2001:14, 42.
15. E.g., Albright 1939.
16. Dever 2003; Faust 2016b.
17. Dever 1994:215–216; Faust 2016a:156, 2018; Sapir-Hen et al. 2013.
18. Gottwald 1979.
19. E.g., Dever 2003.
20. Finkelstein 1992, 1996b; Finkelstein and Na'aman 1994; Finkelstein and Silberman 2001:97–122.
21. Dever 2003; Faust 2016b; Killebrew 2005, 2014.
22. E.g., Dever 2005.
23. For the "house of the father" serving as a "root metaphor" for Israelite society, see Schloen 2001:135–183.
24. Faust 2016b:92–110.
25. For a summary of the evidence for the introduction of the camel, see Walton 2015:281–346. For a general overview of Iron Age Levantine fauna, see Sasson 2010.
26. E.g., Black 2016.
27. Faust 2016a:168.
28. Faust 2016b:131–134.
29. E.g., Finkelstein 1996a, 2010; Mazar 2010.
30. Finkelstein and Silberman 2001:243–246; Geva 2014.
31. Emberling 1997.
32. Finkelstein and Mazar 2007:153–156. Although the Deuteronomistic history (Joshua, Judges, 1 and 2 Samuel, 1 and 2 Kings) probably dates to the Exilic period (see Finkelstein and Silberman 2001:13–14).
33. Finkelstein and Mazar 2007:167.
34. Finkelstein and Silberman 2001:277.
35. Finkelstein and Silberman 2001:3–15.
36. Note that other law codes from ancient Near Eastern history, such as that of Hammurabi, also referenced divine inspiration.
37. Konner 2003:38–43.
38. Black 2016:60.
39. See also Finkelstein and Silberman 2001:296–313.
40. Finkelstein and Silberman 2001:296; Ska 2006:112.
41. Konner 2003:37; Rosenberg 2004.
42. Regarding Iran, Mashkour's (2006) data indicate Iron Age sites generally contained <10 percent pigs. However, pigs represent 12 percent of the main livestock remains at Achaemenid levels at Choga Mish (Lev-Tov et al. 2017) and around 43 percent from Achaemenid Qelich Qoineq, a large military outpost near the Caspian Sea

(Mashkour et al. 2013:570). There are, however, few data from Mesopotamia. Vila (2006) lists seven Mesopotamian sites, all of which are in northern Mesopotamia/Levant—Qatna (Tell Mishrife), Tell Afis, Khirbet Khatuniyeh, Tell Knedig, Shiukh Fawqani, and Horum Höyük.

43. Bertini 2011:109; Redding 2015.
44. Pig bones represent 69 percent of the main livestock taxa from Naukratis (Reese 1997).
45. E.g., Iron I Shiukh Fawqani (6 percent, although this increased to 17 percent in Iron II; Vila 2005) and Iron I–III Tell Nebi Mend (ancient Qadesh; 5–8 percent; Grigson 2015). At Kınık Höyük in Cappadocia, pigs represented around 4 percent of the livestock remains from 6th–4th century BC deposits and were entirely absent from the Achaemenid deposits (Highcock et al. 2015:118–119).
46. E.g., Kaman-Kalehöyük (13–24 percent in various phases from the 12th to 4th century BC; Hongo 1998a); Tell Afis (21 percent in Iron I, 15 percent in Iron II–III, but only 5 percent in Iron III; Wilkens 2000); Gindaris (21 percent; Vila and Dalix 2004); Kinet Höyük (13 percent; Çakırlar 2003). Pig bones made up 29 percent of the livestock remains from the settlement (not the temple) at Ain Dara (Frey and Marean 1999). Pigs typically represent around 15–25 percent of the livestock remains from Iron Age western Anatolia (Çakırlar and Atici 2017:272; Çakırlar et al. 2015).
47. Pig bones represented 14 percent of the livestock remains from the Neo-Assyrian palace and 15 percent from the domestic area Ziyaret Tepe (Greenfield 2015) and from 5 percent up to 21 percent in the Early Iron and Phrygian levels at Gordion (Zeder and Arter 1994).
48. Faust 2018; Faust and Lev-Tov 2011; Hesse 1990; Hesse and Wapnish 1997; Horwitz et al. 2017; Lev-Tov 2000; Sapir-Hen 2018; Sapir-Hen et al. 2013, 2015.
49. Jones 1997; Mac Sweeney 2009.
50. Jaffe et al. 2018; Silliman 2013; Stein 2012b.
51. Bunimovitz and Yasur-Landau 1996; Faust and Lev-Tov 2011; Yasur-Landau 2010.
52. For example, at Tel Miqne-Ekron, pig bones rose from 3 percent of the main livestock species in the pre-Philistine Late Bronze Age to 20 percent in the Iron I (Hesse 1986; Lev-Tov 2000). Similarly, at Ashkelon from Phase 20 (initial Philistine occupation, ca. 1175–1150 BC; 2 percent), Phase 19 (ca. 1150–1100 BC; 9 percent), Phase 18 (ca. 1100–1050 BC; 14 percent), and Phase 17 (ca. 1050–1000 BC; 9 percent; Wapnish and Fulton 2020).
53. Faust and Lev-Tov 2011; Hesse 1990; Lev-Tov 2000.
54. Ben-Shlomo et al. 2008; Faust 2018; Faust and Lev-Tov 2011; Finkelstein 1996b; Lev-Tov 2000; Maher 2017; Mahler-Slasky and Kislev 2010.
55. Horwitz et al. 2017; Sapir-Hen et al. 2013.
56. Faust and Lev-Tov 2011; Horwitz et al. 2017; Wapnish and Fulton 2020.
57. Hesse 1986; Lev-Tov 2000.
58. Hesse et al. 2011.
59. At Tell es-Safi/Gath: 13 percent in the Iron IIA and 16 percent in the Iron IIB periods (Lev-Tov 2012).
60. Faust 2018; Faust and Lev-Tov 2011.
61. Horwitz et al. 2017; Sapir-Hen 2018.

62. E.g., Faust 2016a; Finkelstein 1997.
63. For a discussion, see Fowles (2008) and Politis and Saunders (2002).
64. Politis and Saunders 2002.
65. E.g., Politis and Saunders 2002:118.
66. Fowles 2008:20.
67. Hesse 1990; Sapir-Hen et al. 2013, 2016.
68. Lev-Tov et al. 2011; Sapir-Hen et al. 2013; Vila and Dalix 2004.
69. Bunimovitz and Lederman 2011; Faust 2016a.
70. Sapir-Hen 2018; Sapir-Hen et al. 2013, 2015.
71. Stratum V-A, dating to the late 8th century BC (Angress 1960:166).
72. Angress 1960.
73. Sapir-Hen 2018.
74. Sapir-Hen et al. 2015.
75. Faust 2016b, 2018; Hesse 1986, 1990, 1994; Hesse and Wapnish 1997, 1998; Horwitz, et al. 2017; Meiri et al. 2013, 2017; Sapir-Hen 2018; Sapir-Hen et al. 2013, 2015.
76. Hesse and Wapnish 1997:238.
77. Lipovitch 2006–2007.
78. Ben-Shlomo et al. 2008; Faust and Lev-Tov 2011; Horwitz et al. 2017; Maher 2017.
79. Faust 2018; Faust and Lev-Tov 2011; Hesse 1990; Lev-Tov 2000.
80. Ben-Shlomo et al. 2008.
81. Faust and Lev-Tov 2011; Sapir-Hen et al. 2013, 2015.
82. Faust and Lev-Tov 2011.
83. Faust 2016a.
84. Sapir-Hen 2018.
85. Faust (2017:180) has critiqued Sapir-Hen's hypothesis, arguing that Megiddo, Beth Shean, and other sites with higher frequencies of pig bones had large populations of Canaanites.
86. Finkelstein and Silberman 2001:229–250.
87. Sapir-Hen et al. 2015.
88. Faust 2017; Finkelstein and Silberman 2001:196–295.
89. Finkelstein and Silberman 2001:122.
90. Soler 1979.
91. Milgrom 1998.
92. Houston 1993:237.
93. Houston 1993:122–123.
94. Some of the other taboos (e.g., Leviticus 11: 20) use the Hebrew word *sheketz*. The more general term *to'eba* is also applied (e.g., Deuteronomy 14:3).
95. Hesse and Wapnish 1998:130.
96. Lambert 1996:215. The passage is from the reverse III of tablet VAT 8807, lines 5–165. The tablet is dated to the sixth year of Sargon II's reign, traditionally dated to 716 BC. I have kept Lambert's notation for lacunae and untranslated words.
97. Parayre (2000:167) cites the example of CAD 17/1, which mentions the provisioning of pork to Assurbanipal's (reigned 668–627 BC) cavalry (SAA 13 82).
98. E.g., Briant 2002:232.

99. Parayre 2000:173.
100. KAR 143 (or SAA 3 34; see also SAA 3 35); see Frymer-Kensky 1983.
101. Frymer-Kensky 1983:135.
102. For review, see Hecker 1982; Ikram 1995:29–33; Lobban 1994.
103. Lobban 1994:67. In his book on ethnicity in ancient Egypt, *Wretched Kush*, Stuart Smith argues that the pig taboo "increased [. . .] as a means of differentiating Egyptians in an increasingly cosmopolitan society, and as a means of promoting solidarity in the face of conquest by the Assyrians, Persian, Greeks, and Romans" (Smith 2003:46). While I would argue that this neatly describes the evolution of the pig taboo in Jewish culture, there is no solid evidence that the taboo in Egypt was anything other than one applicable to certain religious contexts. Indeed, Smith (2003:46) continues, "Even in these periods, the pig taboo was primarily religious."
104. E.g., Lobban 1994.
105. For a detailed investigation of the Jewish Diaspora in Egypt, see Barclay 1996.
106. I will not attempt to distinguish magic from religion. The two were intimately connected in the ancient Mediterranean. See sources in Mirecki and Meyer 2002.
107. Betz 1986:97.
108. Betz 1986:97.
109. E.g., De Vaux 1971.
110. Late Bronze and Iron Age faunal data from Lebanon are sparse, but show low numbers of pigs (Chahoud 2014–2015; Lipovitch 2006–2007; Sapir-Hen et al. 2015:9; Vila and Dalix 2004). Pigs were also uncommon at Phoenician-affiliated Tel Dan (<1 percent of Iron Age assemblage; Wapnish and Hesse 1991) and Tel Dor (1 percent of Iron Age assemblage; Raban-Gerstel et al. 2008). The Tel Dor specimens appear to be wild based on size (lower third molar lengths of 37.92 mm and 39.81 mm; Raban-Gerstel et al. 2008:42), as do the ones from Tel Dan (Wapnish and Hesse 1991:57).
111. Cardoso et al. 2016.
112. Docter (2009) indicates that the percentage increased from 4 percent in the Early Punic period to 11 percent in the Middle–Late Punic period. Weinstock (1995) argues that the increase in the proportion of pigs was due to a process of acculturation to Greco-Roman food preferences by the Phoenician inhabitants of Carthage.
113. Larson et al. 2007a.
114. Frantz et al. 2019.
115. Larson et al. 2007a.
116. Meiri et al. 2017; Ottoni et al. 2012; Sapir-Hen et al. 2015.
117. Frantz et al. 2019.
118. Ottoni et al. 2012.
119. Specifically, 2 out of 22 specimens at Middle–Late Bronze Age Lidar Höyük had European genetic signatures (Ottoni et al. 2012).
120. Meiri et al. 2013, 2017.
121. See also Giuffra et al. 2000.
122. Meiri et al. 2013, 2017.
123. Meiri et al. 2013:Table S1.

124. Meiri et al. 2013:4.
125. Ottoni et al. 2012.
126. Lega et al. 2017.
127. Larson et al. 2007a.
128. Cohen 2006:51–75.

Chapter 8

1. E.g., Akkermans and Schwartz 2003:386–392; Burkert 1992; Roosevelt 2012.
2. Cohen 2006:19–36; Mac Sweeney 2009.
3. Ball 2016.
4. Ball 2016:1–27.
5. Barclay 1996.
6. Konner 2003:58–124; Schwartz 2014.
7. Cohen 2006; Schwartz 2014.
8. Cohen 2006:31–35.
9. Black 2016:47–59.
10. Ball 2016:49–62; Barclay 1996; Cohen 2006:19–25; Schwartz 2014.
11. Ball 2016:483–496.
12. Navarette Belda and Saña Seguí 2017; Conolly et al. 2011; Fillios 2006; Trentacoste 2016.
13. Pigs were a major component of urban Mycenaean sites in Greece (20–40 percent; Halstead 2003; Lipovitch 2006–2007). In later periods, sheep and goats tend to dominate Aegean assemblages, but pigs frequently made up 20–40 percent of the livestock taxa and were more common at urban centers. Data summarized by King (1999:Table H) include pig NISPs from a handful of sites dating to the 8th–1st centuries BC: Eleutherna, 15 percent; Isthmia, 10 percent; Kassope, 32 percent; and Messene, 39–45 percent.
14. Whitley 2001:165–194.
15. Trentacoste 2013.
16. Cattle were also important (King 1999:171–172).
17. Trentacoste 2016:308.
18. Trentacoste 2016.
19. King (1999:188–189) argues that the increase in pig production was driven by high-status demand for pork in parts of the empire. In any event, the high proportion of pigs around the Mediterranean continued into the later years of the Roman Empire and the Byzantine period (Kroll 2012).
20. King 1999; MacKinnon 2001.
21. Dalby 2003:360.
22. Dalby 2003:360; Leigh 2015:47.
23. Leigh 2015:47.
24. Ricotti 2015.
25. MacKinnon 2001.

26. Pigs are abundant in the Linear B texts from Pylos, where they account for about 57 percent of the animals recorded as slaughtered. However, they represent only 5 percent of the animals in texts relating to census-taking of live herds. At Knossos, pigs played a more minor role in texts, accounting for only 1 percent of animals recorded as slaughtered and only 0.5 percent of the live animals tallied by the palace officials (Halstead 2003).
27. King 1999:176.
28. Barnish 1987:163; MacKinnon 2001:659. Pigs continued to be taxed into the Byzantine period, especially through the *annona* (tax in kind), which was used to provision the empire's armies and for the dole (Kroll 2012:97).
29. Trentacoste 2016:308.
30. Barnish 1987:160; Essig 2015:71–74.
31. Chandezon 2015:141; Leake 1826:30.
32. Leake 1826:30.
33. Hamilakis and Konsolaki 2004.
34. Especially in *thysia* sacrifices (Ekroth 2007:250). Although these were the main sacrificial animals, the Greeks sacrificed a broad range of species. Wild and domestic animals (e.g., dogs, deer, donkeys, horses, wild boar, and camels) appear in the zooarchaeological record of Greek sanctuary sites; typically 10 percent or less of the entire mammalian fauna are pigs (Ekroth 2007:256–257).
35. Collins 2006:169–171.
36. Cole 2004:138–140.
37. MacKinnon 2001:660.
38. Another epic example is Aeneas's sacrifice of a white sow and her litter of 30 piglets to mark the future location of Alba Longa (Vergil, *Aeneid*, book 8).
39. Cultraro 2004.
40. E.g., Rosenblum 2010a; Weingarten 2007; Wilkins and Nadeau 2015.
41. Anthony King (1999) and Henriette Kroll (2012) have summarized the faunal data from around the Mediterranean in the Roman and Byzantine periods. For Egypt, see summaries by Van Neer (1997) and Redding (2015:331). Even in places where pigs were not abundant, most sites show an increase in the percentage of pig bones over time. For example, Pessinus in the Late Hellenistic (3 percent of livestock) to Early–Late Roman (8–10 percent); see De Cupere 1994; Gruwier and Verlinde 2010.
42. Boessneck and von den Driesch 1985; King 1999:199. As another example, De Cupere et al. (2017) report data from Hellenistic Düzen Tepe (13 percent) and nearby Roman Sagalassos (32 percent).
43. Louise Bertini, personal communication.
44. Forstenpointner et al. 2002. Also Zeugma (20–50 percent in Hellenistic-Byzantine periods; Charles 2013; Rousseau et al. 2008).
45. E.g., the Seleucid fort/military colony at Hacinebi in Anatolia (24 percent; Kathryn Grossman, personal communication); Al Zarqa (85 percent), and Mons Claudianus (79 percent) in Egypt (King 1999; Van Neer 1997). Pigs were not common at all military outposts, e.g., 5 percent in Hellenistic period Jebel Khalid (Steele 2002).

46. Tell Arbid, 16 percent; Tell Beydar, 11 percent. See De Cupere and Van Neer 2014; Kolinski and Piątkowska-Małecka 2008.
47. De Cupere and Van Neer 2014:195.
48. Mashkour 2013.
49. Pigs are 3 percent at Haftavan Tepe in northern Iran from the Median, Persian, and Parthian periods (Mohaseb and Mashkour 2017). In the Persian Gulf, pigs were represented in small numbers (4 percent) at the 4th century AD trading port of Siraf (von den Driesch and Dockner 2002). At the Hellenistic fortress on Failaka Island (Kuwait), pigs represented 20 percent, but no pig remains were found at nearby Tell Akkaz (Monchot 2016).
50. On "Fort 4" and Dasht Qal'eh, see Mashkour et al. 2013.
51. Çakırlar and Marston 2019.
52. De Cupere et al. 2017; Frémondeau et al. 2017; Fuller et al. 2012; Vanpoucke et al. 2007, 2009.
53. De Cupere 2001; Frémondeau et al. 2017; Vanpoucke et al. 2007
54. Frémondeau et al. 2017; Vanpoucke et al. 2007. Third-molar length also decreased over the early Roman to Byzantine periods, possibly indicating increased genetic isolation from wild boar or short-faced breeds (Frémondeau et al. 2017:44–46). Dental microwear suggests pigs consumed soft foods, such as those provided by slop-feeding and foddering (Vanpoucke et al. 2007).
55. Hellenistic δ^{15}N: 5.9 ± 1.7‰; Early Byzantine δ^{15}N 7.8 ± 1.6‰. δ^{13}C remained stable at roughly –19.8‰ (Fuller et al. 2012:5.
56. Frémondeau et al. 2017.
57. This hunger may have been satisfied by the market or through government taxation and distribution. For example, during the Ptolemaic and Roman periods in Egypt, pigs could be collected as tax in kind (Kroll 2012:98; Wallace 1938:145, 194).
58. While the Jewish taboo gained the most notoriety among Roman writers, the Stoic philosopher Epictetus (ca. AD 55–135; *The Teaching of Epictetus* 2.44) noted other pork taboos (perhaps religious) among the Syrians and Egyptians.
59. Cohen 2006:136–146.
60. Cohen 2006:27.
61. Cohen 2006:37–38.
62. Schwartz 2004:20–21. The book of Esther may offer an exception, although its historicity is dubious (Berlin 2001).
63. It is unlikely Alexander visited Jerusalem, although one of his generals may have (Schwartz 2014:31).
64. Ball 2016:xxxi; Schwartz 2014:33.
65. Schwartz 2004:53; 2014:36–39.
66. Honigman (2014:238–250) disputes the historicity of pig sacrifice by Antiochus IV or any other Greek, arguing instead that it was a way to underscore the enormity of Greek oppression (see also Schäfer 1997:66–67).
67. Ball 2016:51.
68. Honigman 2014; Schwartz 2014:44–47.
69. Schwartz 2014:50–55.

70. Ball 2016:49.
71. Schwartz 2004:44–48.
72. The story is related by the early 5th century AD writer Macrobius (*Saturnalia* 2.4:11).
73. Schwartz 2014:63–70.
74. Cohen 1999; Schwartz 2004.
75. Schwartz 2014:80.
76. Ball 2016:60.
77. According to Schwartz's (2014:86) estimate.
78. Schwartz 2004:15.
79. Schwartz 2014:89–97.
80. For example, Herod Agrippa's impassioned speech to the Jews (Josephus, *The Jewish War* 216.4).
81. Schwartz 2004:15.
82. Schwartz (2004) argues that it was this exclusion of Jews by Christians that created the unique form of Rabbinic Judaism.
83. For example, Tel Dor, 18 percent; Maresha, 11 percent; Tel Bet Yerah, 14 percent; and Tel Anafa, 13 percent. See Cope 2006; Lev-Tov 2003; Perry-Gal et al. 2015a. At Tel Dor, pig bones increased from the Persian (<1 percent) to the Hellenistic (18 percent; Sapir-Hen et al. 2014). In general, Persian period sites had <1 percent (Dayan 1999; Hesse 1990:218; Horwitz and Lernau 2003; Sapir-Hen 2017; Sapir-Hen et al. 2014).
84. There are generally fewer pigs (<5 percent) in the more arid regions, including at Petra and other Nabataean settlements. Even Roman military camps (e.g., Lejjun, 3 percent) have few pigs (Horwitz and Studer 2005; Studer 2002, 2007). However, relative abundance increased significantly in the Byzantine period, e.g., 28 percent at Petra (Horwitz and Studer 2005:227).
85. Horwitz and Studer 2005:226.
86. Horwitz and Studer 2005; Kroll 2012.
87. Horwitz and Studer 2005; Perry-Gal et al. 2015a:221. No pig remains were identified in the early Roman layers at Jerusalem (Spiciarich et al. 2017). Pig bones are also <1 percent from 2nd century BC deposits at the Seleucid Acra in Jerusalem (Abra Spiciarich, personal communication).
88. Interestingly, many of these Jewish sites have high proportions of chicken bones (Perry-Gal et al. 2015b).
89. Lev-Tov 2003:21.
90. Cohen 2006:32.
91. Safrai 1994:97.
92. Freidenreich 2011:17–46; Rosenblum 2010b.
93. Rosenblum 2010b:101.
94. Schäfer (1997:81) attributes these satires to the reaction against Jewish proselytism.
95. Schäfer 1997:77–78.
96. Konner 2003:88–89.
97. See Fabre-Vassas 1997.
98. Rosenblum 2010a, 2010b.
99. Goody 1982; Gumerman 1997; Hastorf 2016; Lévi-Strauss 1966 [1962]; Stein 2012b.

100. Rosenblum 2010b:95.
101. Cf. Schäfer 1997:79–81.
102. Rosenblum 2010b:99.
103. Rosenblum 2010b:96.
104. Rosenblum 2010b:104–105.
105. Rosenblum 2010b:109.
106. The parable is discussed on Chabad's website, https://www.chabad.org/library/article_cdo/aid/2376474/jewish/Pigs-Judaism.htm.
107. See also Schwartz 2014.
108. Rosenblum 2010b.
109. See also Rosenblum 2016:38–45. Rosenblum (2016:43–45) notes the curious absence of the weaponization of pork in texts dating to Late Antiquity. Rather than accept this as an indication that such practices ceased, he contends that writers by that time perceived these activities as "business as usual" rather than as something newsworthy.
110. Rosenblum 2010b:102.
111. Nakamura 2017.
112. Rosenblum neglects, or at least underplays, the importance of the feedback in Roman-Jewish relations.
113. An excellent treatment of how this process works with respect to social class identity and education in Britain is *Learning to Labour* (Willis 1977). In some cases, even the recognition of a group of people as a cohesive unit where none existed before can create a sense of identity: e.g., nationalism in the context of 19th–20th century European imperialism (Anderson 2006 [1983]; Emberling 1997; Hobsbowm and Ranger 1983).
114. Cohen 1999:54; Schwartz 2014:106.
115. Kraemer 2007:39–54.
116. Cohen 2006:216.
117. Wilken 2012:6–16.
118. Magness 2011:8–9, 24–25; Wilken 2012:18. But MacCulloch (2010:90) discusses Jesus's flaunting of *halakha* and other "outrageous inversions of normality."
119. Wilken 2012:20–23.
120. Cohen 2006:34–35.
121. MacCulloch 2010:100; Wilken 2012:23.
122. Henderson 1998:40–49; Wilken 2012:37–46.
123. Or "freedom in Christ" (Galatians 2:4), see Wilken 2012:22.
124. Wilken 2012:19–20.
125. Konner 2003:104–105; Wilken 2012:21.
126. Wilken 2012:129.
127. For example, the Epistle to Diognetus (2nd century AD) boasts of the unification of Christians throughout the known world (Wilken 2012:47).
128. Konner 2003:104; Wilken 2012:105–106. Jensen (1996) argues that the prominent role of women in early Christianity represented an early step forward in women's liberation.

129. MacCulloch 2010:97–102.
130. Freidenreich 2011:102.
131. Wilken 2012:37–38.
132. Wilken 2012:39.
133. Similarly, starting with the composition of the Mishnah (AD 200), Jewish food laws grew stricter. In part, this may have been a way to keep Jews separate from their neighbors (Kraemer 2007).
134. Freidenreich 2011:121–122.
135. Freidenreich 2011:205–207.
136. Wilken 2012:122–123. Christian intolerance of alternative ideologies was not focused solely on Judaism; e.g., the destruction of the Greek Magical Papyri (Acts 19:10) in Egypt.
137. Schwartz 2014:135.
138. Schwartz 2014:101,137.
139. Augustine of Hippo 2007:401.
140. Augustine is recycling Philo's (*De Agricultura* 32) rumination hypothesis. Note the similarity to the Babylonian tablet "the pig has no sense" (Lambert 1996:215).
141. Fabre-Vassas 1997:245–246.
142. Root 1988:129.
143. Fabre-Vassas 1997:247.
144. For example, Psalms 80:13 (a boar ruins the fertile fields of Israel); Proverbs 11:22 (jewelry in a pig's snout is compared to an indiscrete woman).
145. Grant 1999:6–7; Magness 2011:51–53.
146. This is perhaps a reference to Proverbs 11:22.
147. Wilken 2012:100–102.
148. Fabre-Vassas 1997:301. Or perhaps skin diseases: pigs and "leprosy" (probably a general term for dermatitis) were allegedly connected in Egyptian thought (e.g., Plutarch, *Moralia, Isis and Osiris* 8.353–354).
149. Thurston and Attwater 1990:108–109.
150. Fabre-Vassas 1997:296.

Chapter 9

1. Ball 2016:65–104.
2. Endress 2002:156–163.
3. Irwin 1996.
4. Although Islam is considerably more orthodoxic than Judaism (Cohen 2006:52–53).
5. Aslan 2011:146.
6. Aslan 2011.
7. Crone 1996:6–8.
8. Several other key Muslim traditions were adopted from Judaism: fasting, honoring the Sabbath (changed from Saturday to Friday), and prayer oriented toward a specific place--Mecca in the case of Muslims (Aslan 2011:101–102).

9. All translations of the Quran are from the Sahih International version, https://quran.com.
10. The ban on animals sacrificed to other gods is made more explicit in the New Testament (Acts 21:25; 1 Corinthians 10:27–31).
11. Freidenreich 2011:135.
12. Freidenreich 2011:131–143.
13. Freidenreich 2011:134.
14. Cohen 1999:54; Schwartz 2014:106.
15. Schimmel 1985:71–73.
16. McCorriston and Martin 2009; Uerpmann et al. 2000. Although note that pigs were raised at Petra in the Nabataean kingdom (Studer 2007).
17. Freidenreich 2011:133; Rodinson 1999.
18. E.g., Monchot 2014.
19. Insoll 1999:5.
20. For example, 12th–13th century Tell Tuneinir in the Middle Khabur, 2 percent (Loyet 2000); Haftavan Tepe in the Fars region, 0 percent (Azadeh Mohaseb and Mashkour 2017); Siraf on the Persian Gulf from the 8th to the 16th centuries, <1 percent (von den Driesch and Dockner 2002); medieval Bastam in northwestern Iran, <1 percent (Boessneck and Kokabi 1988:218). No pig remains are reported from Bahrain, Kuwait, or the Arabian Peninsula (Monchot 2016; Uerpmann 2017).
21. For example, Çadır Höyük, 17 percent in 6th–11th centuries (Steadman et al. 2015:112–116); Amorium, 12 percent in 6th–11th centuries (Silibolatlaz-Baykara 2012); Horum Höyük, 14 percent in 12th–13th centuries (Bartosiewicz 2005); Tell Hadidi, 9 percent, and Ta'as, 4 percent, in the Late Byzantine/Early Islamic periods (Clason and Buitenhuis 1978); Kaman-Kalehöyük, 8 percent in 16th–17th centuries (Hongo 1997). However, pigs were uncommon (<1 percent) at early Islamic levels at Kınık Höyük (Highcock et al. 2015).
22. Stein 1988:327–328. And about 26 percent at nearby Zeugma in the 6th–10th centuries (Charles 2013; Rousseau et al. 2008).
23. Louise Bertini, personal communication. Bertini also reported large numbers of pigs at Kom al-Ahmar. For fauna from 10th–15th century North Africa, see MacKinnon 2017:475.
24. Epstein 1971:330.
25. Cope 1999. Note: Cope provides MNIs but not NISPs. Pigs were also common at Bet She'an, but declined over time: Ummayad, 27 percent; Abbasid,(5 percent; and Mamluk, 4 percent (Manor et al. 1996).
26. Bar-Oz and Raban-Gerstel 2015:100; Brown 2016; Horwitz 1998.
27. Walmsley 1988.
28. Reilly in Walmsley et al. 1993:220–221.
29. Brown 2016.
30. Pines et al. 2017.
31. Epstein 1971:330–331.
32. Raban-Gerstel et al. 2011.

33. Taxel et al. 2017. The authors did not publish measurements of bones, so it is impossible to say whether they came from domestic pigs or wild boar.
34. On Egypt, see Epstein (1971:330), who stated it was "still customary to rear young wild boar together with horses to keep the latter in health," citing Hartmann (1864:226), who referred to "reports from several credible witnesses." On Morocco, see Frazer (1913:31), citing Leared (1876:301), who noted that Moroccan Muslims draw an "affinity between devils and swine."
35. Murray 1935:89.
36. Food and Agriculture Organization of the United Nations 2017.
37. Barak-Erez 2007:4.
38. Louise Bertini, personal communication.
39. Mayton 2009.
40. Iran, 50 million; Turkey, 32 million; Syria, 18 million; Iraq, 6.6 million; Egypt, 5.6 million; Jordan, 3.2 million; Israel, 500,000. Data from Food and Agriculture Organization of the United Nations 2017.
41. Hart 1973; Portes and Haller 2005.
42. There are several other shantytowns where Zabaleen reside (Haynes and El-Hakim 1979). In total, there are about 70,000 Zabaleen families (Fahmi and Sutton 2010a).
43. Fahmi and Sutton 2010a; Haynes and El-Hakim 1979; Miller 1990.
44. Fahmi and Sutton 2010a:1768; Haynes and El-Hakim 1979.
45. Haynes and El-Hakim 1979:103; Miller 1990:127.
46. Estimates vary between 190,000 and 350,000 pigs (Fahmi and Sutton 2010a:1774).
47. Fahmi and Sutton 2010a; Slackman 2009a, 2009b.
48. Fahmi and Sutton 2010a:1775.
49. E.g., El Habachi 2017; Guénard 2013; Kingsley 2014b.
50. Kingsley 2014a.
51. Barak-Erez 2007:33–35.
52. Barak-Erez 2007:60.
53. Barak-Erez 2007:3.
54. Barak-Erez 2007:4.
55. Barak-Erez 2007:69–79.
56. Barak-Erez 2007:12.
57. Food and Agriculture Organization of the United Nations 2017.
58. Barak-Erez 2007:81–105.
59. Traubman 2005.
60. Blunt 1881:124–125.
61. Austen Henry Layard (1903:173–174) recounted a similar story of a wild boar goring his horse during a hunt near Mosul.
62. O'Connor 2017.
63. For the recent history of Egyptian wild boar, see Epstein 1971; Keimer 1932; Manlius and Gautier 1999.
64. Epstein 1971:226.
65. For example, Jordan's recent struggles to control the wild boar populations (Namrouqa 2017).

66. Hattem 2014.
67. Hattem 2014.
68. Levy-Rubin 2018.
69. Grafton 2003:33.
70. Coope 1993.
71. For example, the killing of Christians' pigs by Muslim soldiers after the siege of Qasr Ibrim in Sudan in 1173 (ElMahi 1991:23; Epstein 1971:332); attacks on pork butcher shops in Dar es Salaam in 1993 (Chesworth 2018:398); and the 2009 pig culls in Egypt.
72. Root 1988:129.
73. E.g., Hordes 2005. *Marrano* is a slur—still offensive today—against converted Jews or their descendants. There is some debate about the derivation of this epithet. Although it is spelled and pronounced the same way as Spanish *marrano* (hog), some have argued it derives from other sources, such as *mura'in* (hypocrite in Arabic). It is unclear if the word was originally used by Christians or by other Jews against *conversos* and crypto-Jews (see Hordes 2005:5–7).
74. Fabre-Vassas 1997:112.
75. Clarke-Billings 2016.
76. Fabre-Vassas 1997:99.
77. Fabre-Vassas 1997:103–108.
78. Fabre-Vassas 1997:108; Schacher 1974.
79. Wex 2005:67–68.
80. Wex 2005:67.
81. Wex 2005:100–101.
82. Berger 1979; Talmage 1972.
83. Berger 1979:117–118.
84. Berger 1979:217.
85. Quoted in Lindsay 2005:119.
86. Rubin 1997.
87. Kurban and Tobin 2009:31.
88. Ijzereef 1989.
89. Insoll 1999.
90. Blench 2000; Epstein 1971.
91. ElMahi 1991; Spaulding and Spaulding 1988.
92. E.g., Schorsch 2018:2.
93. E.g., Linfield 2019:142.
94. Schorsch 2018:9–19.
95. The 1885 Pittsburgh Platform declared the food taboos antithetical to modern Jewish practice (Konner 2003:225–243; Sussman 2005).
96. Pew Research Center 2015:88.
97. Sussman 2005.
98. Pew Research Center 2015:88.
99. http://www.goodmuslimbadmuslim.com.

Chapter 10

1. Reich 2018.
2. Diamond 1997.
3. Zeder 2012b.
4. Ervynck et al. 2001.
5. Arbuckle 2013; Price and Arbuckle 2015.
6. Englund 1995.
7. Collins 2006.
8. Murray 1935:89.
9. Coon 1951; Harris 1974.
10. Zeder 1991.
11. See Cline 2014.
12. Freidenreich 2011.
13. Sykes 2015:165–167.
14. Legal Information Institute: https://www.law.cornell.edu/wex/cannibalism.
15. Weber 1992 [1930]:123.

References

Adams, R. M.

1978 Strategies of Maximization, Stability, and Resilience in Mesopotamian Society, Settlement, and Agriculture. *Proceedings of the American Philosophical Society* 122:329–335.

1981 *Heartland of Cities: Surveys of Ancient Settlement and Land Use on the Central Floodplain of the Euphrates*. University of Chicago Press, Chicago.

Akimichi, T.

1998 Pig and Man in Papuan Societies: Two Cases from the Seltaman of the Fringe Highlands and the Gidra of the Lowland. *Senri Ethnological Studies* 47:163–182.

2003 *The Archaeology of Syria*. Cambridge University Press, Cambridge.

Albarella, U., K. Dobney, and P. Rowley-Conwy

2006a The Domestication of the Pig (*Sus scrofa*): New Challenges and Approaches. In *Documenting Domestication: New Genetic and Archaeological Paradigms*, edited by M. A. Zeder, D. G. Bradley, E. Emshwiller, and B. D. Smith, pp. 209–227. University of California Press, Berkeley.

Albarella, U., F. Manconi, and A. Trentacoste

2011 A Week on the Plateau: Pig Husbandry, Mobility and Resource Exploitation in Central Sardinia. In *Ethnozooarchaeology: The Present and Past of Human-Animal Relationships*, edited by U. Albarella and A. Trentacoste, pp. 143–159. Oxbow Books, Oxford.

Albarella, U., F. Manconi, J.-D. Vigne, and P. Rowley-Conwy

2007 Ethnoarchaeology of Pig Husbandry in Sardinia and Corsica. In *Pigs and Humans: 10,000 Years of Interaction*, edited by U. Albarella, K. Dobney, A. Ervynck, and P. Rowley-Conwy, pp. 285–307. Oxford University Press, Oxford.

Albarella, U., M. Rizzetto, H. Russ, K. Vickers, and S. Viner-Daniels (editors)

2017 *The Oxford Handbook of Zooarchaeology*. Oxford University Press, Oxford.

Albarella, U., A. Tagliacozzo, K. Dobney, and P. Rowley-Conwy

2006b Pig Hunting and Husbandry in Prehistoric Italy: A Contribution to the Domestication Debate. *Proceedings of the Prehistoric Society* 72:193–227.

Albert, F. W., M. Somel, M. Carneiro, A. Aximu-Petri, M. Halbwax, O. Thalmann, J. A. Blanco-Aguiar, I. Z. Plyusnina, L. Trut, R. Villafuerte, N. Ferrand, S. Kaiser, P. Jensen, and S. Pääbo

2012 A Comparison of Brain Gene Expression Levels in Domesticated and Wild Animals. *PLoS Genet* 8:e1002962.

Albright, W. F.

1939 The Israelite Conquest of Canaan in the Light of Archaeology. *Bulletin of the American Schools of Oriental Research* 74:11–23.

Algaze, G.

1993 *The Uruk World System*. University of Chicago Press, Chicago.

2008 *Ancient Mesopotamia at the Dawn of Civilization: The Evolution of an Urban Landscape*. University of Chicago Press, Chicago.

Allan, K. and K. Burridge
2006 *Forbidden Words: Taboo and the Censoring of Language*. Cambridge University Press, Cambridge.
Allen, T. G.
1974 *The Book of the Dead or Going Forth by Day*. Oriental Institute of the University of Chicago, Chicago.
Allentuck, A. E.
2013 Human-Livestock Relations in the Early Bronze Age of the Southern Levant. PhD dissertation, Department of Anthropology, University of Toronto, Toronto.
Allentuck, A. E. and H. J. Greenfield
2010 The Organization of Animal Production in an Early Urban Center: The Zooarchaeological Evidence from Early Bronze Age Titriş Höyük, Southeast Turkey. In *Anthropological Approaches to Zooarchaeology: Complexity, Colonialism, and Animal Transformations*, edited by D. Campana, P. J. Crabtree, S. D. deFrance, J. S. Lev-Tov, and A. M. Choyke, pp. 12–29. Oxbow Books, Oxford.
Allentuck, A. E. and A. M. Rosen
2019 The Risky Business of Keeping Pigs During Periods of Climatic Fluctuation: A Case from the Mid-Holocene Near East. *Journal of Archaeological Science: Reports* 24:939–945.
Almond, G. W.
1995 How Much Water Do Pigs Need? *Proceedings of the North Carolina Healthy Hogs Seminar*. North Carlonia Swine Veterinary Group, North Carolina State University. https://projects.ncsu.edu/project/swine_extension/healthyhogs/book1995/almond.htm.
Amberger, G.
1987 Tierknochenfunde vom Tell Abqa'/Irak. *Acta Praehistorica et Archaeologica* 19:111–129.
Anderson, B.
2006 [1983] *Imagined Communities: Reflections on the Origin and Spread of Nationalism*. Verso, London.
Anderson, E. N.
2014 *Everybody Eats: Understanding Food and Culture*. 2nd ed. New York University Press, New York.
Angress, S.
1960 The Pig Skeleton from Area B. In *Hazor II: An Account of the Second Season of Excavations, 1956*, edited by Y. Yadin, Y. Aharoni, R. Amiran, T. Dothan, and I. Dunayevsky, pp. 169. Oxford University Press, Oxford.
Appadurai, A.
1981 Gastro-politics in Hindu South Asia. *American Anthropologist* 8:494–511.
Arbuckle, B. S.
2012a Animals and Inequality in Chalcolithic Central Anatolia. *Journal of Anthropological Archaeology* 31:302–313.
2012b Pastoralism, Provisioning, and Power at Bronze Age Acemhöyük, Turkey. *American Anthropologist* 114:462–476.
2013 The Late Adoption of Cattle and Pig Husbandry in Neolithic Central Turkey. *Journal of Archaeological Science* 40:1805–1815.
2014 Inequality and the Origins of Wool Production in Central Anatolia. In *Animals and Inequality in the Ancient World*, edited by B. S. Arbuckle and S. A. McCarty, pp. 209–229. University of Colorado Press, Boulder.

2015 The Rise of Cattle Cultures in Bronze Age Anatolia. *Journal of Eastern Mediterranean Archaeology and Heritage Studies* 2:277–297.

Arbuckle, B. S. and E. Hammer
2019 The Rise of Pastoralism in the Ancient Near East. *Journal of Archaeological Research* 27:391–449.

Arbuckle, B. S., S. W. Kansa, E. C. Kansa, D. Orton, C. Çakırlar, L. Gourichon, A. Marciniak, J. Mulville, H. Buitenhuis, D. Carruthers, B. De Cupere, A. Demirergi, S. Frame, D. Helmer, L. Martin, J. Peters, N. Pöllath, K. Pawłowska, N. Russell, K. C. Twiss, and D. Würtenberger
2014 Data Sharing Reveals Complexity in the Westward Spread of Domestic Animals Across Neolithic Turkey. *PLoS ONE* 9:e99845.

Arranz-Otaegui, A., S. Colledge, L. Zapata, L. C. Teira-Mayolini, and J. J. Ibáñez
2016 Regional Diversity on the Timing for the Initial Appearance of Cereal Cultivation and Domestication in Southwest Asia. *Proceedings of the National Academy of Sciences* 113:14001–14006.

Aslan, R.
2011 *No god but God: The Origins, Evolution, and Future of Islam.* Random House, New York.

Atici, L.
2009 Implications of Age Structures for Epipaleolithic Hunting Strategies in the Western Taurus Mountains, Southwest Turkey. *Archaeozoologia* 44:13–39.

Augustine of Hippo
2007 *Answer to Faustus, a Manichean.* Translated by R. Teske. New City Press, New York.

Bachhuber, C.
2015 *Citadel and Cemetery in Early Bronze Age Anatolia.* Equinox Publishing Ltd. Monographs in Mediterranean Archaeology, Sheffield.

Bakken, D.
2000 Hunting Strategies of Late Pleistocene Zarzian Populations from Palegawra Cave, Iraq and Warwasi Orck Shelter, Iran. In *Archaeozoology of the Near East*, Vol. IVA, edited by M. Mashkour, A. M. Choyke, H. Buitenhuis, and F. Poplin, pp. 11–17. ARC-Publicaties 32, Groningen.

Balasse, M., A. Evin, C. Tornero, V. Radu, D. Fiorillo, D. Popovici, A. Andreescu, K. Dobney, T. Cucchi, and A. Bălăşescu
2016 Wild, Domestic and Feral? Investigating the Status of Suids in the Romanian Gumelniţa (5th Mil. cal BC) with Biogeochemistry and Geometric Morphometrics. *Journal of Anthropological Archaeology* 42:27–36.

Baldwin, J. A.
1978 Pig Rearing vs. Pig Breeding in New Guinea. *Anthropological Journal of Canada* 16:23–27.

Ball, W.
2016 *Rome in the East: The Transformation of an Empire.* 2nd ed. Routledge, New York.

Bangsgaard, P., L. Yeomans, H. Darabi, K. M. Gregersen, J. Olsen, T. Richter, and P. Mortensen
2019 Feasting on Wild Boar in the Early Neolithic: Evidence from an 11,400 Year Old Placed Deposit at Tappeh Asiab, Central Zagros. *Cambridge Archaeological Journal* 29:443–463.

Banning, E. B.
1998 The Neolithic Period: Triumphs of Architecture, Agriculture and Art. *Near Eastern Archaeology* 61:188–237.

Bar-Matthews, M. and A. Ayalon

2011 Mid-Holocene Climate Variations Revealed by High-Resolution Speleothem Records from Soreq Cave, Israel and Their Correlation with Cultural Changes. *Holocene* 21:163–171.

Bar-Oz, G.

2004 *Epipaleolithic Subsistence Strategies in the Levant: A Zooarchaeological Perspective.* Brill Academic Publishers, Boston.

Bar-Oz, G. and N. Raban-Gerstel

2015 Butcher's Waste: Zooarchaeological Analysis of a Crusader/Ayyubib Bone Deposit from Jerusalem Street, Safed (Zefat). *'Atiqot* 81:99–109.

Bar-Yosef, O.

1998 The Natufian Culture in the Levant, Threshold to the Origins of Agriculture. *Evolutionary Anthropology* 6:159–177.

2001 From Sedentary Foragers to Village Hierarchies: The Emergence of Social Institutions. *Proceedings of the British Academy* 110:1–38.

2002a Natufian: A Complex Society of Foragers. In *Beyond Foraging and Collecting,* edited by B. Fitzhugh and J. Habu, pp. 91–149. Plenum Publishers, New York.

2002b The Upper Paleolithic Revolution. *Annual Review of Anthropology* 31:363–393.

2011 Climatic Fluctuations and Early Farming in West and East Asia. *Current Anthropology* 52:S175–S193.

Bar-Yosef, O. and A. Belfer-Cohen

1988 The Early Upper Paleolithic in Levantine Caves. In *The Early Upper Paleolithic: Evidence from Europe and the Near East,* edited by J. F. Hoffecker and C. Wolf, pp. 23–41. British Archaeological Reports International Series 437, Oxford.

Bar-Yosef, O., B. Vandermeersch, B. Arensburg, A. Belfer-Cohen, P. Goldberg, H. Laville, L. Meignen, Y. Rak, J. D. Speth, E. Tchernov, A.-M. Tillier, and S. Weiner

1992 The Excavations in Kebara Cave, Mt. Carmel. *Current Anthropology* 33:497–550.

Barak-Erez, D.

2007 *Outlawed Pigs: Law, Religion, and Culture in Israel.* University of Wisconsin Press, Madison.

Barclay, J. M. G.

1996 *Jews in the Mediterranean Diaspora: From Alexander to Trajan (323 BCE–117 CE).* University of California Press, Berkeley.

Barnish, S. J. P.

1987 Pigs, Plebeians and Potentates: Rome's Economic Hinterland, C. 300–600 A.D. *Papers of the British School at Rome* 55:157–185.

Barth, F.

1969 Introduction. In *Ethnic Groups and Boundaries: The Social Organization of Culture Difference,* edited by F. Barth, pp. 9–38. Waveland Press, Long Grove, Illinois.

Bartosiewicz, L.

2005 Animal Remains from the Excavations of Horum Höyük, Southeast Anatolia, Turkey. In *Archaeozoology of the Near East,* Vol. VI, edited by H. Buitenhuis, A. M. Choyke, L. Martin, L. Bartosiewicz, and M. Mashkour, pp. 150–162. ARC-Publicaties 123, Groningen.

2010 Herding in Period VI A: Development and Changes from Period VII. In *Economic Centralisation in Formative States: The Archaeological Reconstruction of the Economic System in 4th Millennium Arslantepe,* edited by M. Frangipane, pp. 119–148. Studi di Preistoria Orientale (SPO) Vol. 3. Sapienza Universià di Roma.

Bartosiewicz, L., R. Gillis, L. Girdland-Fink, A. Evin, T. Cucchi, A. R. Hoelzel, U. S. Vidarsdottir, K. Dobney, G. Larson, and U.-D. Schoop

2013 Chalcolithic Pig Remains from Çamlıbel Tarlası, Central Anatolia. In *Archaeozoology of the Near East*, Vol. X, edited by B. De Cupere, V. Linseele, and S. Hamilton-Dyer, pp. 101–120. Ancient Near Eastern Studies Supplement 44. Peeters, Leuven.

Bates, C.

2013 *Masculinity and the Hunt: Wyatt to Spenser*. Oxford University Press, Oxford.

Bazer, F. W., J. J. Ford, and R. S. Kensinger

2001 Reproductive Physiology. In *Biology of the Domestic Pig*, edited by W. G. Pond and H. J. Mersmann, pp. 150–224. Cornell University Press, Ithaca.

Becker, C.

2005 Small Numbers, Large Potential: New Prehistoric Finds of Elephant and Beaver from the Khabur River/Syria. *MUNIBE (Antropologia-Arkeologia)* 57:445–456.

2008 The Faunal Remains from Dur-Katlimmu—Insights into the Diet of the Assyrians. In *Archaeozoology of the Near East*, Vol. XI, edited by E. Vila, L. Gourichon, A. M. Choyke, and H. Buitenhuis, pp. 561–580. Maison de l'Orient et de la Mediterranée, Lyon.

Belyaev, D. K.

1969 Domestication of Animals. *Science Journal (U.K.)* 5:47–92.

Ben-Shlomo, D., A. C. Hill, and Y. Garfinkel

2009 Feasting Between the Revolutions: Evidence from Chalcolithic Tel Tsaf, Israel. *Journal of Mediterranean Archaeology* 22:129–150.

Ben-Shlomo, D., I. Shai, A. Zukerman, and A. Maeir

2008 Cooking Identities: Aegean-Style Cooking Jugs and Cultural Interaction in Iron Age Philistia and Neighboring Regions. *American Journal of Archaeology* 112: 225–246.

Benecke, N.

1993 The Exploitation of *Sus scrofa* (Linné, 1758) on the Crimean Peninsula and in Southern Scandinavia in the Early and Middle Holocene: Two Regions, Two Strategies. In *Exploitation des animaux sauvages a travers le temps*, edited by J. Desse and F. Audoin-Rouzeau, pp. 233–245. Editions APDCA, Juan-les-Pins.

Berger, A.

2018 "Feeding Cities"? Preliminary Notes on the Provisioning of Animal Products at Tel Bet Yerah, Israel. In *Archaeozoology of the Near East XII*, edited by C. Çakırlar, J. Chahoud, R. Berthon, and S. P. Birch, pp. 13–26. Barkhuis, Eelde.

Berger, D.

1979 *The Jewish-Christian Debate in the High Middle Ages: A Critical Edition of the Nizzahon Vetus*. Jewish Publication Society of America, Philadelphia.

Berger, J.-F. and J. Guilaine

2009 The 8200 cal BP Abrupt Environmental Change and the Neolithic Transition: A Mediterranean Perspective. *Quaternary International* 200:31–49.

Berlin, A.

2001 The Book of Esther and Ancient Storytelling. *Journal of Biblical Literature* 120:3–14.

Bernbeck, R.

1995 Lasting Alliances and Emerging Competition: Economic Developments in Early Mesopotamia. *Journal of Anthropological Archaeology* 14:1–25.

Bernbeck, R.

2017 Merging Clay and Fire: Earliest Evidence from the Zagros Mountains. In *The Emergence of Pottery in West Asia*, edited by A. Tsuneki, O. Nieuwenhuyse, and S. Campbell, pp. 97–115. Oxbow Books, Oxford.

Berthon, R.

2011 Animal Exploitation in the Upper Tigris River Valley (Turkey) Between the 3rd and 1st Millennia BC. PhD dissertation, Mathematisch-Naturwissenschaftlichen Fakultät, Christian-Albrechts Universität zu Kiel, Kiel.

Bertini, L.

2011 Changes in Suid and Caprine Husbandry Practices Throughout Dynastic Egypt Using Linear Enamel Hypoplasia (LEH). PhD dissertation, Department of Archaeology, Durham University, Durham.

2016 How Did the Nile Water System Impact Swine Husbandry Practices in Ancient Egypt? In *A History of Water: Water and Food: From Hunter-Gatherers to Global Production in Africa*, edited by T. Tveldt and T. Østigard, pp. 75–100. I. B. Tauris, London.

Bertini, L. and E. Cruz-Rivera

2014 The Size of Ancient Egyptian Pigs: A Biometrical Analysis Using Molar Width. *Bioarchaeology of the Near East* 8:83–107.

Betz, H. D. (editor)

1986 *The Greek Magical Papyri in Translation*. University of Chicago Press, Chicago.

Bieber, C. and T. Ruf

2005 Population Dynamics in Wild Boar *Sus scrofa*: Ecology, Elasticity of Growth Rate and Implications for the Management of Pulsed Resource Consumers. *Journal of Applied Ecology* 42:1203–1213.

Biehl, P. F. and O. P. Nieuwenhuyse (editors)

2016 *Climate and Cultural Change in Prehistoric Europe and the Near East*. State University of New York Press, Albany.

Bigelow, L.

1999 Zooarchaeological Investigation of Economic Organization and Ethnicity at Late Chalcolithic Hacinebi: A Preliminary Report. *Paléorient* 25:83–89.

Binford, L.

1968 Post-Pleistocene Adaptations. In *New Perspectives in Archeology*, edited by S. R. Binford and L. Binford, pp. 313–341. Aldine Publishing Co., Chicago.

Bittman, M.

2014 Christie's Pig-Crate Politics. *New York Times*, 2 December. New York.

Black, A.

2016 *A World History of Ancient Political Thought: Its Significance and Consequences*. Oxford University Press, Oxford.

Blackshaw, J. K., A. W. Blackshaw, F. J. Thomas, and F. W. Newman

1994 Comparison of Behaviour Patterns of Sows and Litters in a Farrowing Crate and a Farrowing Pen. *Applied Animal Behaviour Science* 39:281–295.

Blanton, R. E. and J. Taylor

1995 Patterns of Exchange and the Social Production of Pigs in Highland New Guinea: Their Relevance to Questions About the Origins and Evolution of Agriculture. *Journal of Archaeological Research* 3:113–145.

Bleed, P.

2006 Living in the Human Niche. *Evolutionary Anthropology* 15:8–10.

Bleed, P. and A. Matsui
2010 Why Didn't Agriculture Develop in Japan? A Consideration of Jomon Ecological Style, Niche Construction, and the Origins of Domestication. *Journal of Archaeological Method and Theory* 17:356–370.
Blench, R. M.
2000 A History of Pigs in Africa. In *The Origins and Development of African Livestock: Archaeology, Genetics, Linguistics, and Ethnology*, edited by R. M. Blench and K. C. MacDonald, pp. 355–367. UCL Press, London.
Blunt, A.
1881 *A Pilgrimage to Nejd: The Cradle of the Arab Race*, Vol. 2 John Murray, London.
Böck, B., R. M. Boehmer, J. Boessneck, M. van Ess, G. Meinert, L. Patzelt, and M. Peter-Patzelt
1993 Uruk 39 (1989). *Baghdader Mitteilungen* 24:3–126.
Bocquet-Appel, J.-P.
2009 The Demographic Impact of the Agricultural System in Human History. *Current Anthropology* 50:657–660.
Boessneck, J.
1977 Sonstige Tierknochenfunde aus Išān Bahrīyāt (Isin). In *Isin-Išān Bahrīyāt: I. Die Ergebnisse der Ausgrabungen, 1973–1974*, edited by B. Hrouda, pp. 111–133. Bayerische Akademie der Wissenschaften, Munich.
1978 Tierknochenfunde aus Nippur. In *Excavations at Nippur: Twelfth Season*, edited by M. Gibson, pp. 153–187. Oriental Institute of the University of Chicago, Chicago.
1992 Landschaft, Flora und Fauna: Besprechung der Tierknochen und Molluskenreste von Hassek Höyük. In *Hassek Höyuk: Naturwissenschaftliche Untersuchungen und lithische Industrie*, edited by M. R. Behm-Blanke, pp. 58–74. Ernst Wasmuth Verlag, Tübingen.
Boessneck, J. and A. von den Driesch
1974 The Excavations at Korucutepe, Turkey, 1968–1970: Preliminary Report. Part IX: The Animal Remains. *Journal of Near Eastern Studies* 33:109–112.
1985 *Knochenfunde aus Zisternen in Pergamon*. Institüt für Palaeoanatomie, Domestikationforschung und Geschichte der Teirmedezin de Universität München, Munich.
Boessneck, J. and M. Kokabi
1988 Tierknochenfunde. In *Bastam: 2. Ausgrabungen in den Urartäischen Anlagen, 1977–1978*, edited by W. Kleiss, pp. 175–262. Gebr. Mann Verlag, Berlin.
Bogaard, A.
2005 "Garden Agriculture" and the Nature of Early Farming in Europe and the Near East. *World Archaeology* 37:177–196.
Bogaard, A. and V. Isaakidou
2010 From Mega-Sites to Farmsteads: Community Size, Ideology and the Nature of Early Farming Landscapes in Western Asia and Europe. In *Landscapes in Transition*, edited by B. Finlayson and G. Warren, pp. 192-207. Oxbow Books, Oxford.
Bogucki, P.
1993 Animal Traction and Household Economies in Neolithic Europe. *Antiquity* 67:492–503.
Bökönyi, S.
1973 The Fauna of Umm Dabaghiyah: A Preliminary Report. *Iraq* 35:9–11.

1977 *The Animal Remains from Four Sites in the Kermanshah Valley, Iran: Asiab, Sarab, Dehsavar, and Siahib*. British Archaeological Reports Supplementary Series 34, Oxford.

1978a The Animal Remains of the 1970–1972 Excavation Seasons at Tell el-Der: A Preliminary Report. In *Tell ed-Der II: Progress Reports (First Series)*, edited by L. De Meyer, pp. 185–189. Peeters, Leuven.

1978b Environmental and Cultural Differences as Reflected in the Animal Bones from Five Early Neolithic Sites in Southwest Asia. In *Approaches to Faunal Analysis in the Middle East*, edited by R. H. Meadow and M. A. Zeder, pp. 57–62. Peabody Museum Bulletin 2. Peabody Museum, Cambridge.

Bottero, J.

2004 *The Oldest Cuisine in the World: Cooking in Mesopotamia*. University of Chicago Press, Chicago.

Bouchud, J.

1987 Les Mammiferes et la petite faune du gisement Natoufien de Mallaha (Eynan), Israel. In *La Faune du gisement Natoufien de Mallaha (Eynan), Israel*, edited by J. Bouchud, pp. 13–114. Association Paleorient, Paris.

Boyd, D. J.

1985 "We Must Follow the Fore": Pig Husbandry Intensification and Ritual Diffusion among the Irakia Awa, Papua New Guinea. *American Ethnologist* 12:119–136.

Brellas, D. A. J.

2016 By the Rivers of Babylon: Patterns of Heterarchy, Sustainable Wetland Agroecology, and Urban Dynamics in Old Babylonian Mashkan-Shapir. PhD dissertation, Department of Archaeology, Boston University, Boston.

Breniquet, C. and C. Michel (editors)

2014 *Wool Economy in the Ancient Near East and the Aegean: From the Beginnings of Sheep Husbandry to Institutional Textile Industry*. Oxbow Books, Oxford.

Briant, P.

2002 *From Cyrus to Alexander: A History of the Persian Empire*. Translated by P. T. Daniels. Eisenbrauns, Winona Lake, Indiana.

Bridault, A., R. Rabinovich and T. Simmons

2008 Human Activities, Site Location and Taphonomic Process: A Relevant Combination for Understanding the Fauna of Eynan (Ain Mallaha), Level Ib (Final Natufian, Israel). In *Archaeozoology of the Near East*, Vol. VIII, edited by E. Vila, L. Gourichon, A. M. Choyke, and H. Buitenhuis, pp. 99–117. Maison de l'Orient de la Méditerranée, Lyon.

Brown, R.

2016 Faunal Distributions from the Southern Highlands of Transjordan: Regional and Historical Perspectives on the Representations and Roles of Animals in the Middle Islamic Period. In *Landscapes of the Islamic World: Archaeology, History, and Ethnography*, edited by S. McPhillips and P. D. Wordsworth, pp. 71–93. University of Pennsylvania Press, Philadelphia.

Buccellati, G. and M. Kelly-Buccellati

2005 Urkesh as a Hurrian Religious Center. *Studi Micenei ed Egeo-Anatolici* 2005:27–59.

Budiansky, S.

1992 *The Covenant of the Wild: Why Animals Chose Domestication*. Yale University Press, New Haven.

Buitenhuis, H.
1985 Animal Remains from Tell Sweyhat, Syria. *Palaeohistoria* 25:131–144.
Bulliet, R. W.
2005 *Hunters, Herders, and Hamburgers: The Past and Future of Human-Animal Relationships*. Columbia University Press, New York.
Bulmer, R.
1967 Why Is the Cassowary Not a Bird? A Problem of Zoological Taxonomy among the Karam of the New Guinea Highlands. *Man* 2:5–25.
Bunimovitz, S. and Z. Lederman
2011 Canaanite Resistance: The Philistines and Beth-Shemesh: A Case Study from Iron Age I. *Bulletin of the American Schools of Oriental Research* 364:37–51.
Bunimovitz, S. and A. Yasur-Landau
1996 Philistine and Israelite Pottery: A Comparative Approach to the Question of Pots and People. *Tel Aviv* 23:88–101.
Burkert, W.
1992 *The Orientalizing Revolution: Near Eastern Influence on Greek Culture in the Early Archaic Age*. Harvard University Press, Cambridge.
Burrin, D. G.
2001 Nutrient Requirements and Metabolism. In *Biology of the Domestic Pig*, edited by W. G. Pond and H. J. Mersmann, pp. 309–389. Cornell University Press, Ithaca.
Bywater, K. A., M. Apollonio, N. Cappai, and P. A. Stephens
2010 Litter Size and Latitude in a Large Mammal: The Wild Boar *Sus scrofa*. *Mammal Review* 40:212–220.
Çakırlar, C.
2003 Animal Exploitation at Kinet Höyük (Hatay,Turkey) During the First Half of the Late Iron Age. MA thesis, Department of History and Archaeology, American University of Beirut, Beirut.
2012 The Evolution of Animal Husbandry in Neolithic CentralWest Anatolia: The Zooarchaeological Record from Ulucak Höyük (c. 7040–5660 cal. BC, Izmir, Turkey). *Anatolian Studies* 62:1–33.
2016 Early Bronze Age Foodways in the Aegean: Social Archaeozoology on the Eastern Side. In *Early Bronze Age Troy: Chronology, Cultural Development, and Interregional Contacts*, edited by E. Pernicka, S. Ünlüsöy, and S. W. E. Blum, pp. 291–305. Verlag Dr. Rudolf Habelt GMBH, Bonn.
Çakırlar, C. and L. Atici
2017 Animal Exploitation in Western Turkey. In *The Oxford Handbook of Zooarchaeology*, edited by U. Albarella, M. Rizzetto, H. Russ, K. Vickers, and S. Viner-Daniels, pp. 266–279. Oxford University Press, Oxford.
Çakırlar, C., R. Breider, Y. Ersoy, and E. Koparal
2015 Klazomenai'de Zooarkeoloji Çalışmaları. *Arkeometri Sonuçları Toplantısı* 31:189–206.
Çakırlar, C., L. Gourichon, S. Pilaar Birch, and R. Berthon
2014 Provisioning an Urban Center Under Foreign Occupation: Zooarchaeological Insights in the Hittite Presence in Late Fourteenth-Century BCE Alalakh. *Journal of Eastern Mediterranean Archaeology and Heritage Studies* 2:259–276.
Çakırlar, C. and J. M. Marston
2019 Rural Agricultural Economies and Military Provisioning at Roman Gordion (Central Turkey). *Environmental Archaeology* 24:91–105.

Caliebe, A., A. Nebel, C. Makarewicz, M. Krawczak, and B. Krause-Kyora
2017 Insights into Early Pig Domestication Provided by Ancient DNA Analysis. *Scientific Reports* 7:44550.
Campbell, S.
2000 The Burnt House at Arpachiyah: A Reexamination. *Bulletin of the American Schools of Oriental Research*: 1–40.
2007–2008 The Dead and the Living in Late Neolithic Mesopotamia. In *Sepolti tra i vivi: Evidenza ed interpretazione di contesti funerari in abitato*, edited by G. Bartoloni and M. G. Benedettini, pp. 125–140. Universita degli studi di Roma, Rome.
Campbell, S., S. W. Kansa, R. Bichener, and H. Lau
2014 Burying Things: Practices of Cultural Disposal at Late Neolithic Domuztepe, Southeast Turkey. In *Remembering the Dead in the Ancient Near East*, edited by B. W. Porter and A. T. Boutin, pp. 27–60. University Press of Colorado, Boulder.
Cardoso, J. L., J. L. L. Castro, A. Ferjaoui, A. M. Martín, V. M. Hahnmüller, and I. B. Jerbania
2016 What the People of Utica (Tunisia) Ate at a Banquet in the 9th Century BCE: Zooarchaeology of a North African Early Phoenician Settlement. *Journal of Archaeological Science: Reports* 8:314–322.
Çevik, Ö.
2007 The Emergence of Different Social Systems in Early Bronze Age Anatolia: Urbanisation versus Centralisation. *Anatolian Studies* 57:131–140.
Chahoud, J.
2014–2015 L'Aspect archéozoologique des pratiques funéraires à l'Âge du bronze au Liban. *Archaeology and History in the Lebanon* 40–41:274–290.
Chandezon, C.
2015 Animals, Meat, and Alimentary By-products: Patterns of Production and Consumption. In *A Companion to Food in the Ancient World*, edited by J. Wilkins and R. Nadeau, pp. 135–146. Wiley-Blackwell, New York.
Charles, B.
2013 Faunal Remains. In *Excavations at Zeugma Conducted by Oxford Archaeology*, edited by W. Aylward, pp. 399–410. Packard Humanities Institute, Los Altos, California.
Chase, P. G. and H. L. Dibble
1987 Middle Paleolithic Symbolism: A Review of Current Evidence and Interpretations. *Journal of Anthropological Archaeology* 6:263–296.
Chesworth, J.
2018 Christians and Muslims in Sub-Saharan Africa. In *Routledge Handbook on Christian–Muslim Relations*, edited by D. Thomas, pp. 393–401. Routledge, New York.
Choquenot, D. and W. A. Ruscoe
2003 Landscape Complementation and Food Limitation of Large Herbivores: Habitat-Related Constraints on the Foraging Efficiency of Wild Pigs. *Journal of Animal Ecology* 72:14–26.
Clark, D.
2004 The Raw and the Rotten: Punk Cuisine. *Ethnology* 43:19–31.
Clarke-Billings, L.
2016 Man Throws Rotten Pork Meat at Mosque in Hate Crime. *Newsweek*, 4 July, https://www.newsweek.com/man-throws-rotten-pork-meat-mosque-post-brexit-hate-crime-477449.

Clason, A.
1979–1980 The Animal Remains from Tell es-Sin Compared with Those from Bouqras. *Anatolica* 7:35–48.
Clason, A. and H. Buitenhuis
1978 A Preliminary Report on the Faunal Remains of Nahr el Homr, Hadidi and Ta'as in the Tabqa Dam Region in Syria. *Journal of Archaeological Science* 5:75–83.
Cline, E. H.
2014 *1177 BC: The Year Civilization Collapsed*. Princeton University Press, Princeton.
Clutton-Brock, J.
1979 The Mammalian Remains from the Jericho Tell. *Proceedings of the Prehistoric Society* 45:135–157.
1992 The Process of Domestication. *Mammal Review* 22:79–85.
Cohen, S.
1999 *The Beginnings of Jewishness: Boundaries, Varieties, Uncertainties*. University of California Press, Berkeley.
2006 *From the Maccabees to the Mishnah*. 2nd ed. Westminster John Knox Press, Louisville.
Cole, S. G.
2004 *Landscapes, Gender, and Ritual Space: The Ancient Greek Experience*. University of California Press, Berkeley.
Collins, B. J.
2002 Necromancy, Fertility and the Dark Earth: The Use of Ritual Pits in Hittite Cult. In *Magic and Ritual in the Ancient World*, edited by P. Mirecki and M. Meyer, pp. 224–241. Brill, Leiden.
2004 A Channel to the Underworld in Syria. *Near Eastern Archaeology* 67:53–56.
2006 Pigs at the Gate: Hittite Pig Sacrifice in Its Eastern Mediterranean Context. *Journal of Near Eastern Religions* 6:155–188.
Colonna d'Istria, L.
2014 Wool Economy in the Royal Archive of Mari During the Sakkanakku Period. In *Wool Economy in the Ancient Near East and the Aegean: From the Beginnings of Sheep Husbandry to Institutional Textile Industry*, edited by C. Breniquet and C. Michel, pp. 169–201. Oxbow Books, Oxford.
Comaroff, J. and J. Comaroff
1992 *Ethnography and the Historical Imagination*. Westview Press, Boulder.
Conolly, J., S. Colledge, K. Dobney, J.-D. Vigne, J. Peters, B. Stopp, K. Manning, and S. Shennan
2011 Meta-analysis of Zooarchaeological Data from SW Asia and SE Europe Provides Insight into the Origins and Spread of Animal Husbandry. *Journal of Archaeological Science* 38:538–545.
Coogan, M. D.
2017 *The Old Testament: A Historical and Literary Introduction to the Hebrew Scriptures*. 4th ed. Oxford University Press, Oxford.
Coon, C. S.
1951 *Caravan: The Story of the Middle East*. Henry Holt, New York.
Coope, J. A.
1993 Religious and Cultural Conversion in Ninth-Century Umayya Córdoba. *Journal of World History* 4:47–68.

Cooper, J. S.
1996 Magic and M(is)use: Poetic Promiscuity in Mesopotamian Ritual. In *Mesopotamian Poetic Language: Sumerian and Akkadian*, edited by M. E. Vogelzang and H. L. J. Vanstiphout, pp. 47–57. Styx Publications, Groningen.

Cope, C. R.
1999 Faunal Remains and Butchery Practices from Byzantine and Islamic Contexts (1993–94 Seasons). In *Caesarea Papers: 2. Herod's Temple, the Provincial Govenor's Praetorium and Granaries, the Later Harbor, a Gold Coin Hoard, and Other Studies*, edited by K. G. Holum, A. Raban, and J. Patrich, pp. 405–417. Journal of Roman Archaeology Supplementary Series 35, Portsmouth.
2006 The Fauna: Preliminary Results. In *The Tel Bet Yerah Excavations, 1994–1995*, edited by N. Getzov, pp. 169–174. Israel Antiquities Authority, Jerusalem.

Cowgill, G. L.
2004 Origins and Development of Urbanism: Archaeological Perspectives. *Annual Review of Anthropology* 33:525–549.

Crabtree, P.
1990 Zooarchaeology and Complex Societies: Some Uses of Faunal Analysis for the Study of Trade, Social Status, and Ethnicity. *Archaeological Method and Theory* 2:155–205.

Crone, P.
1996 The Rise of Islam in the World. In *The Cambridge Illustrated History of the Islamic World*, edited by F. Robinson, pp. 2–31. Cambridge University Press, Cambridge.

Cucchi, T., A. Hulme-Beaman, J. Yuan, and K. Dobney
2011 Early Neolithic Pig Domestication at Jiahu, Henan Province, China: Clues from Molar Shape Analyses Using Geometric Morphometric Approaches. *Journal of Archaeological Science* 38:11–22.

Cultraro, M.
2004 Exercise of Dominance: Boar Hunting in Mycenaean Religion and Hittite Royal Rituals. In *Offizielle Religion, Locale Kulte und Individuelle Religiosität*, edited by M. Hutter and S. Hutter-Braunsar, pp. 117–135. Ugarit-Verlag, Münster.

D'Altroy, T. N. and T. K. Earle
1985 Staple Finance, Wealth Finance, and Storage in the Inka Political Economy. *Current Anthropology* 26:187–206.

D'Anna, M. B.
2012 Between Inclusion and Exclusion: Feasting and Redistribution of Meals at Late Chalcolithic Arslantepe (Malatya, Turkey). *eTopoi Journal for Ancient Studies* Special Volume 2:97–123.

Dahl, J.
2006 Early Swine Herding. In *De la domestication au tabou*, edited by B. Lion and C. Michel, pp. 31–38. De Boccard, Paris.

Dalby, A.
2003 *Food in the Ancient World from A to Z*. Routledge, New York.

Dalix, A.-S. and E. Vila
2006 Wild Boar Hunting in the Eastern Mediterranean from the 2nd to the 1st Millennium BC. In *Pigs and Humans: 10,000 Years of Interaction*, edited by U. Albarella, K. Dobney, A. Ervynck, and P. Rowley-Conwy, pp. 359–372. Oxford University Press, Oxford.

Dalix Meier, A.-S.
2006 Ba'al et les sangliers dans CAT 1.12. *Historiae* 3:35–68.
Darabi, H., T. Richter, and P. Mortensen
2018 New Excavations at Tappeh Asiab, Kermanshah Province, Iran. *Antiquity* 92:e2.
Darwin, C.
1868 *The Variation of Plants and Animals Under Domestication.* John Murray, London.
Davis, S. J. M.
1981 The Effects of Temperature Change and Domestication on the Body Size of Late Pleistocene to Holocene Mammals of Israel. *Paleobiology* 7:101–114.
Davis, S. J. M., R. Rabinovich, and N. Goren-Inbar
1988 Quaternary Extinctions and Population Increase in Western Asia: The Animal Remains from Biq'at Quneitra. *Paléorient* 14:95–105.
Davis, S. J. M. and F. F. Valla
1978 Evidence for Domestication of the Dog 12,000 Years Ago in the Natufian of Israel. *Nature* 276:608–610.
Dayan, T.
1999 Faunal Remains: Areas A–G. In *Tel 'Ira: A Stronghold in the Biblical Negev*, edited by I. Beit-Arieh, pp. 480–487. Monograph Series 15. Sonia and Marco Nadler Institute of Archaeology, Tel Aviv University, Tel Aviv.
De Beauvoir, S.
2011 [1949] *The Second Sex.* Vintage, New York.
De Cupere, B.
1994 Report of the Faunal Remains from Trench K (Roman Pessinus, Central Anatolia). *Archaeofauna* 3:63–75.
2001 *Animals at Ancient Sagalassos: Evidence of the Faunal Remains.* Brepols, Turnhout.
De Cupere, B. and R. Duru
2003 Faunal Remains from Neolithic Höyücek (SW-Turkey) and the Presence of Early Domestic Cattle in Anatolia. *Paléorient* 29:107–120.
De Cupere, B., D. Frémondeau, E. Kaptijn, E. Marinova, J. Poblome, R. Vandam, and W. Van Neer
2017 Subsistence Economy and Land Use Strategies in the Burdur Province (SW Anatolia) from Prehistory to the Byzantine Period. *Quaternary International* 436:4–17.
De Cupere, B. and W. Van Neer
2014 Consumption Refuse, Carcasses and Ritual Deposits at Tell Beydar (Northeastern Syria). In *Paleonutrition and Food Practices in the Ancient Near East: Towards a Multidisciplinary Approach*, edited by L. Milano, pp. 187–213. S.A.R.G.O.N. Editrice e Libreria, Padova.
De Vaux, R.
1971 *The Bible and the Ancient Near East.* Translated by D. McHugh. Doubleday and Company, New York.
DeBoer, W. R.
1987 You Are What You Don't Eat: Yet Another Look at Food Taboos in Amazonia. In *Ethnicity and Culture*, edited by R. Auger, M. F. Glass, S. MacEachern, and P. H. McCartney, pp. 45–54. University of Calgary Press, Calgary.
Deckers, K. and H. Pessin
2010 Vegetation Development in the Middle Euphrates and Upper Jazirah (Syria/Turkey) During the Bronze Age. *Quaternary Research* 74:216–226.

Desse, J.

1983 Les faunes du gisement obéidien final de Tell el 'Oueili. In *Larsa et 'Ouelli: Traveau de 197po8o8.981*, edited by J.-L. Huot, pp. 193–199. Editions Recherche sur les Civilisations, Paris.

Dever, W. G.

1994 From Tribe to Nation: State Formation Processes in Ancient Israel. In *Nuove fondazioni nel vicino Oriente antico: Realtà e ideologia*, edited by S. Mazzoni, pp. 213–239. Giradini, Pisa.

2003 *Who Were the Early Israelites and Where Did They Come From?* Eerdmans, Grand Rapids, Michigan.

2005 *Did God Have a Wife? Archaeology and Folk Religion in Ancient Israel.* Eerdmans, Grand Rapids, Michigan.

Diamond, J.

1997 *Guns, Germs, and Steel: The Fate of Human Societies.* W. W. Norton & Co., New York.

2005 *Collapse: How Societies Choose to Fail or Succeed.* Viking, New York.

Diener, P. and E. E. Robkin

1978 Ecology, Evolution, and the Search for Cultural Origins: The Question of Islamic Pig Prohibition. *Current Anthropology* 19:493–540.

Dietler, M.

2001 Theorizing the Feast: Rituals of Consumption, Commensal Politics, and Power in African Contexts. In *Feasts: Archaeological and Ethnographic Perspectives on Food, Politics, and Power*, edited by M. Dietler and B. Hayden, pp. 65–114. Smithsonian Institution Press, Washington, DC.

Dietler, M. and B. Hayden

2001 Digesting the Feast: Good to Eat, Good to Drink, Good to Think. In *Feasts: Archaeological and Ethnographic Perspectives on Food, Politics, and Power*, edited by M. Dietler and B. Hayden, pp. 1–19. Smithsonian Institution Press, Washington, DC.

Dietrich, O., M. Heun, J. Notroff, K. Schmidt, and M. Zarnkow

2012 The Role of Cult and Feasting in the Emergence of Neolithic Communities: New Evidence from Göbekli Tepe, South-eastern Turkey. *Antiquity* 86:674–695.

Dobney, K. and A. Ervynck

2000 Interpreting Developmental Stress in Archaeological Pigs: The Chronology of Linear Enamel Hypoplasia. *Journal of Archaeological Science* 27:597–607.

Dobney, K., A. Ervynck, U. Albarella, and P. Rowley-Conwy

2007 The Transition from Wild Boar to Domestic Pig in Eurasia, Illustrated by a Tooth Development Defect and Biometrical Data. In *Pigs and Humans: 10,000 Years of Interaction*, edited by U. Albarella, K. Dobney, A. Ervynck, and P. Rowley-Conwy, pp. 57–82. Oxford University Press, Oxford.

Dobney, K., D. Jaques, and W. Van Neer

2003 Diet, Economy and Status: Evidence from the Animal Bones. In *Excavations at Tell Brak: 4. Exploring an Upper Mesopotamian Regional Centre, 1994–1996*, edited by R. Matthews, pp. 417–430. British School of Archaeology in Iraq, Oxford.

Docter, R. F.

2009 Carthage and Its Hinterland. *Iberia Archaeologica* 13:179–189.

Dolansky, S.

2013 Syria-Canaan. In *The Cambridge Companion to Ancient Mediterranean Religions*, edited by B. S. Spaeth, pp. 55–75. Cambridge University Press, Cambridge.

Doll, M.
2010 Meat, Traction, Wool: Urban Livestock in Tell Mozan. In *Development of the Environment, Subsistence, and Settlement of the City of Urkeš and Its Region*, edited by K. Deckers, M. Doll, P. Pfälzner, and S. Riehl, pp. 191–360. Harrassowitz, Wiesbaden.
Dothan, T.
1998 Initial Philistine Settlement: From Migration to Coexistence. In *Mediterranean Peoples in Transition, Thirteenth to Early Tenth Centuries BCE*, edited by S. Gitin, A. Mazar, and E. Stern, pp. 148–161. Israel Exploration Society, Jerusalem.
Douglas, M.
1975 *Implicit Meanings: Essays in Anthropology*. Routledge and Kegan Paul, London.
1999 *Leviticus as Literature*. Oxford University Press, Oxford.
2002 [1966] *Purity and Danger: An Analysis of Concepts of Pollution and Taboo*. Routledge Classics, New York.
Ducos, P.
1967 Les Débuts de l'élevage en Palestine. *Syria* 44:375–400.
1968 *L'Origine des animaux domestiques au Palestine*. Publications de l'Institut de Préhistoire de l'Université de Bordeaux, Bordeaux.
1978 Domestication Defined and Methodological Approaches to Its Recognition in Faunal Assemblages. In *Approaches to Faunal Analysis in the Middle East*, edited by R. H. Meadow and M. A. Zeder, pp. 53–56. Peabody Museum Bulletin No. 2. Peabody Museum, Cambridge.
Düring, B. S.
2013 Breaking the Bond: Investigating the Neolithic Expansion in Asia Minor in the Seventh Millennium BC. *Journal of World Prehistory* 26:75–100.
Durkheim, E.
1995 [1912] *The Elementary Forms of the Religious Life*. Translated by K. E. Fields. The Free Press, New York.
Dwyer, P. D.
1996 Boars, Barrows, and Breeders: The Reproductive Status of Domestic Pig Populations in Mainland New Guinea. *Journal of Anthropological Research* 52:481–500.
Dwyer, P. D. and M. Minnegal
2005 Person, Place or Pig: Animal Attachments and Human Transactions in New Guinea. In *Animals in Person: Cultural Perspectives on Human-Animal Relations*, edited by J. Knight, pp. 37–60. Berg, Oxford.
Dyson-Hudson, R. and N. Dyson-Hudson
1980 Nomadic Pastoralism. *Annual Review of Anthropology* 9:15–61.
Ehituv, S.
1978 Economic Factors in the Egyptian Conquest of Canaan. *Israel Exploration Journal* 28:93–105.
Ekroth, G.
2007 Meat in Ancient Greece: Sacrificial, Sacred or Secular? *Food and History* 5:249–272.
El Habachi, M.
2017 Cairo's Treasures of Trash. *Newsweek Middle East*. 24 May: http://newsweekme.com/cairos-treasures-trash/.
El Mahi, A. T.
1991 Pigs and Politics in Medieval Sudan: The Origin of the Sennar Pig. *Dirasat Ifriqiyya* 8:19–33.

Emberling, G.
1997 Ethnicity in Complex Societies: Archaeological Perspectives. *Journal of Archaeological Research* 5:295–344.
Emerton, J. A.
2004 The Date of the Yahwist. In *In Search of Pre-Exhilic Israel*, edited by J. Day, pp. 107–129. T & T Clark, London.
Endress, G.
2002 *Islam: An Historical Introduction.* 2nd ed. Columbia University Press, New York.
Englund, R. K.
1995 Late Uruk Pigs and Other Herded Animals. In *Beiträge zur Kulturgeschichte Vorderasiens*, edited by U. Finkbeiner, R. Dittmann, and H. Hauptmann, pp. 121–133. Philipp von Zabern, Mainz.
Epstein, H.
1971 *The Origin of the Domestic Animals of Africa*, Vol. 2. Africana Publishing, New York.
Ervynck, A. and K. Dobney
1999 Lining Up on the M1: A Tooth Defect as a Bio-Indicator for Environment and Husbandry in Ancient Pigs. *Environmental Archaeology* 4:1–8.
Ervynck, A., K. Dobney, H. Hongo, and R. H. Meadow
2001 Born Free? New Evidence for the Status of *Sus scrofa* at Neolithic Çayönü Tepesi (Southeastern Anatolia, Turkey). *Paléorient* 27:47–73.
Essig, M.
2015 *Lesser Beasts: A Snout-to-Tail History of the Humble Pig.* Basic Books, New York.
Evershed, R. P., S. Payne, A. Sherratt, M. S. Copley, J. Coolidge, D. Urem-Kotsu, K. Kotsakis, M. Özdoğan, A. E. Özdoğan, O. Nieuwenhuyse, P. M. M. G. Akkermans, D. Bailey, R.-R. Andeescu, S. Campbell, S. Farid, I. Hodder, N. Yalman, M. Özbaşaran, E. Bıçakcı, Y. Garfinkel, T. E. Levy, and M. M. Burton
2008 Earliest Date for Milk Use in the Near East and Southeastern Europe Linked to Cattle Herding. *Nature* 455:528–531.
Evin, A., T. Cucchi, A. Cardini, U. S. Vidarsdottir, G. Larson, and K. Dobney
2013 The Long and Winding Road: Identifying Pig Domestication through Molar Size and Shape. *Journal of Archaeological Science* 40:735–743.
Evin, A., L. Girdland Flink, A. Bălăşescu, D. Popovici, A. Andreescu, D. Bailey, P. Mirea, C. Lazăr, A. Boroneanţ, C. Bonsall, U. Strand Vidarsdottir, S. Brehard, A. Tresset, T. Cucchi, G. Larson, and K. Dobney
2015 Unravelling the Complexity of Domestication: A Case Study Using Morphometrics and Ancient DNA Analyses of Archaeological Pigs from Romania. *Philosophical Transactions of the Royal Society B* 370:20130616.
Evins, M. A.
1982 The Fauna from Shanidar Cave: Mousterian Wild Goat Exploitation in Northeastern Iraq. *Paléorient* 8:37–58.
Fabre-Vassas, C.
1997 *The Singular Beast: Jews Christians, and the Pig.* Translated by C. Volk. Columbia University Press, New York.
Fagan, B.
2015 *The Intimate Bond: How Animals Shaped Human History.* Bloomsbury Press, New York.

Fahmi, W. and K. Sutton
2010 Cairo's Contested Garbage: Sustainable Solid Waste Management and the Zabaleen's Right to the City. *Sustainability* 2010:1765–1783.
Falconer, S. E.
1995 Rural Responses to Early Urbanism: Bronze Age Household and Village Economy at Tell el-Hayyat, Jordan. *Journal of Field Archaeology* 22:399–419.
Fall, P. L., L. Lines, and S. E. Falconer
1998 Seeds of Civilization: Bronze Age Rural Economy and Ecology in the Southern Levant. *Annals of the Association of American Geographers* 88:107–125.
Fang, M., G. Larson, H. Soares Ribeiro, N. Li, and L. Andersson
2009 Contrasting Mode of Evolution at a Coat Color Locus in Wild and Domestic Pigs. *PLoS Genet* 5:e1000341.
Faust, A.
2005 The Canaanite Village: Social Structure of Middle Bronze Age Rural Communities. *Levant* 37:105–125.
2016a The Emergence of Israel and Theories of Ethnogenesis. In *The Wiley Blackwell Companion to Ancient Israel*, edited by S. Niditch, pp. 155–173. John Wiley and Sons, New York.
2016b *Israel's Ethnogenesis: Settlement, Interaction, Expansion and Resistance*. London, Routledge.
2017 An All-Israelite Identity: Historical Reality or Biblical Myth? In *The Wide Lens in Archaeology: Honoring Brian Hesse's Contributions to Anthropological Archaeology*, edited by J. Lev-Tov, P. Hesse, and A. Gilbert, pp. 169–190. Lockwood Press, Atlanta.
2018 Pigs in Space (and Time): Pork Consumption and Identity Negotiations in the Late Bronze and Iron Ages of Ancient Israel. *Near Eastern Archaeology* 81:276–299.
Faust, A. and J. S. Lev-Tov
2011 The Constitution of Philistine Identity: Ethnic Dynamics in Twelfth to Tenth Century Philistia. *Oxford Journal of Archaeology* 30:13–31.
Fessler, D. M. T. and C. D. Navarrete
2003 Meat Is Good to Taboo: Dietary Proscriptions as a Product of the Interaction of Psychological Mechanisms and Social Processes. *Journal of Cognition and Culture* 3:1–40.
Fiddes, N.
1991 *Meat: A Natural Symbol*. Routledge, New York.
Fillios, M. A.
2006 Measuring Complexity in Early Bronze Age Greece: The Pig as a Proxy Indicator of Socio-Economic Structures. PhD dissertation, Department of Anthropology, University of Minnesota, Minneapolis.
Finkelstein, I.
1988 *The Archaeology of the Israelite Settlement*. Israel Exploration Society, Jerusalem.
1992 Pastoralism in the Highlands of Canaan in the Third and Second Millennia B.C.E. In *Pastoralism in the Levant*, edited by O. Bar-Yosef and A. Khazanov, pp. 133–142. Prehistory Press, Madison.
1996a The Archaeology of the United Monarchy: An Alternative View. *Levant* 28:177–187.
1996b Ethnicity and Origin of the Iron I Settlers in the Highlands of Canaan: Can the Real Israel Stand Up? *Biblical Archaeologist* 59:198–212.

1997 Pots and People Revisited: Ethnic Boundaries in the Iron Age I. In *The Archaeology of Israel: Constructing the Past, Interpreting the Present*, edited by N. A. Silberman and D. B. Small, pp. 216–237. Journal for the Study of the Old Testament Supplement Series 237. Sheffield Academic Press, Sheffield.

2010 A Great United Monarchy: Archaeological and Historical Perspectives. In *One God—One Cult—One Nation: Archaeological and Biblical Perspectives*, edited by R. G. Kratz and H. Spieckermann, pp. 3–28. Walter de Gruyter, Berlin.

Finkelstein, I. and A. Mazar

2007 *The Quest for the Historical Israel: Debating Archaeology and the History of Early Israel*. SBL Archaeology and Biblical Studies 17. Society of Biblical Literature, Atlanta.

Finkelstein, I. and N. Na'aman (editors)

1994 *From Nomadism to Monarchy*. Yad Ben-Zvi, Jerusalem.

Finkelstein, I. and N. A. Silberman

2001 *The Bible Unearthed: Archeology's New Vision of Ancient Israel and the Origin of Its Sacred Texts*. Free Press, New York.

Fisher, L. R.

2013 *The Eloquent Peasant*. Cascade Books, Eugene, Oregon.

Flad, R. K., J. Yuan, and S. Li

2007 Zooarcheological Evidence for Animal Domestication in Northwest China. *Developments in Quaternary Sciences* 9:167–203.

Flannery, K. V.

1969 Origins and Ecological Effects of Early Domestication in Iran and the Near East. In *The Domestication of Plants and Animals*, edited by P. J. Ucko and G. W. Dimbleby, pp. 73–100. Aldine, Chicago.

1983 Early Pig Domestication in the Fertile Crescent: A Retrospective Look. In *The Hilly Flanks: Essays on the Pre-History of Southwestern Asia*, edited by T. C. Young, P. E. L. Smith, and P. Mortensen, pp. 163–188. Studies in Ancient Oriental Civilization 36. Oriental Institute of the University of Chicago, Chicago.

Fleming, D. E.

2004 *Democracy's Ancient Ancestors: Mari and Early Collective Governance*. Cambridge University Press, Cambridge.

2012 *The Legacy of Israel in Judah's Bible: History, Politics, and the Reinscribing Tradition*. Cambridge University Press, Cambridge.

Foer, J. S.

2009 Let Them Eat Dog: A Modest Proposal for Tossing Fido in the Oven. *Wall Street Journal*, 31 October. New York.

Food and Agriculture Organization of the United Nations

2017 FAO Statistics. Livestock Processed. http://www.fao.org/faostat.

Forstenpointner, G., G. Weissengruber, and A. Galik

2002 Banquets at Ephesos: Archaeozoological Evidence of Well Stratified Greek and Roman Kitchen Waste. In *Archaeozoology of the Near East*, Vol. V, edited by H. Buitenhuis, A. Choyke, M. Mashkour, and A. H. al-Shiyab, pp. 282–304. Rijksuniversitit, Groningen.

Foster, B. R.

2006 Everything Except the Squeal: Pigs in Early Mesopotamia. In *De la domestication au tabou*, edited by B. Lion and C. Michel, pp. 283–291. De Boccard, Paris.

2014 Wool in the Economy of Sargonic Mesopotamia. In *Wool Economy in the Ancient Near East and the Aegean: From the Beginnings of Sheep Husbandry to Institutional Textile Industry*, edited by C. Breniquet and C. Michel, pp. 115–123. Oxbow Books, Oxford.

2016 *The Age of Agade: Inventing Empire in Ancient Mesopotamia*. Routledge, New York.

Fowles, S.

2008 Steps Toward an Archaeology of Taboo. In *Religion, Archaeology, and the Material World*, edited by L. Fogelin, pp. 15–37. Occasional Paper No. 36. Center for Archaeological Investigations, Southern Illinois University, Carbondale.

Frangipane, M.

2000 The Development of Administration from Collective to Centralized Economies in the Mesopotamian World: The Transformation of an Institution from "System-Serving" to "Self-Serving." In *Cultural Evolution: Contemporary Viewpoints*, edited by G. Feinman and L. R. Manzanilla, pp. 215–232. Kluwer Academic / Plenum, New York.

2007 Different Types of Egalitarian Societies and the Development of Inequality in Early Mesopotamia. *World Archaeology* 39:151–176.

2010 Different Models of Power Structuring at the Rise of Hierarchical Societies in the Near East: Primary Economy versus Luxury and Defence Management. In *The Development of Pre-State Communities in the Ancient Near East*, edited by D. Bolger and L. C. Maguire, pp. 79–86. Oxbow Books, Oxford.

Frangipane, M., C. Alvora, F. Balossi, and G. Siracusano

2002 The 2000 Campaign at Zeytinli Bahçe Höyük. In *Salvage Project of the Archaeological Heritage of the Ilisu and Carchemish Dam Reservoirs: Activities in 2000*, edited by N. Tuna, J. Greenhalgh, and J. Velibeyoglu, pp. 41–99. Middle East Technical University (METU), Ankara.

Frantz, L. A. F., E. Meijaard, J. Gongora, J. Haile, M. A. M. Groenen, and G. Larson

2016 The Evolution of Suidae. *Annual Review of Animal Biosciences* 4:61–85.

Frantz, L. A. F., J. Haile, A. T. Lin, A. Scheu, C. Geörg, N. Benecke, M. Alexander, A. Linderholm, V. E. Mullin, K. G. Daly, V. M. Battista, M. Price, R.-M. Arbogast, B. Arbuckle, A. Bălăşescu, R. Barnett, L. Bartosiewicz, G. Baryshnikov, C. Bonsall, D. Borić, A. Boronean, J. Bulatović, C. Çakırlar, J.-M. Carretero, J. Chapman, M. Church, R. Crooijmans, B. D. Cupere, C. Detry, V. Dimitrijevic, V. Dumitrascu, C. Edwards, M. Erek, A. Erim-Özdoğan, A. Ervynck, D. Fulgione, M. Gligor, A. Götherström, L. Gourichon, M. Groenen, D. Helmer, H. Hongo, L. K. Horwitz, E. K. Irving-Pease, O. Lebrasseur, J. Lesur, C. Malone, N. Manaseryan, A. Marciniak, H. Martlew, M. Mashkour, R. Matthews, G. Matuzeviciute, S. Maziar, E. Meijaard, T. McGovern, H-J. Megens, R. Miller, A. Mohaseb, J. Orschiedt, D. Orton, A. Papathanasiou, M. P. Pearson, R. Pinhasi, D. Radmanović, F.-X. Ricaut, M. Richards, R. Sabin, L. Sarti, W. Schier, S. Sheikhi, E. Stephan, J. R. Stewart, S. Stoddart, A. Tagliacozzo, N. Tasić, K. Trantalidou, A. Tresset, C. Valdiosera, Y. van den Hurk, S. V. Poucke, J.-D. Vigne, A. Yanevich, A. Zeeb-Lanz, M. Thomas, P. Gilbert, J. Schibler, M. Zeder, J. Peters, T. Cucchi, D. G. Bradley, K. Dobney, J. Burger, A. Evin, L. Girdland Flink, and G. Larson 2019 Ancient Pig Genomes Reveal a Near Complete Turnover Following Their Introduction to Europe. *Proceedings of the National Academy of Sciences* 116:17231–17238.

Frantz, L. A. F., J. G. Schraiber, O. Madsen, H.-J. Megens, A. Cagan, M. Bosse, Y. Paudel, R. P. M. A. Crooijmans, G. Larson, and M. A. M. Groenen 2015 Evidence of Long-Term Gene Flow and Selection During Domestication from Analyses of Eurasian Wild and Domestic Pig Genomes. *Nature Genetics* 47:1141–1148.

Frazer, J. G.
1911 *The Golden Bough: A Study in Magic and Religion: 3. Taboo and the Perils of the Soul.* Macmillan and Co., London.
1912 *The Golden Bough: A Study in Magic and Religion: 8. Spirits of the Corn and of the Wild (Part 2)* Macmillan and Co., London.
1913 *The Golden Bough: A Study in Magic and Religion: 6. The Scapegoat.* Macmillan and Co., London.
Freidenreich, D. M.
2011 *Foreigners and Their Food: Constructing Otherness in Jewish, Christian, and Islamic Law.* University of California Press, Berkeley.
Frémondeau, D., T. Cucchi, F. Casabianca, J. Ughetto-Monfrin, M.-P. Horard-Herbin, and M. Balasse
2012 Seasonality of Birth and Diet of Pigs from Stable Isotope Analyses of Tooth Enamel (d18O, d13C): A Modern Reference Data Set from Corsica, France. *Journal of Archaeological Science* 39:2023–2035.
Frémondeau, D., B. De Cupere, A. Evin, and W. Van Neer
2017 Diversity in Pig Husbandry from the Classical-Hellenistic to the Byzantine Periods: An Integrated Dental Analysis of Düzen Tepe and Sagalassos Assemblages (Turkey). *Journal of Archaeological Science: Reports* 11:38–52.
Freud, S.
1918 *Totem and Taboo: Resemblances Between the Psychic Lives of Savages and Neurotics.* Translated by A. A. Brill. Moffat, Yard and Company, New York.
Frey, C. J. and C. W. Marean
1999 Mammal Remains. In *The Iron Age Settlement of 'Ain Dara, Syria: Survey and Surroundings,* edited by E. C. Stone and P. E. Zimansky, pp. 123–137. British Archaeological Reports International Series 786, Oxford.
Friedman, R. E.
2017 *The Exodus: How It Happened and Why It Matters.* Harper Collins Books, New York.
Frymer-Kensky, T.
1983 The Tribulations of Marduk: The So-Called "Marduk Ordeal Text." *Journal of the American Oriental Society* 103:131–141.
Fuchs, E.
1921 *Die Juden in der Karikatur: Ein Beitrag zur Kulturgeschichte.* Albert Langen, Munich.
Fuller, B. T., B. De Cupere, E. Marinova, W. Van Neer, M. Waelkens, and M. P. Richards
2012 Isotopic Reconstruction of Human Diet and Animal Husbandry Practices During the Classical-Hellenistic, Imperial, and Byzantine Periods at Sagalassos, Turkey. *American Journal of Physical Anthropology* 149:157–171.
Fuller, D. Q., R. G. Allaby, and C. J. Stevens
2010 Domestication as Innovation: The Entanglement of Techniques, Technology and Chance in the Domestication of Cereal Crops. *World Archaeology* 42:13–28.
Fuller, D. Q. and C. J. Stevens
2017 Open for Competition: Domesticates, Parasitic Domesticoids and the Agricultural Niche. *Archaeology International* 20:112–123.
Fuller, D. Q., G. Willcox, and R. G. Allaby
2011 Cultivation and Domestication Had Multiple Origins: Arguments Against the Core Area Hypothesis for the Origins of Agriculture in the Near East. *World Archaeology* 43:628–652.

Gabunia, L., A. Vekua, D. Lordkipanidze, C. C. Swisher, R. Ferring, A. Justus, M. Nioradze, M. Tvalchrelidze, S. C. Antón, G. Bosinski, O. Jöris, M.-A. de Lumley, G. Majsuradze, and A. Mouskhelishvili

2000 Earliest Pleistocene Hominid Cranial Remains from Dmanisi, Republic of Georgia: Taxonomy, Geological Setting, and Age. *Science* 288:1019–1025.

Galili, E., D. J. Stanley, J. Sharvit, and M. Weinstein-Evron

1997 Evidence for Earliest Olive-Oil Production in Submerged Settlements off the Carmel Coast, Israel. *Journal of Archaeological Science* 24:1141–1150.

Garfinkel, Y.

2017 The Ethnic Identification of Khirbet Qeiyafa: Why It Matters. In *The Wide Lens in Archaeology: Honoring Brian Hesse's Contributions to Anthropological Archaeology*, edited by J. Lev-Tov, P. Hesse, and A. Gilbert, pp. 149–167. Lockwood Press, Atlanta.

Geiger, M., M. R. Sánchez-Villagra, and A. K. Lindholm

2018 A Longitudinal Study of Phenotypic Changes in Early Domestication of House Mice. *Royal Society Open Science* 5:172099.

Geller, M. J.

1991 Akkadian Medicine in the Babylonian Talmud. In *A Traditional Quest: Essays in Honour of Louis Jacobs*, edited by D. Cohn-Sherbok, pp. 102–112. Journal for the Study of the Old Testament Supplement Series 114. University of Sheffield Press, Sheffield.

George, A. R.

2015 On Babylonian Lavatories and Sewers. *Iraq* 77:75–106.

Geva, H.

2014 Jerusalem's Population in Antiquity: A Minimalist View. *Tel Aviv* 41:131–160.

Giménez-Anaya, A., J. Herrero, C. Rosell, S. Couto, and A. García-Serrano

2008 Food Habitats of Wild Boars (*Sus scrofa*) in a Mediterranean Coastal Wetland. *Wetlands* 28:197–203.

Girdland-Fink, L. and G. Larson

2011 Archaeological, Morphological and Genetic Approaches to Pig Domestication. In *The Origins and Spread of Domestic Animals in Southwest Asia and Europe*, edited by S. Colledge, J. Conolly, K. Dobney, K. Manning, and S. Shennan, pp. 27–36. Routledge, London.

Giuffra, E., J. M. H. Kijas, V. Amarger, Ö. Carlborg, J.-T. Jeon, and L. Andersson

2000 The Origin of the Domestic Pig: Independent Domestication and Subsequent Introgression. *Genetics* 154:1785–1791.

Gladfelter, B. G.

1997 The Ahmarian Tradition of the Levantine Upper Paleolithic: The Environment of the Archaeology. *Geoarchaeology* 12:363–393.

Gongora, J., R. E. Cuddahee, F. F. do Nascimento, C. J. Palgrave, S. Lowden, S. Y. W. Ho, D. Simond, C. S. Damayanti, D. J. White, W. T. Tay, E. Randi, H. Klingel, C. J. Rodrigues-Zarate, K. Allen, C. Moran, and G. Larson

2011 Rethinking the Evolution of Extant Sub-Saharan African Suids (Suidae, Artiodactyla). *Zoologica Scripta* 40:327–335.

Goody, J.

1982 *Cooking, Cuisine and Class: A Study in Comparative Sociology*. Cambridge University Press, Cambridge.

Goossaert, V.
2005 The Beef Taboo and the Sacrificial Structure of Late Imperial Chinese Society. In *Of Tripod and Palate: Food, Politics, and Religion in Traditional China*, edited by R. Sterckx, pp. 237–248. Palgrave Macmillan, New York.
Goring-Morris, A. N. and A. Belfer-Cohen
2002 Symbolic Behaviour from the Epipalaeolithic and Early Neolithic of the Near East: Preliminary Observations on Continuity and Change. In *Magic Practices and Ritual in the Near Eastern Neolithic*, edited by H. G. K. Gebel, B. D. Hermansen, and C. H. Jensen, pp. 67–79. Studies in Early Near Eastern Production, Subsistence, and Environment 8. Ex Oriente, Berlin.
Gottwald, N. K.
1979 *The Tribes of Yahweh: A Sociology of the Religion of Liberated Israel, 1250–1050 BCE*. Sheffield Academic Press, Sheffield.
Grafton, D. D.
2003 *The Christians of Lebanon: Political Rights in Islamic Law*. Tauris Academic Studies, London.
Graham, P. J.
2011 Ubaid Period Agriculture at Kenan Tepe, Southeastern Turkey. PhD dissertation, Department of Anthropology, University of Connecticut, Storrs.
Grant, R. M.
1999 *Early Christians and Animals*. Routledge, New York.
Greenfield, H. J.
2002 Faunal Remains from the Early Bronze Age Site of Titriş Höyük, Turkey. In *Archaeozoology of the Near East V*, edited by H. Buitenhuis, A. Choyke, M. Mashkour, and A. H. al-Shiyab, pp. 252–261. Archaeological Research and Consultancy Publication 62. Rijksuniversitit, Groningen.
2010 The Secondary Products Revolution: The Past, the Present and the Future. *World Archaeology* 42:29–54.
2014 *Animal Secondary Products: Domestic Animal Exploitation in Prehistoric Europe, the Near East and the Far East*. Oxbow Books, Oxford.
Greenfield, T. L.
2015 The Palace versus the Home: Social Status and Zooarchaeology at Tušḫan (Ziyaret Tepe), a Neo-Assyrian Administrative Provincial Capital in Southeastern Turkey. *Journal of Eastern Mediterranean Archaeology and Heritage Studies* 3:1–26.
Grigson, C.
1982 Porridge and Pannage: Pig Husbandry in Neolithic England. In *Archaeological Aspects of Woodland Ecology*, edited by M. Bell and S. Limbrey, pp. 297–314. British Archaeological Reports British Series 109, Oxford.
2007 Culture, Ecology, and Pigs from the 5th to the 3rd Millennium BC around the Fertile Crescent. In *Pigs and Humans: 10,000 Years of Interaction*, edited by U. Albarella, K. Dobney, A. Ervynck, and P. Rowley-Conwy, pp. 83–108. Oxford University Press, Oxford.
2015 The Fauna of Tell Nebi Mend (Syria) in the Bronze and Iron Age—A Diachronic Overview. Part 1: Stability and Change—Animal Husbandry. *Levant* 47:5–29.
Grosman, L., N. D. Munro, I. Abadi, E. Boaretto, D. Shaham, A. Belfer-Cohen, and O. Bar-Yosef
2016 Nahal Ein Gev II, a Late Natufian Community at the Sea of Galilee. *PLoS ONE* 11:e0146647.

Grossman, K. M.
2013 Early Bronze Age Hamoukar: A Settlement Biography. PhD dissertation, Department of Near Eastern Languages and Civilizations, University of Chicago, Chicago.
Groves, C.
1981 *Ancestors for the Pigs: Taxonomy and Phylogeny of the Genus Sus.* Australian National University, Department of Prehistory Technical Bulletin No. 3, Canberra.
2007 Current Views on Taxonomy and Zoogeography of the Genus *Sus.* In *Pigs and Humans: 10,000 Years of Interaction*, edited by U. Albarella, K. Dobney, A. Ervynck, and P. Rowley-Conwy, pp. 15–29. Oxford University Press, Oxford.
Gruwier, B. and A. Verlinde
2010 Preliminary Archaeozoological Report on Sectors B6 and B6D in the Sanctuary Area (Pessinus, Ballıhisar, Turkey). *Anatolia Antiqua* 18:157–162.
Guénard, M.
2013 Cairo Puts Its Faith in Ragpickers to Manage the City's Waste Problem. *Guardian*, 19 November. London.
Gumerman IV, G.
1997 Food and Complex Societies. *Journal of Archaeological Method and Theory* 4:105–139.
Gündem, C.
2010 Animal Based Subsistence Economy of Emar During the Bronze Age. In *Emar After the Closure of the Tabqa Dam: The Syrian-German Excavations, 1996–2002: I. Late Roman and Medieval Cemeteries and Environmental Studies*, edited by U. Finkbeiner and F. Sakal, pp. 125–176. Subartu 25. Brepols, Turnhout.
Haak, W., O. Balanovsky, J. J. Sanchez, S. Koshel, V. Zaporozhchenko, C. J. Adler, C. S. I. D. Sarkissian, G. Brandt, C. Schwarz, N. Nicklisch, V. Dresely, B. Fritsch, E. Balanovska, R. Villems, H. Meller, K. W. Alt, and A. Cooper
2010 Ancient DNA from European Early Neolithic Farmers Reveals Their Near Eastern Affinities. *PLoS ONE* 8:e1000536.
Haber, A. and T. Dayan
2004 Analyzing the Process of Domestication: Hagoshrim as a Case Study. *Journal of Archaeological Science* 31:1587–1601.
Hadjikoumis, A.
2012 Traditional Pig Herding Practices in Southwest Iberia: Questions of Scale and Zooarchaeological Implications. *Journal of Anthropological Archaeology* 31:353–364.
Halstead, P.
2003 Texts and Bones: Contrasting Linear B and Archaeozoological Evidence for Animal Exploitation in Mycenaean Southern Greece. *British School at Athens Studies* 9:257–261.
2014 *Two Oxen Ahead: Pre-Mechanized Farming in the Mediterranean.* Wiley-Blackwell, Hoboken.
Halstead, P. and V. Isaakidou
2011a Revolutionary Secondary Products: The Development and Significance of Milking, Animal-Traction and Wool-Gathering in Later Prehistoric Europe and the Near East. In *Interweaving Worlds: Systemic Interactions in Eurasia, 7th to 1st Millennia BC*, edited by T. Wilkinson, S. Sherratt, and J. Bennet, pp. 61–76. Oxbow Books, Oxford.
2011b A Pig Fed by Hand Is Worth Two in the Bush: Ethnoarchaeology of Pig Husbandry in Greece and Its Archaeological Implications. In *Ethnozooarchaeology: The*

Present and Past of Human-Animal Relationships, edited by U. Albarella and A. Trentacoste, pp. 160–174. Oxbow Books, Oxford.
Hamilakis, Y. and E. Konsolaki
2004 Pigs for the Gods: Burnt Animal Sacrifices as Embodied Rituals at a Mycenaean Sanctuary. *Oxford Journal of Archaeology* 23:135–151.
Hamilton, J., R. E. M. Hedges, and M. Robinson
2009 Rooting for Pigfruit: Pig Feeding in Neolithic and Iron Age Britain Compared. *Antiquity* 83:998–1011.
Haran, M.
1981 Behind the Scenes of History: Determining the Date of the Priestly Source. *Journal of Biblical Literature* 100:321–333.
Harper, R. F.
1904 *The Code of Hammurabi*. University of Chicago Press, Chicago.
Harris, M.
1974 *Cows, Pigs, Wars, and Witches*. Vintage Books, New York.
1985 *The Sacred Cow and the Abominable Pig: Riddles of Food and Culture*. Simon & Schuster, New York.
Hart, K.
1973 Informal Income Opportunities and Urban Employment in Ghana. *Journal of Modern African Studies* 11:61–89.
Hartman, G., O. Bar-Yosef, A. Brittingham, L. Grosman, and N. D. Munro
2016 Hunted Gazelles Evidence Cooling, but Not Drying, During the Younger Dryas in the Southern Levant. *Proceedings of the National Academy of Sciences* 113:3997–4002.
Hartmann, R.
1864 Die Haussäugetiere der Nilländer. *Annalen der Lanwirtschaft* 44:7–38 and 208–230.
Harvey, L. P.
2005 *Muslims in Spain, 1500 to 1614*. University of Chicago Press, Chicago.
Hastorf, C. A.
2016 *The Social Archaeology of Food: Thinking about Eating from Prehistory to the Present*. Cambridge University Press, Cambridge.
Hattem, B.
2014 The Wild Boar and Feces Epidemic in Palestine. *Vice News*, 3 January. New York.
Haynes, K. E. and S. M. El-Hakim
1979 Appropriate Technology and Public Policy: The Urban Waste Management System in Cairo. *Geographical Review* 69:101–108.
Hecker, H. M.
1982 A Zooarchaeological Inquiry into Pork Consumption in Egypt from Prehistoric to New Kingdom Times. *Journal of the American Research Center in Egypt* 19:59–71.
1984 Preliminary Report on the Faunal Remains from the Workmen's Village. In *Amarna Reports*, Vol. I, edited by B. J. Kemp, pp. 154–177. Egypt Exploration Society, London.
Heisen, W. M. H.
1891 A Case of Trichinosis. *Journal of Materia Medica* 29:51.
Helmer, D.
2008 Révision de la faune de Cafer Höyük (Malatya Turquie): Apports des méthodes de l'analyse des mélanges et de l'analyse de kernel à la mise en évidence de la Domestication.

In *Archaeozoology of the Near East VIII*, edited by E. Vila, L. Gourichon, A. M. Choyke, and H. Buitenhuis, pp. 169–195. Maison de l'Orient et de la Mediterranée, Lyon.

Helmer, D. and L. Gourichon

2008 Premières données sur les modalités de subsistance dans les niveaux récents de Tell Aswad (Damascène, Syrie): Fouilles, 2001–2005. In *Archaeozoology of the Near East VIII*, edited by E. Vila, L. Gourichon, A. M. Choyke, and H. Buitenhuis, pp. 119–151. Maison de l'Orient et de la Mediterranée, Lyon.

2017 The Fauna of Tell Aswad (Damascus, Syria), Early Neolithic Levels: Comparison with Northern and Southern Levant Sites. In *Archaeozoology of the Near East*, Vol. IX, edited by M. Mashkour and M. Beech, pp. 23–38. Oxbow Books, Oxford.

Helwing, B.

2003 Feasts as a Social Dynamic in Prehistoric Western Asia: Three Case Studies from Syria and Anatolia. *Paléorient* 29:63–85.

Hellwing, S. and R. Gophna

1984 The Animal Remains from the Early and Middle Bronze Ages at Tel Aphek and Tel Dalit: A Comparative Study. *Tel Aviv* 11:48–59.

Hemmer, H.

1990 *Domestication: The Decline of Environmental Appreciation*. Cambridge University Press, Cambridge.

Henderson, J. B.

1998 *The Construction of Orthodoxy and Heresy: Neo-Confucian, Islamic, Jewish, and Early Christian Patterns*. State University of New York Press, Albany.

Hendrickx, S.

2011 Hunting and Social Complexity in Predynastic Egypt. *Bulletin des Séances Mededelingen der Zittingen* 57:237–263.

Henrich, J. and N. Henrich

2010 The Evolution of Cultural Adaptations: Fijian Food Taboos Protect against Dangerous Marine Toxins. *Proceedings of the Royal Society B* 277:3715–3724.

Herrero, J., A. Garcia-Serrano, S. Couto, V. M. Ortuño, and R. Garcîa-González

2006 Diet of Wild Boar *Sus scrofa* L. and Crop Damage in an Intensive Agroecosystem. *European Journal of Wildlife Research* 52:245–250.

Hesse, B.

1986 Animal Use at Tel Miqne-Ekron in the Bronze Age and Iron Age. *Bulletin of the American Schools of Oriental Research* 264:17–27.

1990 Pig Lovers and Pig Haters: Patterns of Palestinian Pork Production. *Journal of Ethnobiology* 10:195–225.

1994 Husbandry, Dietary Taboos, and the Bones of Ancient Near East: Zooarchaeology in the Post-Processual World. In *Methods in the Mediterranean: Historical and Archaeological Views on Texts and Archaeology*, edited by D. B. Small, pp. 197–232. Brill, Leiden.

Hesse, B., D. N. Fulton, and P. Wapnish

2011 Animal Remains. In *Final Reports of the Leon Levy Expedition to Ashkelon: 3. The Seventh Century B.C.*, edited by L. E. Stager, D. M. Master, and J. D. Schloen, pp. 615–643. Eisenbrauns, Winona Lake, Indiana.

Hesse, B. and P. Wapnish

1997 Can Pig Remains Be Used for Ethnic Diagnosis in the Ancient Near East? In *The Archaeology of Israel: Constructing the Past, Interpreting the Present*, edited by N. A.

Silberman and D. B. Small, pp. 238–270. Journal for the Study of the Old Testament Supplement Series 237. Sheffield Academic Press, Sheffield.

1998 Pig Use and Abuse in the Ancient Levant: Ethnoreligious Boundary-Building and Swine. In *Ancestors for the Pigs*, edited by S. Nelson, pp. 123–135. MASCA Research Papers in Science and Archaeology 15. University of Pennsylvania Museum of Archaeology and Anthropology, Philadelphia.

2001 Commodities and Cuisine: Animals in the Early Bronze Age of Northern Palestine. In *Studies in the Archaeology of Israel and Neighboring Lands in Memory of Douglas L. Esse*, edited by S. R. Wolff, pp. 251–282. Studies in Ancient Oriental Civilization No. 59. Oriental Institute of the University of Chicago, Chicago.

Hessler, P.

2014 Letters from Cairo: Tales of the Trash. *New Yorker*, 13 October: 90–99.

Hide, R.

2003 *Pig Husbandry in New Guinea: A Literature Review and Bibliography*. Australian Centre for International Agricultural Research, Canberra.

Highcock, N., P. Crabtree, D. V. Campana, M. Capardoni, A. Lanaro, A. Matessi, N. F. Miller, P. Strosahal, A. Trameri, and L. D'Alfonso

2015 Kınık Höyük, Niğde: A New Archaeological Project in Southern Cappadocia. In *The Archaeology of Anatolia: Recent Discoveries (2011–2014)*, Vol. 1, edited by S. R. Steadman and G. McMahon, pp. 98–127. Cambridge Scholars Publishing, Newcastle upon Tyne.

Hill, A. C.

2011 Specialized Pastoralism and Social Stratification: Analysis of the Fauna from Chalcolithic Tel Tsaf, Israel. PhD dissertation, Department of Anthropology, University of Connecticut, Storrs.

Hill, A. C., M. D. Price, and Y. M. Rowan

2016 Feasting at Marj Rabba, an Early Chalcolithic Site in the Galilee. *Oxford Journal of Archaeology* 35:127–140.

Hilzheimer, L.

1941 *Animal Remains from Tell Asmar*. Studies in Ancient Oriental Civilization 20. Oriental Institute of the University of Chicago, Chicago.

Hobsbowm, E. and T. Ranger (editors)

1983 *The Invention of Tradition*. Cambridge University Press, Cambridge.

Homer

1996 *The Odyssey*. Translated by R. Fagles. Penguin Book, New York.

Hongo, H.

1997 Patterns of Animal Husbandry, Environment, and Ethnicity in Central Anatolia in the Ottoman Empire Period: Faunal Remains from Islamic Layers at Kaman-Kalehöyük. *Japan Review* 8:275–307.

1998a Patterns of Animal Husbandry in Central Anatolia from the Second and First Millennia BC: Faunal Remains from Kaman-Kalehöyük, Turkey. In *Archaeozoology of the Near East III*, edited by H. Buitenhuis, L. Bartosiewicz, and A. Choyke, pp. 255–275. ARC-Publicaties 18, Groningen.

1998b Patterns of Animal Husbandry in Central Anatolia from the Second Millennium BC through the Middle Ages: Faunal Remains from Kaman-Kalehöyük, Turkey. PhD dissertation, Department of Anthropology, Harvard University, Cambridge.

Hongo, H., T. Anezaki, K. Yamazaki, O. Takahashi, and H. Sugawara
2007 Hunting or Management? The Status of *Sus* in the Jomon Period in Japan. In *Pigs and Humans: 10,000 Years of Interaction*, edited by U. Albarella, K. Dobney, A. Ervynck, and P. Rowley-Conwy, pp. 109–130. Oxford University Press, Oxford.
Hongo, H. and R. H. Meadow
1998a Faunal Remains from Prepottery Neolithic Levels at Çayönü, Southeastern Turkey: A Preliminary Report Focusing on Pigs (*Sus* sp.). In *Archaeozoology of the Near East*, Vol. IVA, edited by M. Mashkour, A. M. Choyke, H. Buitenhuis, and F. Poplin, pp. 121–140. ARC-Publicaties 32, Groningen.
1998b Pig Exploitation at Neolithic Çayönü Tepesi (Southeastern Anatolia). In *Ancestors for the Pigs*, edited by S. Nelson, pp. 77–98. MASCA Research Papers in Science and Archaeology 15. University of Pennsylvania Museum of Archaeology and Anthropology, Philadelphia.
Hongo, H., J. Pearson and B. Öksüz
2009 The Process of Ungulate Domestication at Çayönü, Southeastern Turkey: A Multidisciplinary Approach Focusing on *Bos* sp. and *Cervus elaphus*. *Anthropozoologica* 44:63–78.
Honigman, S.
2014 *Tales of High Priests and Taxes: The Books of the Maccabees and the Judean Rebellion against Antiochus IV*. University of California Press, Berkeley.
Hordes, S. M.
2005 *To the End of the Earth: A History of the Crypto-Jews of New Mexico*. Columbia University Press, New York.
Horwitz, L. K.
1997 Faunal Remains. In *Yiftaḥ'el: Salvage and Rescue Excavations at a Prehistoric Village in the Lower Galilee*, edited by E. Braun, pp. 155–172. Israel Antiquities Authority, Jerusalem.
1998 Animal Exploitation During the Early Islamic Period in the Negev: The Fauna from Elat-Elot. *'Atiqot* 36:27–38.
2003 Early Bronze Age Animal Exploitation at Qiryat Ata. In *Salvage Excavations at the Early Bronze Age Site of Qiryat Ata*, edited by A. Golani, pp. 225–241. Israel Antiquities Authority, Jerusalem.
2009 Terrestrial Fauna. In *Aphek-Antipatris: II. The Remains on the Acropolis—The Moshe Kochavi and Pirhiya Beck Excavations*, edited by Y. Gadot and Y. Yadin, pp. 526–561. Monograph Series of the Institute of Archaeology of Tel Aviv University No. 27. Institute of Archaeology of Tel Aviv University, Tel Aviv.
Horwitz, L. K., A. Gardeisen, A. M. Maeir, and L. A. Hitchcock
2017 A Contribution to the Iron Age Philistine Pig Debate. In *The Wide Lens in Archaeology: Honoring Brian Hesse's Contributions to Anthropological Archaeology*, edited by J. Lev-Tov, P. Hesse, and A. Gilbert, pp. 93–116. Lockwood Press, Atlanta.
Horwitz, L. K. and H. Lernau
2005 Faunal and Malacological Remains from the Middle Bronze, Late Bronze and Iron Age levels at Tel Yoqne'am. In *Yoqne'am III: The Middle and Late Bronze Ages: Final Report of the Archaeological Excavations (1977–1988)*, edited by A. Ben-Tor, D. Ben-Ami, and A. Livneh, pp. 395–435. Qedem Reports 7. Institute of Archaeology, Hebrew University, Jerusalem.

Horwitz, L. K. and O. Lernau

2003 Fauna from Tel Qashish. In *Tel Qashish: A Village in the Jezreel Valley—Final Report of the Archaeological Excavations (1978–1987)*, edited by A. Ben-Tor, R. Bonfil, and S. Zuckerman, pp. 427–443. Qedem Reports 5. Institute of Archaeology, Hebrew University, Jerusalem.

Horwitz, L. K. and H. Monchot

2002 Représentation squelettique au Paléolithique inférieur, le site d'Holon (Israël). *Paléorient* 28:71–85.

2007 *Sus, Hippopotamus, Bos,* and *Gazella.* In *Holon: A Lower Paleolithic Site in Israel,* edited by M. Chazan and L. K. Horwitz, pp. 91–109. Peabody Museum, Cambridge.

Horwitz, L. K. and J. Studer

2005 Pig Production During the Classical Periods in the Southern Levant. In *Archaeozoology of the Near East*, Vol. VI, edited by H. Buitenhuis, A. M. Choyke, L. Martin, L. Bartosiewicz, and M. Mashkour, pp. 222–239. Arc-Publicaties 123, Groningen.

Horwitz, L. K. and E. Tchernov

1989 Animal Exploitation in the Early Bronze Age of the Southern Levant: An Overview. In *L'Urbanisation de la Palestine à l'Âge du bronze ancien: Bilan et perspectives des recherches actuelles*, edited by P. D. Miroschedji, pp. 279–296. British Archaeological Reports International Series 527, Oxford.

Houston, W. J.

1993 *Purity and Monotheism: Clean and Unclean Animals in Biblical Law*. Sheffield Academic Press, Sheffield.

Howell, S.

2012 Knowledge, Morality, and Causality in a "Luckless Society": The Case of the Chewong in the Malaysian Rain Forest. *Social Analysis* 56:133–147.

Hurvitz, A.

1988 Dating the Priestly Source in Light of the Historical Study of Biblical Hebrew: A Century after Wellhausen. *Zeitschrift für die Alttestamentliche Wissenschaft* 100:88–100.

Ijzereef, F. G.

1989 Social Differentiation from Animal Bone Studies. In *Diet and Crafts in Towns*, edited by D. Serjeantson and T. Waldron, pp. 41–53. British Archaeological Reports International Series 199, Oxford.

Ikram, S.

1995 *Choice Cuts: Meat Production in Ancient Egypt*. Uitgeverij Peeters en Departement Oosterse Studies, Leuven.

2008 Food and Funerals: Sustaining the Dead for Eternity. *Polish Archaeology in the Mediterranean* 20:361–371.

Ilgezdi, G.

2008 The Domestication Process in Southeastern Turkey: The Evidence of Mezraa-Teleilat. PhD dissertation, Geowissenschaftlichen Fakultät, Universität Tübingen, Tübingen.

Ingold, T.

1988 Introduction. In *What Is an Animal?*, edited by T. Ingold, pp. 1–16. Unwin Hyman, London.

Insoll, T.

1999 *The Archaeology of Islam*. Blackwell Publishers, Oxford.

Irwin, R.
1996 The Emergence of the Islamic World System: 1000–1500. In *The Cambridge Illustrated History of the Islamic World*, edited by F. Robinson, pp. 32–61. Cambridge University Press, Cambridge.
Izadi, E.
2015 "Ignorant Act": Severed Pig's Head Thrown at Philadelphia Mosque Door. *Washington Post*, 9 December. Washington, DC.
Jaffe, Y., Q. Wei, and Y. Zhao
2018 Foodways and the Archaeology of Colonial Contact: Rethinking the Western Zhou Expansion in Shandong. *American Anthropologist* 120:55–71.
Jensen, A.
1996 *God's Self-Confident Daughters: Early Christianity and the Liberation of Women*. Translated by O. C. Dean, Jr. Westminster John Knox Press, Louisville.
Jones, S.
1997 *The Archaeology of Ethnicity*. London, Routledge.
Kagira, J. M., P. W. N. Kanyari, N. Maingi, S. M. Githigia, J. C. Ng'ang'a, and J. W. Karuga
2010 Characteristics of the Smallholder Free-Range Pig Production System in Western Kenya. *Tropical Animal Health and Production* 42:865–873.
Kaniewski, D., E. Van Campo, J. Guiot, S. Le Burel, T. Otto, and C. Baeteman
2013 Environmental Roots of the Late Bronze Age Crisis. *PLoS One* 8:e71004.
Kansa, S. W.
2004 Animal Exploitation at Early Bronze Age Ashqelon, Afridar: What the Bones Tell Us—Initial Analysis of the Animal Bones from Areas E, F and G. *'Atiqot* 45: 279–297.
Kansa, S. W., S. C. Gauld, S. Campbell, and E. Carter
2009a Whose Bones Are Those? Preliminary Comparative Analysis of Fragmented Human and Animal Bones in the "Death Pit" at Domuztepe, a Late Neolithic Settlement in Southeastern Turkey. *Anthropozoologica* 44:159–172.
Kansa, S. W., A. Kennedy, S. Campbell, and E. Carter
2009b Resource Exploitation at Late Neolithic Domuztepe. *Current Anthropology* 50:897–914.
Kapoor Sharma, R.
2002 Un maiale fra i rifiuti. *Slow* 26:44–49.
Kees, H.
1961 *Ancient Egypt: A Cultural Topography*. University of Chicago Press, Chicago.
Keimer, L.
1932 Le Sanglier égyptien. *Chronique D'Égypte* 10:26–33.
Kelley, C. P., S. Mohtadi, M. A. Cane, R. Seager, and Y. Kushnir
2015 Climate Change in the Fertile Crescent and Implications of the Recent Syrian Drought. *Proceedings of the National Academy of Sciences* 112:3241–3246.
Kelly, R. C.
1988 Etoro Suidology: A Reassessment of the Pig's Role in the Prehistory and Comparative Ethnology of New Guinea. In *Mountain Papuans: Historical and Comparative Perspectives from New Guinea Fringe Highlands Societies*, edited by J. F. Weiner, pp. 111–186. University of Michigan Press, Ann Arbor.
Kelly, R. L.
1995 *The Foraging Spectrum*. Smithsonian Institution Press, Washington, DC.

Kelly-Buccellati, M.
2005 Introduction to the Archeo-Zoology of the *Abi*. *Studi Micenei ed Egeo-Anatolici* 47:61–66.
Kemp, B. J. (editor)
1984 *Amarna Reports I*. Egypt Exploration Society, London.
Kennedy, M. A.
2016 The End of the 3rd Millennium BC in the Levant: New Perspectives and Old Ideas. *Levant* 48:1–32.
Killebrew, A.
2005 *Biblical Peoples and Ethnicity: An Archaeological Study of Egyptians, Canaanites, Philistines, and Early Israel, 1300–1100 BCE*. Society of Biblical Literature, Atlanta.
2014 Hybridity, Hapiru, and the Archaeology of Ethnicity in Second Millennium BCE Western Asia. In *A Companion to Ethnicity in the Ancient Mediterranean*, edited by J. McInerney, pp. 142–157. Wiley Blackwell, New York.
King, A.
1999 Diet in the Roman World: A Regional Inter-Site Comparison of the Mammal Bones. *Journal of Roman Archaeology* 12:168–202.
Kingsley, P.
2014a Morsi's Overthrow Helps Egypt's Pork Farmers Get Their Sizzle Back. *Guardian*, 24 March. London.
2014b Waste Not: Egypt's Refuse Collectors Regain Role at Heart of Cairo Society. *Guardian*, 27 March. London.
Klein, R. G.
2009 *The Human Career: Human Biological and Cultural Origins*. 3rd ed. University of Chicago Press, Chicago.
Knapp, A. B.
2010 Cyprus's Earliest Prehistory: Seafarers, Foragers and Settlers. *Journal of World Prehistory* 23:79–120.
Kolinski, R.
2012 "The Mountain Sheep are Sweeter . . . ". In *Looking North: The Socioeconomic Dynamics of Northern Mesopotamian and Anatolian Regions During the Late Third and Early Second Millennium BC*, edited by N. Laneri, P. Pfälzner, and S. Valentino, pp. 237–251. Harrassowitz, Wiesbaden.
Kolinski, R. and J. Piątkowska-Małecka
2008 Animals in the Steppe: Patterns of Animal Husbandry as a Reflection of Changing Environmental Conditions in the Khabur Triangle. In *Proceedings of the 4th International Congress of the Archaeology of the Ancient Near East*, edited by H. Kühne, R. M. Czichon, and F. J. Kreppner, pp. 115–128. Harrassowitz, Wiesbaden.
Konner, M.
2003 *Unsettled: An Anthropology of the Jews*. Viking Compass, New York.
Kornum, B. R. and G. M. Knudsen
2011 Cognitive Testing of Pigs (*Sus scrofa*) in Translational Biobehavioral Research. *Neuroscience & Biobehavioral Reviews* 35:437–451.
Kozlowski, S. K. and O. Aurenche
2005 *Territories, Boundaries and Cultures in the Near East*. British Archaeological Reports International Series 1362, Oxford.
Kraemer, D.
2007 *Jewish Eating and Identity Through the Ages*. Routledge, London.

Krause-Kyora, B., C. Makarewicz, A. Evin, L. Girdland Flink, K. Dobney, G. Larson, S. Hartz, S. Schreiber, C. von Carnap-Bornheim, N. von Wurmb-Schwark, and A. Nebel

2013 Use of Domesticated Pigs by Mesolithic Hunter-Gatherers in Northwestern Europe. *Nature Communications* 4:2348.

Kroll, H.

2012 Animals in the Byzantine Empire: An Overview of the Archaeozoological Evidence. *Archeologia Medievale* 39:93–121.

Kruska, D. C. T.

2005 On the Evolutionary Significance of Encephalization in Some Eutherian Mammals: Effects of Adaptive Radiation, Domestication, and Feralization. *Brain, Behavior and Evolution* 65:73–108.

Kuhn, S. L., M. C. Stiner, E. Güleç, I. Özer, H. Yılmaz, I. Baykara, A. Açıkkol, P. Goldberg, K. Martínez Molina, E. Ünay, and F. Suata-Alpaslan

2009 The Early Upper Paleolithic Occupations at Üçagızlı Cave (Hatay, Turkey). *Journal of Human Evolution* 56:87–113.

Kuijt, I. and N. Goring-Morris

2002 Foraging, Farming, and Social Complexity in the Pre-Pottery Neolithic of the Southern Levant: A Review and Synthesis. *Journal of World Prehistory* 16:361–440.

Kurban, F. and J. Tobin

2009 "They Don't Like Us": Reflections of Turkish Children in a German Preschool. *Contemporary Issues in Early Childhood* 10:24–34.

Kuşatman, B. 1991 The Origins of Pig Domestication with Particular Reference to the Near East. PhD dissertation, Institute of Archaeology, University College London, London.

Laffer, J. P. 1983 The Faunal Remains from Banahilk. In *Prehistoric Archeology Along the Zagros Flanks*, edited by L. S. Braidwood, R. J. Braidwood, B. Howe, C. A. Reed, and P. J. Watson, pp. 629–648. Oriental Institute of the University of Chicago, Chicago.

Lambert, W. G.

1996 *Babylonian Wisdom Literature*. Eisenbrauns, Winona Lake, Indiana.

Landau, E.

2010 *Ha-sefer ha-lavan* [In Hebrew; The White Book]. Keter Publishing House, Jerusalem.

Laneri, N.

2011 Connecting Fragments: A Sensorial Approach to the Materialization of Religious Beliefs in Rural Mesopotamia at the Beginning of the Second Millennium BC. *Cambridge Archaeological Journal* 21:77–94.

Langgut, D., I. Finkelstein, and T. Litt

2013 Climate and the Late Bronze Collapse: New Evidence from the Southern Levant. *Tel Aviv* 40:149–175.

Larsen, M. T.

2015 *Ancient Kanesh: A Merchant Colony in Bronze Age Anatolia*. Cambridge University Press, Cambridge.

Larson, G., U. Albarella, K. Dobney, P. Rowley-Conwy, J. Schibler, A. Tresset, J.-D. Vigne, C. J. Edwards, A. Schlumbaum, A. Dinu, A. Bălăçsescu, G. Dolman, A. Tagliacozzo, N. Manaseryan, P. Miracle, L. V. Wijngaarden-Bakker, M. Masseti, D. G. Bradley, and A. Cooper

2007a Ancient DNA, Pig Domestication, and the Spread of the Neolithic into Europe. *Proceedings of the National Academy of Sciences* 104:15276–15281.

Larson, G., T. Cucchi, M. Fujita, E. Matisoo-Smith, J. Robins, A. Anderson, B. Rolett, M. Spriggs, G. Dolman, T.-H. Kim, N. T. D. Thuy, E. Randi, M. Doherty, R. A. Due, R. Bollt, T. Djubiantono, B. Griffin, M. Intoh, E. Keane, P. Kirch, K.-T. Li, M. Morwood, L. M. Pedriña, P. J. Piper, R. J. Rabett, P. Shooter, G. Van den Bergh, E. West, S. Wickler, J. Yuan, A. Cooper, and K. Dobney
2007b Phylogeny and Ancient DNA of *Sus* Provides Insights into Neolithic Expansion in Island Southeast Asia and Oceania. *Proceedings of the National Academy of Sciences* 104:4834–4839.

Larson, G., K. Dobney, U. Albarella, M. Fang, E. Matisoo-Smith, J. Robins, S. Lowden, H. Finlayson, T. Brand, E. Willerslev, P. Rowley-Conwy, L. Andersson, and A. Cooper
2005 Worldwide Phylogeography of Wild Boar Reveals Multiple Centers of Pig Domestication. *Science* 307:1618–1621.

Larson, G., R. Liu, X. Zhao, J. Yuan, D. Q. Fuller, L. Barton, K. Dobney, Q. Fan, Z. Gu, X.-H. Liu, Y. Luo, P. Lv, L. Andersson, and N. Li
2010 Patterns of East Asian Pig Domestication, Migration, and Turnover Revealed by Modern and Ancient DNA. *Proceedings of the National Academy of Sciences* 107:7686–7691.

Lawrence, D., G. Philip, H. Hunt, L. Snape-Kennedy, and T. Wilkinson
2016 Long Term Population, City Size and Climate Trends in the Fertile Crescent: A First Approximation. *PloS ONE* 11:e0152563.

Lawrence, D. and T. Wilkinson
2015 Hubs and Upstarts: Pathways to Urbanism in the Northern Fertile Crescent. *Antiquity* 89:328–344.

Layard, A. H.
1903 *Sir A. Henry Layard, G.C.B, D.C.L: Autobiography and Letters from His Childhood until His Appointment as H. M. Ambassador at Madrid*, Vol. 2. John Murray, London.

Leahy, M. A.
1985 The Hieratic Labels. In *Amarna Reports II*, edited by B. J. Kemp, pp. 65–109. Egypt Exploration Society, London.

Leake, W. M.
1826 *An Edict of Diocletian, Fixing a Maximum of Prices Throughout the Roman Empire, A.D. 303.* John Murray, London.

Leared, A.
1876 *Morocco and the Moors.* Sampson Low, Marston, Searle, and Rivington, Limited, London.

Leduc, C., A. Bridault, and C. Cupillard
2015 Wild Boar (*Sus scrofa scrofa*) Hunting and Exploitation Strategies During the Mesolithic at Les Cabônes (Ranchot Jura, France), Layer 3. *Journal of Archaeological Science: Reports* 2:473–484.

Lega, C., D. Fulgione, A. Genovese, L. Rook, M. Masseti, M. Meiri, A. Cinzia Marra, F. Carotenuto, and P Raia
2017 Like a Pig Out of Water: Seaborne Spread of Domestic Pigs in Southern Italy and Sardinia During the Bronze and Iron Ages. *Heredity* 118:154–159.

Legal Information Institute
Cannibalism. Cornell Law School. https://www.law.cornell.edu/wex/cannibalism.

Legge, A. J. and P. Rowley-Conwy
2000 The Exploitation of Animals. In *Village on the Euphrates: Foraging to Farming at Abu Hureyra*, edited by A. M. T. Moore, G. C. Hillman, and A. J. Legge, pp. 423–471. Oxford University Press, Oxford.

Leigh, M.
2015 Food in Latin Literature. In *A Companion to Food in the Ancient World*, edited by J. Wilkins and R. Nadeau, pp. 43–52. Wiley-Blackwell, New York.
Lemoine, X.
2012 Pig (*Sus scrofa*) Exploitation at Hallan Çemi, Southeastern Anatolia: Proposing an Alternative Model. BA thesis, Department of Anthropology, Portland State University, Portland.
Lemoine, X., M. A. Zeder, K. J. Bishop, and S. J. Rufolo
2014 A New System for Computing Dentition-Based Age Profiles in *Sus scrofa*. *Journal of Archaeological Science* 47:179–193.
Lemonnier, P.
2002 Pigs as Ordinary Wealth: Technical Logic, Exchange and Leadership in New Guinea. In *Technological Choices: Transformation in Material Cultures Since the Neolithic*, edited by P. Lemonnier, pp. 126–156. Routledge, New York.
2012 *Mundane Objects: Materiality and Non-Verbal Communication*. Left Coast Press, Walnut Creek, California.
Lev-Tov, J. S.
2000 Pigs, Philistines, and the Ancient Animal Economy of Ekron from the Late Bronze Age to the Iron Age II. PhD dissertation, Department of Anthropology, University of Tennessee, Knoxville.
2003 "Upon What Meat Doth This Our Caesar Feed . . .?"A Dietary Perspective on Hellenistic and Roman Influence in Palestine. In *Zeichen aus Text und Stein: Studien auf dem Weg zu einer Archäologie des Neuen Testaments*, edited by S. Alkier and J. K. Zangenberg, pp. 420–446. Francke, Tübingen.
2010 A Plebeian Perspective on Empire Economies: Faunal Remains from Tel Miqne-Ekron, Israel. In *Anthropological Approaches to Zooarchaeology: Complexity, Colonialism, and Animal Transformations*, edited by D. V. Campana, P. Crabtree, and S. D. deFrance, pp. 90–104. Oxbow Books, Oxford.
2012 A Preliminary Report on the Late Bronze and Iron Age Faunal Assemblage from Tell es-Safi/Gath. In *Tell es-Safi/Gath: I. Report on the 1996–2005 Seasons*, edited by A. M. Maeir, pp. 589–612. Ägypten und Altes Testament 69. Harrassowitz, Wiesbaden.
Lev-Tov, J. S., S. W. Kansa, L. Atici, and J. C. Wheeler
2017 New Light on Faunal Remains from Choga Mish, Iran. In *The Wide Lens in Archaeology: Honoring Brian Hesse's Contributions to Anthropological Archaeology*, edited by J. Lev-Tov, P. Hesse, and A. Gilbert, pp. 443–475. Lockwood Press, Atlanta.
Lev-Tov, J. S. and K. McGeough
2007 Examining Feasting in Late Bronze Age Syro-Palestine through Ancient Texts and Bones. In *The Archaeology of Food and Identity*, edited by K. C. Twiss, pp. 85–111. Occasional Paper No. 34. Center for Archaeological Investigations, Southern Illinois University, Carbondale.
Lev-Tov, J. S., B. Porter, and B. E. Routledge
2011 Measuring Local Diversity in Early Iron Age Animal Economies: A View from Khirbat al-Mudayna al-ʿAliya (Jordan). *Bulletin of the American Schools of Oriental Research* 361:67–93.
Lévi-Strauss, C.
1966 [1962] *The Savage Mind*. University of Chicago Press, Chicago.
Levy-Rubin, M.
2018 The Pact of Umar. In *Routledge Handbook on Christian–Muslim Relations*, edited by D. Thomas, pp. 69–74. Routledge, New York.

Liebmann, M.
2015 The Mickey Mouse Kachina and Other "Double Objects": Hybridity in the Material Culture of Colonial Encounters. *Journal of Social Archaeology* 15:319–341.
Lindsay, J. E.
2005 *Daily Life in the Medieval Islamic World.* Greenwood Press, Westport, Connecticut.
Linfield, S.
2019 *The Lion's Den: Zionism and the Left from Hannah Arendt to Noam Chomsky.* Yale University Press, New Haven.
Linseele, V., W. Van Neer, S. Thys, R. Phillipps, R. Cappers, W. Wendrich, and S. Holdaway
2014 New Archaeozoological Data from the Fayum "Neolithic" with a Critical Assessment of the Evidence for Early Stock Keeping in Egypt. *PLoS ONE* 9:e108517.
Lion, B. and C. Michel
2006 L'Élevage des porcs en Haute Mésopotamie, Syrie et Transtigrine au debut du IIe millénaire. In *De la domestication au tabou*, edited by B. Lion and C. Michel, pp. 89–101. De Boccard, Paris.
Lipovitch, D. R.
2006–2007 Modeling a Mycenaean Menu: Can Aegean Populations Be Defined in Near Eastern Contexts Based on Their Diet? 28 (2015). *Scripta Mediterranea* 27–28:147–159.
Liverani, M.
1987 The Collapse of the Near Eastern Regional System at the End of the Bronze Age: The Case of Syria. In *Centre and Periphery in the Ancient World*, edited by M. J. Rowlands, M. Larsen, and K. Kristiansen, pp. 66–73. Cambridge Univeristy Press, Cambridge.
2017 *Assyria: The Imperial Mission.* Eisenbrauns, Winona Lake, Indiana.
Lobban, R.
1994 Pigs and Their Prohibition. *Journal of Middle East Studies* 26:57–75.
Loyet, M. A.
2000 The Potential for Within-Site Variation of Faunal Remains: A Case Study from the Islamic Period Urban Center of Tell Tuneinir, Syria. *Bulletin of the American Schools of Oriental Research* 320:23–48.
Lupo, K. D. and D. N. Schmitt
2002 Upper Paleolithic Net-Hunting, Small Prey Exploitation, and Women's Work Effort: A View From the Ethnographic and Ethnoarchaeological Record of the Congo Basin. *Journal of Archaeological Method and Theory* 9:147–179.
Mac Sweeney, N.
2009 Beyond Ethnicity: The Overlooked Diversity of Group Identities. *Journal of Mediterranean Archaeology* 22:101–126.
MacCulloch, D.
2010 *Christianity: The First Three Thousand Years.* Penguin, London.
MacKinnon, M.
2001 High on the Hog: Linking Zooarchaeological, Literary, and Artistic Data for Pig Breeds in Roman Italy. *American Journal of Archaeology* 105:649–673.
2017 Animals, Acculturation, and Colonization in Ancient and Islamic North Africa. In *The Oxford Handbook of Zooarchaeology*, edited by U. Albarella, M. Rizzetto, H. Russ, K. Vickers, and S. Viner-Daniels, pp. 466–478. Oxford University Press, Oxford.

Maeir, A. M. and L. A. Hitchcock
2017 The Appearance, Formation and Transformation of Philistine Culture: New Perspectives and New Finds. In *"Sea Peoples" Up-to-Date: New Research on Transformations in the Eastern Mediterranean in the 13th–11th Centuries BCE*, edited by P. M. Fischer and T. Bürge, pp. 149–162. Austrian Academy of Sciences, Vienna.
Magness, J.
2011 *Stone and Dung, Oil and Spit: Jewish Daily Life in the Time of Jesus*. Grand Rapids, Michigan.
Maher, E.
2017 Flair of the Dog: The Philistine Consumption of Canines. In *The Wide Lens in Archaeology: Honoring Brian Hesse's Contributions to Anthropological Archaeology*, edited by J. Lev-Tov, P. Hesse, and A. Gilbert, pp. 117–148. Lockwood Press, Atlanta.
Mahler-Slasky, Y. and M. E. Kislev
2010 Lathyrus Consumption in Late Bronze and Iron Age Sites in Israel: An Aegean affinity. *Journal of Archaeological Science* 37:2477–2485.
Makarewicz, C.
2012 The Younger Dryas and Hunter-Gatherer Transitions to Food Production in the Near East. In *Hunter-Gatherer Behavior: Human Response During the Younger Dryas*, edited by M. I. Erin, pp. 195–230. Left Coast Press, Walnut Creek, California.
2016 Caprine Husbandry and Initial Pig Management East of the Jordan Valley: Animal Exploitation at Neolithic Wadi Shu'eib, Jordan. *Paléorient* 42:151–168.
Makarewicz, C. and B. Finlayson
2018 Constructing Community in the Neolithic of Southern Jordan: Quotidian Practice in Communal Architecture. *PLoS ONE* 13:e0193712.
Malcolmson, R. and S. Mastoris
1998 *The English Pig: A History*. Hambledom Press, London.
Malinowski, B.
1932 *Argonauts of the Western Pacific*. George Routledge and Sons, New York.
Manlius, N. and A. Gautier
1999 Le sanglier en Égypte. *Comptes Rendus de l'Académie des Sciences—Series III—Sciences de la Vie* 322:577.
Manning, S. W., C. McCartney, B. Kromer, and S. T. Stewart
2010 The Earlier Neolithic in Cyprus: Recognition and Dating of a Pre-Pottery Neolithic A Occupation. *Antiquity* 84:693–706.
Manor, M., R. Rabinovich, and L. K. Horwitz
1996 Diachronic Changes in Urban Diet: The Historical Periods at Bet She'an, Israel. *Archaeozoologia* 8:89–104.
Manunza, A., A. Zidi, S. Yeghoyan, V. A. Balteanu, T. C. Carsai, O. Scherbakov, O. Ramírez, S. Eghbalsaied, A. Castelló, A. Mercadé, and M. Amills
2013 A High Throughput Genotyping Approach Reveals Distinctive Autosomal Genetic Signatures for European and Near Eastern Wild Boar. *PLoS ONE* 8:e55891.
Marom, N. and G. Bar-Oz
2009 "Man Made Oases": Neolithic Patterns of Wild Ungulate Exploitation & Their Consequences for the Domestication of Pigs & Cattle. *Before Farming* 2009:1–12.
2013 The Prey Pathway: A Regional History of Cattle (*Bos taurus*) and Pig (*Sus scrofa*) Domestication in the Northern Jordan Valley, Israel. *PLoS ONE* 8:e55958.

Marom, N., A. Yasur-Landau, S. Zuckerman, E. H. Cline, A. Ben-Tor, and G. Bar-Oz
2014 Shepherd Kings? A Zooarchaeological Investigation of Elite Precincts in Middle Bronze Age Tel Hazor and Tel Kabri. *Bulletin of the American Schools of Oriental Research* 371:59–82.

Marom, N. and S. Zuckerman
2012 The Zooarchaeology of Exclusion and Expropriation: Looking Up from the Lower City in Late Bronze Age Hazor. *Journal of Anthropological Archaeology* 31:573–585.

Martin, L.
1999 Mammal Remains from the Eastern Jordanian Neolithic, and the Nature of Caprine Herding in the Steppe. *Paléorient* 25:87–104.

Mashkour, M.
2006 Boars and Pigs: A View from the Iranian Plateau. In *De la domestication au tabou*, edited by B. Lion and C. Michel, pp. 155–167. De Boccard, Paris.
2013 Animal Remains at Hatra, City of the Sun: First Archaeozoological Evidence for the Parthian Period. In *Animals, Gods and Men from East to West: Papers on Archaeology and History in Honour of Roberta Venco Ricciardi*, edited by A. Peruzzetto, F. D. Metzger, and L. Dirven, pp. 135–146. British Archaeological Reports International Series 2516, Oxford.

Mashkour, M., V. Radu, and R. Thomas
2013 Animal Bones. In *Persia's Imperial Power in Late Antiquity: The Great Wall of Gorgan and Frontier Landscapes of Sasanian Iran*, edited by E. W. Sauer, H. O. Rekavandi, T. J. Wilkinson, and J. Nokandeh, pp. 539–580. British Institute of Persian Studies, Archaeological Monograph Series II. Oxbow Books, Oxford.

Masseti, M.
2007 The Economic Role of *Sus* in Early Human Fishing Communities. In *Pigs and Humans: 10,000 Years of Interaction*, edited by U. Albarella, K. Dobney, A. Ervynck, and P. Rowley-Conwy, pp. 156–170. Oxford University Press, Oxford.

Matthews, W., J. N. Postgate, S. Payne, M. P. Charles, and K. Dobney
1994 The Imprint of Living in an Early Mesopotamian City: Questions and Answers. In *Whither Environmental Archaeology?*, edited by R. Luff and P. Rowley-Conwy, pp. 171–212. Oxbow Monograph 38. Oxbow Books, Oxford.

Matthiae, P. and N. Marchetti (editors)
2003 *Ebla and Its Landscape: Early State Formation in the Ancient Near East*. Left Coast Press, Walnut Creek, California.

Mayton, J.
2009 Pig Slaughter Lands Cairo in the Mire. *Guardian*, 8 October. London.

Mazar, A.
2010 Archaeology and the Biblical Narrative: The Case of the United Monarchy. In *One God—One Cult—One Nation: Archaeological and Biblical Perspectives*, edited by R. G. Kratz and H. Spieckermann, pp. 29–58. Walter de Gruyter, Berlin.

McAnany, P. A. and N. Yoffee (editors)
2010 *Questioning Collapse: Human Resilience, Ecological Vulnerability, and the Aftermath of Empire*. Cambridge University Press, Cambridge.

McCartney, C., S. W. Manning, D. Sewell, and S. T. Stewart
2007 Elaborating Early Neolithic Cyprus (EENC): Preliminary Report on the 2007 Field Season—Excavations and Regional Field Survey at Agia Varvara-Asprokremmos. *Report of the Department of Antiquities Cyprus* 2008:67–86.

McCorriston, J.
1997 The Fiber Revolution: Textile Extensification, Alienation, and Social Stratification in Ancient Mesopotamia. *Current Anthropology* 38:517–535.
McCorriston, J. and L. Martin
2009 Southern Arabia's Early Pastoral Population History: Some Recent Evidence. In *The Evolution of Human Populations in Arabia*, edited by M. D. Petraglia and J. I. Rose, pp. 237–250. Springer, New York.
McGlone, J. J. and S. E. Curtis
1985 Behavior and Performance of Weanling Pigs in Pens Equipped with Hide Areas. *Journal of Animal Science* 60:20–24.
McInerney, J.
2010 *The Cattle of the Sun: Cows and Culture in the World of the Ancient Greeks.* Princeton University Press, Princeton.
McMahon, A., A. Sołtysiak, and J. A. Weber
2011 Late Chalcolithic Mass Graves at Tell Brak, Syria, and Violent Conflict During the Growth of Early City-states. *Journal of Field Archaeology* 36:201–220.
Meadow, R. H.
1983 Appendix G: The Vertebrate Faunal Remains from Hasanlu Period X at Hajji Firuz. In *Hasanlu Excavation Reports: 2. Hajji Firuz Tepe, Iran: The Neolithic Settlement*, edited by M. Voigt, pp. 369–422. University Museum, University of Pennsylvania, Philadelphia.
1984 Animal Domestication in the Middle East: A View from the Eastern Margin. In *Animals in Archaeology: 3. Early Herders and Their Flocks*, edited by J. Clutton-Brock and C. Grigson, pp. 309–337. British Archaeological Reports International Series 202, Oxford.
Meek, T. J.
1969 The Middle Assyrian Laws. In *Ancient Near Eastern Texts Relating to the Old Testament with Supplement*, edited by J. B. Pritchard, pp. 180–188. Princeton University Press, Princeton.
Meiri, M., D. Huchon, G. Bar-Oz, E. Boaretto, L. K. Horwitz, A. M. Maeir, L. Sapir-Hen, G. Larson, S. Weiner, and I. Finkelstein
2013 Ancient DNA and Population Turnover in Southern Levantine Pigs: Signature of the Sea Peoples Migration? *Scientific Reports* 3:3035.
Meiri, M., P. W. Stockhammer, N. Marom, G. Bar-Oz, L. Sapir-Hen, P. Morgenstern, S. Macheridis, B. Rosen, D. Huchon, J. Maran, and I. Finkelstein
2017 Eastern Mediterranean Mobility in the Bronze and Early Iron Ages: Inferences from Ancient DNA of Pigs and Cattle. *Scientific Reports* 7:701.
Meyer-Rochow, V. B.
2009 Food Taboos: Their Origins and Purposes. *Journal of Ethnobiology and Ethnomedicine* 5.
Michel, C.
2006 Les Suidés dans la documentation de Kaniš au debut du IIe millénaire avant J.-C. In *De la domestication au tabou*, edited by B. Lion and C. Michel, pp. 169–180. De Boccard, Paris.
Milgrom, J.
1998 *Leviticus 1–16*. Anchor Bible, New York.
Miller, N. F.
2008 Sweeter than Wine? The Use of the Grape in Early Western Asia. *Antiquity* 82:937–946.
Miller, R.
1990 Hogs and Hygiene. *Journal of Egyptian Archaeology* 76:125–140.

Miller, S.
2004 *Disgust: The Gatekeeper Emotion*. Routledge, New York.
Minniti, C.
2013 Exploiting Animals: The Zooarchaeological Evidence through the Early Bronze Age. In *Ebla and Its Landscape: Early State Formation in the Ancient Near East*, edited by P. Matthiae and N. Marchetti, pp. 413–430. Left Coast Press, Walnut Creek, California.
Minniti, C. and L. Peyronel
2005 Symbolic or Functional Astragali from Tell Mardikh-Ebla (Syria). *Archaeofauna* 2005:7–26.
Mintz, S. W. and C. M. Du Bois
2002 The Anthropology of Food and Eating. *Annual Review of Anthropology* 31:99–119.
Mirecki, P. and M. Meyer (editors)
2002 *Magic and Ritual in the Ancient World*. Brill, Leiden.
Mizelle, B.
2011 *Pig*. Reaktion Books, London.
Mohaseb, F. A. and M. Mashkour
2017 Animal Exploitation from the Bronze Age to the Early Islamic Period in Haftavan Tepe (Western Azerbaijan-Iran). In *Archaeozoology of the Near East*, Vol. IX, edited by M. Mashkour and M. Beech, pp. 146–170. Oxbow Books, Oxford.
Monahan, B. H.
2000 The Organization of Domestication at Gritille, a Pre-Pottery Neolithic B Site in Southeastern Turkey. PhD dissertation, Department of Anthropology, Northwestern University, Evanston.
Monchot, H.
2014 Camels in Saudi Oasis During the Last Two Millennia: The Examples of Dûmat al-Jandal (Al-Jawf Province) and al-Yamâma (Riyadh Province). *Anthropozoologica* 49:195–206.
2016 Tell Said, the Hellenistic Settlement: The Archaeological Material—The Faunal Remains. In *French-Kuwaiti Archaeological Missions in Failaka: 2016 Campaign—Preliminary Report*, edited by J. Bonnéric and M. Gelin, pp. 127–144. Institut Français du Proche-Orient, Amman.
Moorey, P. R. S.
1986 The Emergence of the Light, Horse-Drawn Chariot in the Near-East c. 2000–1500 B.C. *World Archaeology* 18:196–215.
Moran, W. L.
1992 *The Amarna Letters*. Johns Hopkins University Press, Baltimore.
Mount, L. E.
1968 *The Climatic Physiology of the Pig*. Arnold, London.
Mudar, K.
1982 Early Dynastic III Animal Utilization in Lagash: A Report on the Fauna from Tell Al-Hiba. *Journal of Near Eastern Studies* 41: 23–34.
Müller, W.
2005 The Domestication of the Wolf: The Inevitable First? In *The First Steps of Animal Domestication: New Archaeological Approaches*, edited by J.-D. Vigne, J. Peters, and D. Helmer. Oxbow Books, Oxford.

Munro, N. D.
2004 Zooarchaeological Measures of Hunting Pressure and Occupation Intensity in the Natufian. *Current Anthropology* 45:S5–S33.

Munro, N. D., G. Bar-Oz, J. S. Meier, L. Sapir-Hen, M. C. Stiner, and R. Yeshurun
2018 The Emergence of Animal Management in the Southern Levant. *Scientific Reports* 8:9279.

Munro, N. D. and L. Grosman
2010 Early Evidence (ca. 12,000 B.P.) for Feasting at a Burial Cave in Israel. *Proceedings of the National Academy of Sciences* 107:15362–15366.

Murray, G. W.
1935 *Sons of Ishmael: A Study of the Egyptian Bedouin*. Routledge, Abingdon.

Nakamura, D.
2017 Trump Recycles Discredited Islamic Pigs' Blood Tale After Terrorist Attack in Barcelona. *Washington Post*, 17 August. Washington, DC.

Namrouqa, H.
2017 Jordan Valley Farmers Say Wild Boars Wreaking Havoc. *Jordan Times*, 30 January. Amman.

Nannoni, E., G. Martelli, M. Cecchini, G. Vignola, M. Giammarco, G. Zaghini, and L. Sardi
2013 Water Requirements of Liquid-Fed Heavy Pigs: Effect of Water Restriction on Growth Traits, Animal Welfare and Meat and Ham Quality. *Livestock Science* 151:21–28.

Navarette Belda, V. and M. Saña Seguî
2017 Size Changes in Wild and Domestic Pig Populations Between 10,000 and 800 cal. BC in the Iberian Peninsula: Evaluation of Natural versus Social Impacts in Animal Populations During the First Domestication Stages. *Holocene* 27:1526–1539.

Neghina, R., R. Moldovan, I. Marincu, C. L. Calma, and A. M. Neghina
2012 The Roots of Evil: The Amazing History of Trichinellosis and Trichinella Parasites. *Parasitology Research* 110:503–508.

Nemeth, D. J.
1998 Privy-Pigs in Prehistory? A Korean Analog for Neolithic Chinese Subsistence Practices. In *Ancestors for the Pigs*, edited by S. Nelson, pp. 11–25. MASCA Research Papers in Science and Archaeology 15. University of Pennsylvania Museum of Archaeology and Anthropology, Philadelphia.

Neumann, E. J., A. Ramirez, and K. J. Schwartz (editors)
2010 *Swine Disease Manual.* 4th ed. American Association of Swine Veterinarians, Perry, Iowa.

New York Times
1859 The Offal and Piggery Nuisances. *New York Times*, 27 July. New York.

Nieuwenhuyse, O., P. M. M. G. Akkermans, and J. van der Plicht
2010 Not So Coarse, Nor Always Plain: The Earliest Pottery of Syria. *Antiquity* 84:71–85.

Nieuwenhuyse, O., M. Roffet-Salque, R. P. Evershed, P. M. M. G. Akkermans, and A. Russell
2015 Tracing Pottery Use and the Emergence of Secondary Product Exploitation through Lipid Residue Analysis at Late Neolithic Tell Sabi Abyad (Syria). *Journal of Archaeological Science* 64:54–66.

NOAA [National Oceanic and Atmospheric Administration]
2019 Climate at a Glance. http://www.ncdc.noaa.gov/cag/.

O'Brien, M. J. and K. N. Laland
2012 Genes, Culture, and Agriculture: An Example of Human Niche Construction. *Current Anthropology* 53:434–470.

O'Connor, T.
2017 ISIS Fighters in Iraq Killed by Wild Boars Before They Can Ambush Locals. *Newsweek*, 25 April, https://www.newsweek.com/isis-fighters-iraq-killed-wild-boars-ambush-civilians-589816.

Oates, J.
1969 Choga Mami, 1967–68: A Preliminary Report. *Iraq* 31:115–152.
1978 Religion and Ritual in Sixth-Millennium B.C. Mesopotamia. *World Archaeology* 10:117–124.

Oates, J., A. McMahon, P. Karsgaard, S. Al-Quntar, and J. A. Ur
2007 Early Mesopotamian Urbanism: A New View from the North. *Antiquity* 81:585–600.

Oren, E. D. (editor)
2000 *The Sea Peoples and Their World: A Reassessment*. University of Pennsylvania Museum of Archaeology and Anthropology, Philadelphia.

Orliac, M. J., P.-O. Antoine, and S. Ducrocq
2010 Phylogenetic Relationships of the Suidae (Mammalia, Cetartiodactyla): New Insights on the Relationships Within Suoidea. *Zoologica Scripta* 39:315–330.

Ottoni, C., L. Girdland Flink, A. Evin, C. Geörg, B. De Cupere, W. Van Neer, L. Bartosiewicz, A. Linderholm, R. Barnett, J. Peters, R. Decorte, M. Waelkens, N. Vanderheyden, F.-X. Ricaut, C. Çakırlar, O. Çevik, A. R. Hoelzel, M. Mashkour, A. F. M. Karimlu, S. S. Seno, J. Daujat, F. Brock, R. Pinhasi, H. Hongo, M. Perez-Enciso, M. Rasmussen, L. Frantz, H.-J. Megens, R. Crooijmans, M. Groenen, B. Arbuckle, N. Benecke, U. S. Vidarsdottir, J. Burger, T. Cucchi, K. Dobney, and G. Larson
2012 Pig Domestication and Human-Mediated Dispersal in Western Eurasia Revealed through Ancient DNA and Geometric Morphometrics. *Molecular Biology and Evolution* 30:824–832.

Özdoğan, A., N. Karul, and E. Özdoğan
2011–2012 Mezraa-Teleilat Excavations. In *Salvage Project of the Archaeological Heritage of the Ilısu and Carchemish Dam Reservoirs Activities in 2002*, Vol. 1, edited by N. Tuna and O. Doonan, pp. 35–96. Middle East Technical University, Ankara.

Panagiotakopulu, E.
1999 An Examination of Biological Materials from Coprolites from XVIII Dynasty Amarna, Egypt. *Journal of Archaeological Science* 26:547–551.

Parayre, D.
2000 Les Suidés dans le monde Syro-Mésopotamien aux époques historiques. *Topoi Orient-Occident Supplément* 2:141–206.

Parsons, J. J.
1962 The Acorn-Hog Economy of the Oak Woodlands of Southwestern Spain. *Geographical Review* 52:211–235.

Paulette, T.
2016 Grain, Storage, and State Making in Mesopotamia (3200–2000 BC). In *Storage in Ancient Complex Societies: Administration, Organization, and Control*, edited by L. R. Manzanilla and M. Rothman, pp. 85–109. Routledge, New York.

Payne, S.

1973 Kill-off Patterns in Sheep and Goats: The Mandibles from Aşvan Kale. *Anatolian Studies* 23:281–303.

1988 Animal Bones from Tell Rubeideh. In *Tell Rubeideh: An Uruk Village in the Jebel Hamrin*, edited by R. G. Killick, pp. 98–135. British School of Archaeology in Iraq, Baghdad.

Payne, S. and G. Bull

1988 Components of Variation in Measurements of Pig Bones and Teeth, and the Use of Measurements to Distinguish Wild from Domestic Pig Remains. *Archaeozoologia* 2:27–66.

Peasnall, B. L., R. Redding, M. R. Nesbitt, and M. Rosenberg

1998 Hallan Çemi, Pig Husbandry, and Post-Pleistocene Adaptations along the Taurus-Zagros Arc (Turkey). *Paléorient* 24:25–41.

Peltenburg, E.

2000 From Nucleation to Dispersal: Late 3rd Millennium BC Settlement Pattern Transformations in the Near East and Aegean. In *La Djéziré et l'Euphrate Syrien de le protohistoire à la fin du second millénaire av. J.C.: Tendences dans l'interprétation des données nouvelles*, edited by O. Rouault and M. Wäfler, pp. 183–206. Brepols, Turnhout.

Perkins, D.

1964 Prehistoric Fauna from Shanidar, Iraq. *Science* 144:1565–1566.

Perry-Gal, L., G. Bar-Oz and A. Erlich

2015a Livestock Animal Trends in Idumaean Maresha: Preliminary Analysis of Cultural and Economic Aspects. *Aram* 27:213–216.

Perry-Gal, L., A. Erlich, A. Gilboa, and G. Bar-Oz

2015b Earliest Economic Exploitation of Chicken Outside East Asia: Evidence from the Hellenistic Southern Levant. *Proceedings of the National Academy of Sciences* 112:9849–9854.

Peters, J., D. Helmer, A. von den Driesch, and M. Saña Seguí

1999 Early Animal Husbandry in the Northern Levant. *Paléorient* 25:27–47.

Peters, J. and K. Schmidt

2004 Animals in the Symbolic World of Pre-Pottery Neolithic Göbekli Tepe, South-Eastern Turkey: A Preliminary Assessment. *Anthropozoologica* 39:179–218.

Peters, J., A. von den Driesch, and D. Helmer

2005 The Upper Euphrates-Tigris Basin: Cradle of Agropastoralism? In *First Steps of Animal Domestication: New Archaeozoological Approaches*, edited by J.-D. Vigne, J. Peters, and D. Helmer, pp. 96–124. Oxbow Books, Oxford.

Pew Research Center

2015 U.S. Becoming Less Religious. https://www.pewforum.org/2015/11/03/u-s-public-becoming-less-religious/.

Piątkowska-Małecka, J.

2015 Different Types of Animal Economy at Bakr Awa, Iraq, in the Bronze Age. *Bioarchaeology of the Near East* 9:1–19.

Piątkowska-Małecka, J. and A. Smogorzewska

2010 Animal Economy at Tell Arbid, North-East Syria, in the Third Millennium BC. *Bioarchaeology of the Near East* 4:25–43.

2013 Animal Bone Remains from Tell Arbid (Season 2009): Archaeozoological Analysis. *Polish Archaeology in the Mediterranean* 22:438–449.

Pimental, D.
2007 Environmental and Economic Costs of Vertebrate Species Invasions into the United States. *Managing Vertebrate Invasive Species* 38:2–8.
Pines, M., L. Sapir-Hen, and O. Tal
2017 Crusader Diet in Times of War and Peace: Arsur (Israel) as a Case Study. *Oxford Journal of Archaeology* 36:307–328.
Polcaro, A., J. Muñiz, V. Alvarez, and S. Mogliazza
2014 Dolmen 317 and Its Hidden Burial: An Early Bronze Age I Megalithic Tomb from Jebel al-Mutawwaq (Jordan). *Bulletin of the American Schools of Oriental Research* 372:1–17.
Politis, G. G.
2007 *Nukak: Ethnoarchaeology of an Amazonian People*. Translated by B. Alberti. Left Coast Press, Walnut Creek, California.
Politis, G. G. and N. J. Saunders
2002 Archaeological Correlates of Ideological Activity: Food Taboos and Spirit-Animals in an Amazonian Hunter-Gatherer Society. In *Consuming Passions and Patterns of Consumption*, edited by C. P. T. Miracle and N. Milner, pp. 113–130. McDonald Institute for Archaeological Research, Cambridge.
Pollock, S. M.
1999 *Ancient Mesopotamia: The Eden That Never Was*. Cambridge University Press, Cambridge.
Pond, W. G. and H. J. Mersmann
2001 General Characteristics. In *Biology of the Domestic Pig*, edited by W. G. Pond and H. J. Mersmann, pp. 1–40. Cornell University Press, Ithaca.
Porter, A.
2002 The Dynamics of Death: Ancestors, Pastoralism, and the Origins of a Third-Millennium City in Syria. *Bulletin of the American Schools of Oriental Research* 325:1–36.
Portes, A. and W. Haller
2005 The Informal Economy. In *The Handbook of Economic Sociology*, edited by N. J. Smelser and R. Swedberg, pp. 403–425. Princeton University Press, Princeton.
Postgate, N.
1992 *Early Mesopotamia: Society and Economy at the Dawn of History*. Routledge, London.
Postgate, N., T. Wang, and T. Wilkinson
1995 The Evidence for Early Writing: Utilitarian or Ceremonial? *Antiquity* 69:459–480.
Price, M. D.
2016 Pigs and Power: Pig Husbandry in Northern Mesopotamia During the Emergence of Social Complexity (6500–2000 BC). PhD dissertation, Department of Anthropology, Harvard University, Cambridge.
In press Pigs in Between: Pig Husbandry in the Late Neolithic in Northern Mesopotamia. In *Archaeozoology of the Near East*, Vol. XIII, edited by J. Daujat, A. Hadjikoumis, R. Berthon, J. Chahoud, L. Kassianidou, and J.-D. Vigne. Lockwood Press, Atlanta.
Price, M. D. and B. S. Arbuckle
2015 Early Domestic Pigs in the Zagros Flanks: Reanalysis of the Fauna from Neolithic Jarmo, Northern Iraq. *International Journal of Osteoarchaeology* 25:441–453.

Price, M. D., M. Buckley, Y. M. Rowan, and M. Kersel
2013 Animal Management Strategies During the Chalcolithic in the Lower Galilee: New Data from Marj Rabba. *Paléorient* 39:183–200.
Price, M. D., M. S. Chesson, and C. Makarewicz
2018 Domestic Animal Production and Consumption at Tall al-Handaquq South (Jordan) in the Early Bronze III. *Paléorient* 44:75–91.
Price, M. D. and A. Evin
2019 Long-Term Morphological Changes and Evolving Human-Pig Relations in the Northern Fertile Crescent from 11,000 to 2000 cal. BC. *Archaeological and Anthropological Sciences* 11:237–251.
Price, M. D., K. Grossman, and T. Paulette
2017 Pigs and the Pastoral Bias: The Other Animal Economy in Northern Mesopotamia (3000–2000 BCE). *Journal of Anthropological Archaeology* 48:46–62.
Price, M. D. and H. Hongo
In press The Archaeology of Pig Domestication: Methods, Models, and Case Studies. *Journal of Archaeological Research*, https://doi.org/10.1007/s10814-019-09142-9.
Pugliese, C., G. Madonia, V. Chiofalo, S. Margiotta, A. Acciaioli, and G. Gandini
2003 Comparison of the Performances of Nero Siciliano Pigs Reared Indoors and Outdoors: 1. Growth and Carcass Composition. *Meat Science* 65:825–831.
Raban-Gerstel, N. and G. Bar-Oz
2010 The Faunal Remains. In *An Early Pottery Neolithic Occurrence at Beisamoun, the Hula Valley, Northern Israel*, edited by D. Rosenberg, pp. 97–104. British Archaeological Reports International Series 2095, Oxford.
Raban-Gerstel, N., G. Bar-Oz, and Y. Tepper
2011 The Bone and Horn Industry in Late Ottoman Nazareth: The Evidence from Shihad ad-Din. *'Atiqot* 67:61–80.
Raban-Gerstel, N., I. Zohar, G. Bar-Oz, I. Sharon, and A. Gilboa
2008 Early Iron Age Dor (Israel): A Faunal Perspective. *Bulletin of the American Schools of Oriental Research* 349:25–59.
Rabinovich, R.
2003 The Levantine Upper Palaeolithic Faunal Record. In *More than Meets the Eye: Studies on Upper Palaeolithic Diversity in the Near East*, edited by A. N. Goring-Morris and A. Belfer-Cohen, pp. 33–48. Oxbow Books, Oxford.
Rabinovich, R. and R. Biton
2011 The Early-Middle Pleistocene Faunal Assemblages of Gesher Benot Ya'aqov: Inter-Site Variability. *Journal of Human Evolution* 60:357–374.
Rabinovich, R. and D. Nadel
2005 Broken Mammal Bones: Taphonomy and Food Sharing at the Ohalo II Submerged Prehistoric Camp. In *Archaeolozoology of the Near East*, Vol. VI, edited by H. Buitenhuis, A. M. Choyke, L. Martin, L. Bartosiewicz, and M. Mashkour, pp. 34–50. ARC-Publicaties 123, Groningen.
Rabinovich, R. and E. Tchernov
1995 Chronological, Paleoecological and Taphonomical Aspects of the Middle Paleolithic Site of Qafzeh, Israel. In *Archaeozoology of the Near East*, Vol. II, edited by H. Buitenhuis and H.-P. Uerpmann, pp. 5–44. Backhuys, Leiden.
Radcliffe-Brown, A. R.
1939 *Taboo*. Cambridge University Press, Cambridge.

Ramos-Onsins, S. E., W. Burgos-Paz, A. Manunza, and M. Amills
2014 Mining the Pig Genome to Investigate the Domestication Process. *Heredity* 113:471–484.
Rappaport, R. A.
1968 *Pigs for the Ancestors*. Yale University Press, New Haven.
Reade, J. E.
1973 Tell Taya (1972–1973): Summary Report. *Iraq* 35:155–188.
Redding, R.
1991 The Role of the Pig in the Subsistence System of Ancient Egypt: A Parable of Faunal Data. In *Animal Use and Culture Change*, edited by P. J. Crabtree and K. A. Ryan, pp. 20–30. MASCA Research Papers in Science and Archaeology 8. University of Pennsylvania Museum of Archaeology and Anthropology, Philadelphia.
1992 Egyptian Old Kingdom Patterns of Animal Use and the Value of Faunal Data in Modeling Socioeconomic Systems. *Paléorient* 18:99–107.
2010 Status and Diet at Giza. In *Anthropological Approaches to Zooarchaeology: Complexity, Colonialism, and Animal Transformations*, edited by D. Campana, P. J. Crabtree, S. D. deFrance, J. S. Lev-Tov, and A. M. Choyke, pp. 65–75. Oxbow Books, Oxford.
2015 The Pig and the Chicken in the Middle East: Modeling Human Subsistence Behavior in the Archaeological Record Using Historical and Animal Husbandry Data. *Journal of Archaeological Research* 23:325–368.
Redding, R. and M. Rosenberg
1998 Ancestral Pigs: A New (Guinea) Model for Pig Domestication in the Middle East. In *Ancestors for the Pigs*, edited by S. Nelson, pp. 65–76. MASCA Research Papers in Science and Archaeology 15. University of Pennsylvania Museum of Archaeology and Anthropology, Philadelphia.
Reese, D. S.
1997 The Animal Bones and Shells. In *Ancient Naukratis: Excavations at a Greek Emporium in Egypt: I. The Excavations at Kom Ge'if*, edited by A. Leonard, Jr., pp. 347–349. American Schools of Oriental Research, Boston.
Reich, D.
2018 *Who We Are and How We Got Here: Ancient DNA and the New Science of the Human Past*. Pantheon Books, New York.
Reichel-Dolmatoff, G.
1985 Tapir Avoidance in the Colombian Northwest Amazon. In *Animal Myths and Metaphors in South America*, edited by G. Urton, pp. 107–143. University of Utah Press, Salt Lake City.
Rhyder, J.
2019 *Centralizing the Cult: The Holiness Legislation in Leviticus 17-26*. Mohr Siebeck, Tübingen.
Richardson, S.
2007 Death and Dismemberment in Mesopotamia: Discorporation between the Body and Body Politic. In *Performing Death: Social Analysis of Funerary Traditions in the Ancient Near East and Mediterranean*, edited by N. Laneri, pp. 189–208. Oriental Institute of the University of Chicago, Chicago.
Ricotti, E. S. P.
2015 *Meals and Recipes from Ancient Greece*. Translated by R. A. Lotero. J. Paul Getty Museum, Los Angeles.

Riehl, S.
2008 Climate and Agriculture in the Ancient Near East: A Synthesis of the Archaeobotanical and Stable Carbon Isotope Evidence. *Vegetation History and Archaeobotany* 17:S43–S51.
Rindos, D.
1984 *The Origins of Agriculture: An Evolutionary Perspective*. Academic Press, New York.
Ristvet, L.
2012 The Development of Underdevelopment? Imperialism, Economic Exploitation and Settlement Dynamics on the Khabur Plains, ca. 2300–2200 BC. In *Seven Generations Since the Fall of Akkad*, edited by H. Weiss, pp. 241–260. Harrassowitz, Wiesbaden.
2014 *Ritual, Performance, and Politics in the Ancient Near East*. Cambridge University Press, Cambridge.
Rodinson, M.
1999 Ghida. In *Encyclopedia of Islam*, 2nd ed., edited by P. Bearman, T. Bianquis, C. E. Bosworth, E. van Donzel, and W. P. Heinrichs, pp. 1057–1072. Brill, Leiden.
Rollefson, G. O.
1989 The Aceramic Neolithic of the Southern Levant: The View From 'Ain Ghazal. *Paléorient* 15:135–140.
Rollefson, G. O. and I. Kohler-Rollefson
1992 Early Neolithic Exploitation Patterns in the Levant: Cultural Impact on the Environment. *Population and Environment* 13:243–254.
1993 PPNC Adaptations in the First Half of the 6th Millennium B.C. *Paléorient* 19:33–42.
Romanes, G. J.
1883 *Animal Intelligence*. D. Appleton and Company, New York.
Rooijakkers, C. T.
2012 Spinning Animal Fibres at Late Neolithic Tell Sabi Abyad, Syria? *Paléorient* 38:93–109.
Roosevelt, C. H.
2012 *The Archaeology of Lydia, from Gyges to Alexander*. Cambridge University Press, Cambridge.
Root, D.
1988 Speaking Christian: Orthodoxy and Difference in Sixteenth-Century Spain. *Representations* 23:118–134.
Rosen, A. M.
2007 *Civilizing Climate: Social Responses to Climate Change in the Ancient Near East*. Altamira Press, Lanham, Maryland.
2010 Natufian Plant Exploitation: Managing Risk and Stability in an Environment of Change. *Eurasian Prehistory* 7:113–127.
Rosenberg, M.
1994 Hallan Çemi Tepesi: Some Further Observations Concerning Stratigraphy and Material Culture. *Anatolica* 20:121–141.
Rosenberg, M. and M. Davis
1992 Hallan Çemi Tepesi, an Early Aceramic Neolithic Site in Eastern Anatolia: Some Preliminary Observations Concerning Material Culture. *Anatolica* 18: 1–18.

Rosenberg, M. and R. Redding
1998 Early Pig Husbandry in Southwestern Asia and Its Implications for Modeling the Origins of Food Production. In *Ancestors for the Pigs*, edited by S. Nelson, pp. 55–64. MASCA Research Papers in Science and Archaeology 15. University of Pennsylvania Museum of Archaeology and Anthropology, Philadelphia.
2000 Hallan Çemi and Early Village Organization in Eastern Anatolia. In *Life in Neolithic Farming Communities: Social Organization, Identity and Differentiation*, edited by I. Kuijt, pp. 39–61. Kluwer Academic / Plenum Publishers, New York.
Rosenberg, S. G.
2004 The Jewish Temple at Elephantine. *Near Eastern Archaeology* 67:4–13.
Rosenblum, J. D.
2010a *Food and Identity in Early Rabbinic Judaism*. Cambridge University Press, Cambridge.
2010b "Why Do You Refuse to Eat Pork?": Jews, Food, and Identity in Roman Palestine. *Jewish Quarterly Review* 100:95–110.
2016 *The Jewish Dietary Laws in the Ancient World*. Cambridge Univeristy Press, Cambridge.
Rossel, S., F. Marshall, J. Peters, T. Pilgram, M. D. Adams, and D. O'Connor
2008 Domestication of the Donkey: Timing, Processes, and Indicators. *Proceedings of the National Academy of Sciences* 105:3715–3720.
Roth, M. T.
1980 The Scholastic Exercise "Laws About Rented Oxen." *Journal of Cuneiform Studies* 32:127–146.
Rousseau, G., C. Guintard, and C. Abadie-Reynal
2008 La Gestion des animaux à Zeugma (Turquie): Étude des restes fauniques du Chantier 9 (Époques hellénistique, romaine, byzantine et islamique). *Revue de Médecine Vétérinaire* 159:251–275.
Rowan, Y. M. and J. Golden
2009 The Chalcolithic Period of the Southern Levant: A Synthetic Review. *Journal of World Prehistory* 22:1–92.
Rowley-Conwy, P.
2011 Westward Ho! The Spread of Agriculture from Central Europe to the Atlantic. *Current Anthropology* 52:S431–S451.
Rowley-Conwy, P., U. Albarella, and K. Dobney
2012 Distinguishing Wild Boar from Domestic Pigs in Prehistory: A Review of Approaches and Recent Results. *Journal of World Prehistory* 25:1–44.
Rowley-Conwy, P. and K. Dobney
2007 Wild Boar and Domestic Pigs in Mesolithic and Neolithic Southern Scandinavia. In *Pigs and Humans: 10,000 Years of Interaction*, edited by U. Albarella, K. Dobney, A. Ervynck, and P. Rowley-Conwy. Oxford University Press, Oxford.
Rozin, P., J. Haidt, C. McCauley, and S. Imada
1997a Disgust: The Cultural Evolution of a Food-Based Emotion. In *Food Preference and Taste: Continuity and Change*, edited by H. Macbeth, pp. 65–82. Berghahn Books, New York.
Rozin, P., M. Markwith, and C. Stoess
1997b Moralization and Becoming a Vegetarian: The Transformation of Preferences into Values and the Recruitment of Disgust. *Psychological Science* 8:67–73.

Rubin, U.
1997 Apes, Pigs, and the Islamic Identity. In *Dhimmis and Others: Jews and Christians in the World of Classical Islam*, edited by U. Rubin and D. J. Wasserstein, pp. 89–105. Eisenbrauns, Winona Lake, Indiana.
Rufolo, S. J.
2011 Specialized Pastoralism and Urban Process in Third Millennium BC Northern Mesopotamia. PhD dissertation, Department of Near Eastern Studies, Johns Hopkins University, Baltimore.
Russell, A.
2010 Retracing the Steppes: A Zooarchaeological Analysis of Changing Subsistence Patterns in the Late Neolithic at Tell Sabi Abyad, Northern Syria, c. 6900 to 5900 BC. PhD dissertation, Department of Archaeology, Leiden University, Leiden.
Russell, N.
2012 *Social Zooarchaeology: Humans and Animals in Prehistory*. Cambridge University Press, Cambridge.
Safrai, Z.
1994 *The Economy of Roman Palestine*. Routledge, New York.
Sallaberger, W.
2014 The Value of Wool in Early Bronze Age Mesopotamia: On the Control of Sheep and the Handling of Wool in the Presargonic to the Ur III Periods (c. 2400–2000 BC). In *Wool Economy in the Ancient Near East and the Aegean: From the Beginnings of Sheep Husbandry to Institutional Textile Industry*, edited by C. Breniquet and C. Michel, pp. 94–114. Oxbow Books, Oxford.
Salleberger, W. and A. Pruß
2015 Home and Work in Early Bronze Age Mesopotamia: "Ration Lists" and "Private Houses" at Tell Beydar/Nabada. In *Labor in the Ancient World*, edited by P. Steinkeller and M. Hudson, pp. 69–136. ISLET, Dresden.
Sapir-Hen, L.
2017 Pax Assyriaca and the Animal Economy in the Southern Levant: Regional and Local-Scale Imperial Contacts. In *Rethinking Israel: Studies in the History and Archaeology of Ancient Israel in Honor of Israel Finkelstein*, edited by O. Lipschits, Y. Gadot, and M. J. Adams, pp. 341–353. Eisenbrauns, Winona Lake, Indiana.
2018 Food, Pork Consumption, and Identity in Ancient Israel. *Near Eastern Archaeology* 81:52–59.
Sapir-Hen, L., G. Bar-Oz, Y. Gadot, and I. Finkelstein
2013 Pig Husbandry in Iron Age Israel and Judah: New Insights Regarding the Origin of the 'Taboo.' *Zeitschrift des Deutschen Palästine-Vereins* 129: 1–20
Sapir-Hen, L., G. Bar-Oz, I. Sharon, A. Gilboa, and T. Dayan
2014 Food, Economy, and Culture at Tel Dor, Israel: A Diachronic Study of Faunal Remains from 15 Centuries of Occupation. *Bulletin of the American Schools of Oriental Research* 371:83–101.
Sapir-Hen, L., Y. Gadot, and I. Finkelstein
2016 Animal Economy in a Temple City and Its Countryside: Iron Age Jerusalem as a Case Study. *Bulletin of the American Schools of Oriental Research* 375:103–118.
Sapir-Hen, L., M. Meiri, and I. Finkelstein
2015 Iron Age Pigs: New Evidence on Their Origin and Role in Forming Identity Boundaries. *Radiocarbon* 57:307–315.

Sapir-Hen, L., A. Sasson, A. Kleiman, and I. Finkelstein
2016 Social Stratification in the Late Bronze and Early Iron Ages: An Intra-Site Investigation at Megiddo. *Oxford Journal of Archaeology* 35:47–73.
Sasson, A.
2010 *Animal Husbandry in Ancient Israel.* Equinox, London.
Scarborough, J.
1982 Beans, Pythagoras, Taboos, and Ancient Dietetics. *Classical World* 75:355–358.
Schacher, I.
1974 *The "Judensau": A Medieval Anti-Jewish Motif and Its History.* Warburg Institute, University of London, London.
Schäfer, P.
1997 *Judeophobia: Attitudes Toward the Jews in the Ancient World.* Harvard University Press, Cambridge.
Schieffelin, E. L.
2005 [1976] *The Sorrow of the Lonely and the Burning of the Dancers.* 2nd ed. Palgrave MacMillan, New York.
Schimmel, A.
1985 *And Muhammad Is His Messenger: The Veneration of the Prophet in Islamic Piety.* University of North Carolina Press, Chapel Hill.
Schley, L., M. Dufrêne, A. Krier, and A. C. Frantz
2008 Patterns of Crop Damage by Wild Boar (*Sus scrofa*) in Luxembourg over a 10-Year Period. *European Journal of Wildlife Research* 54:589–599.
Schley, L. and T. J. Roper
2003 Diet of Wild Boar *Sus scrofa* in Western Europe, with Particular Reference to Consumption of Agricultural Crops. *Mammal Review* 33:43–56.
Schloen, J. D.
2001 *The House of the Father as Fact and Symbol: Patrimonialism in Ugarit and the Ancient Near East.* Eisenbrauns, Winona Lake, Indiana.
Schmidt, K.
2000 Göbekli Tepe, Southeastern Turkey: A Preliminary Report on the 1995–1999 Excavations. *Paléorient* 26:45–54.
Schorsch, J.
2018 *The Food Movement, Culture, and Religion: A Tale of Pigs, Christians, Jews, and Politics.* New York, Springer.
Schwartz, G. M.
2013 Memory and Its Demolition: Ancestors, Animals and Sacrice at Umm el-Marra, Syria. *Cambridge Archaeological Journal* 23:495–522.
Schwartz, G. M., C. D. Brinker, A. T. Creekmore, M. H. Feldman, A. Smith, and J. A. Weber
2017 Excavations at Kurd Qaburstan, a Second Millennium B.C. Urban Site on the Erbil Plain. *Iraq* 79:213–255.
Schwartz, G. M., H. H. Curvers, and B. Stuart
2000 A Third Millennium B.C. Elite Tomb from Tell Umm el-Marra, Syria. *Antiquity* 74:771–772.
Schwartz, S.
2004 *Imperialism and Jewish Society: 200 B.C.E. to 640 C.E.* Princeton University Press, Princeton.
2014 *The Ancient Jews from Alexander to Muhammad.* Cambridge University Press, Cambridge.

Schwemer, D.
2007 Witchcraft and War: The Ritual Fragment Ki 1904-10-9, 18 (BM 98989). *Iraq* 69:29–42.
Scott, J. C.
1998 *Seeing Like a State: How Certain Schemes to Improve the Human Condition Have Failed.* Yale University Press, New Haven.
2017 *Against the Grain: A Deep History of the Earliest States.* Yale University Press, New Haven.
Scurlock, J.
2002 Animals in Ancient Mesopotamian Religion. In *A History of the Animal World in the Ancient Near East*, edited by B. J. Collins, pp. 361–387. Brill, Leiden.
Seeskin, K.
2017 Maimonides. In *Stanford Encyclopedia of Philosophy*, edited by E. N. Zalta. https://plato.stanford.edu/archives/spr2017/entries/maimonides/.
Serpell, J.
1996 [1986] *In the Company of Animals: A Study of Human-Animal Relationships.* Cambridge University Press, Cambridge.
Shaw, I.
1984 Report on the 1983 Excavations: The Animal Pens (Building 400). In *Amarna Reports I*, edited by B. J. Kemp, pp. 40–56. Egypt Exploration Society, London.
Shea, J. J.
2006 The Origins of Lithic Projectile Point Technology: Evidence from Africa, the Levant, and Europe. *Journal of Archaeological Science* 33:823–846.
Sherratt, A.
1981 Plough and Pastoralism: Aspects of the Secondary Products Revolution. In *Patterns of the Past: Studies in Honour of David Clarke*, edited by I. Hodder, G. Isaac, and N. Hammond, pp. 261–305. Cambridge University Press, Cambridge.
1983 The Secondary Products Revolution of Animals in the Old World. *World Archaeology* 15:90–104.
Shipman, P.
2010 The Animal Connection and Human Evolution. *Current Anthropology* 51 519–538.
Shirres, M. P.
1982 Tapu. *Journal of the Polynesian Society* 91:29–51.
Silibolatlaz-Baykara, D.
2012 Faunal Studies on Byzantine City of the Amorium. *Ankara University Journal of the Faculty of Letters* 52:71–82.
Silliman, S. W.
2013 What, Where and When Is Hybridity? In *The Archaeology of Hybrid Material Culture*, edited by J. Card, pp. 486–500. Occasional Paper No. 39. Center for Archaeological Investigations, Southern Illinois University, Carbondale.
Sillitoe, P.
2007 Pigs in the New Guinea Highlands: An Ethnographic Example. In *Pigs and Humans: 10,000 Years of Interaction*, edited by U. Albarella, K. Dobney, A. Ervynck, and P. Rowley-Conwy, pp. 330–358. Oxford University Press, Oxford.
Simmons, A.
1988 Extinct Pygmy Hippopotamus and Early Man in Cyprus. *Nature* 333:554–557.

Simoons, F. J.
1994 *Eat Not This Flesh: Food Avoidances in the Old World.* 2nd ed. University of Wisconsin Press, Madison.
Ska, J. L.
2006 *Introduction to Reading the Pentateuch.* Translated by P. Dominique. Eisenbrauns, Winona Lake, Indiana.
Slackman, M.
2009a Belatedly, Egypt Spots Flaws in Wiping Out Pigs. *New York Times*, 19 September. New York.
2009b Cleaning Cairo, but Taking a Livelihood. *New York Times*, 24 May. New York.
Slim, F., C. Çakırlar, C. H. Roosevelt 2020 Pigs in Sight: Late Bronze Age Pig Husbandries in the Aegean and Anatolia. *Journal of Field Archaeology* 45:315–333.
Smith, B. D. 2007 The Ultimate Ecosystem Engineers. *Science* 315:1797–1798.
2011 General Patterns of Niche Construction and the Management of "Wild" Plant and Animal Resources by Small-Scale Pre-Industrial Societies. *Philosophical Transactions of the Royal Society B* 366:836–848.
Smith, S. T.
2003 *Wretched Kush: Ethnic Identities and Boundaries in Egypt's Nubian Empire.* Routledge, New York.
Snir, A., D. Nadel, I. Groman-Yaroslavski, Y. Melamed, M. Sternberg, O. Bar-Yosef, E. Weiss
2015 The Origin of Cultivation and Proto-Weeds Long Before Neolithic Farming. *PLoS ONE* 10:e0131422.
Soffer, O.
2004 Recovering Perishable Technologies through Use Wear on Tools: Preliminary Evidence for Upper Paleolithic Weaving and Net Making. *Current Anthropology* 45:407–413.
Soler, J.
1979 The Semiotics of Food in the Bible. In *Food and Drink in History: Selections from the Annales, Économies, Sociétés, Civilisations*, Vol. 5, edited by R. Forster and O. Ranum, pp. 126–138. Translated by E. Forster and P. M. Ranum. Johns Hopkins University Press, Baltimore.
Spaulding, J. and J. L. Spaulding
1988 The Democratic Philosophers of the Medieval Sudan: The Pig. *Sprache und Geschichte in Afrika* 9:247–268.
Speth, J. D.
2009 *The Paleoanthropology and Archaeology of Big-Game Hunting: Protein, Fat, or Politics?* Springer, New York.
2012 Middle Palaeolithic Subsistence in the Near East: Zooarchaeological Perspectives—Past, Present and Future. *Before Farming* 2012:1–45.
Speth, J. D. and E. Tchernov
1998 The Role of Hunting and Scavenging in Neanderthal Procurement Strategies: New Evidence from Kebara Cave (Israel). In *Neanderthals and Modern Humans in Western Asia*, edited by T. Akazawa, A. Kenichi, and O. Bar-Yosef, pp. 223–240. Kluwer Academic, New York.
Spiciarich, A., Y. Gadot, and L. Sapir-Hen
2017 The Faunal Evidence from Early Roman Jerusalem: The People Behind the Garbage. *Tel Aviv* 44:98–117.

Spinka, M.
2009 Behaviour of Pigs. In *The Ethology of Domestic Animals*, 2nd ed., edited by P. Jensen, pp. 177–191. CAB International, Oxfordshire.
Stager, L. E.
1985 The Archaeology of the Family in Ancient Israel. *Bulletin of the American Schools of Oriental Research* 260:1–35.
Stampfli, H. R.
1983 The Fauna of Jarmo with Notes on Animal Bones from Matarrah, the Amuq, and Karim Shahir. In *Prehistoric Archeology Along the Zagros Flanks*, edited by L. S. Braidwood, R. J. Braidwood, B. Howe, C. A. Reed, and P. J. Watson, pp. 431–484. Oriental Institute of the University of Chicago, Chicago.
Starkovich, B. M. and M. C. Stiner
2009 Hallan Çemi Tepesi: High-Ranked Game Exploitation Alongside Intensive Seed Processing at the Epipaleolithic-Neolithic Transition in Southeastern Turkey. *Paléorient* 4:41–61.
Steadman, S. R., G. McMahon, J. C. Ross, M. Cassis, T. E. Şerifoğlu, B. S. Arbuckle, S. E. Adcock, S. A. Roodenberg, M. von Baeyer, and A. J. Lauricella
2015 The 2013 and 2014 Excavation Seasons at Çadır Höyük on the Anatolian North Central Plateau. *Anatolica* 41:87–124.
Steele, D.
2002 Faunal Remains. In *Jebel Khalid on the Euphrates: Report on Excavations, 1986–1996*, edited by G. W. Clarke, P. J. O'Connor, L. Crewe, B. Frohlich, H. Jackson, J. Littleton, C. E. V. Nixon, M. O'Hea, and D. Steele, pp. 125–146. Mediterranean Archaeology Supplement 5. Meditarch, Sydney.
Stein, G. J.
1988 Pastoral Production in Complex Societies: Mid-Late Third Millennium BC and Medieval Faunal Remains from Gritille Höyük in the Karababa Basin, Southeast Turkey. PhD dissertation, Department of Anthropology, University of Pennsylvania, Philadelphia.
1999 *Rethinking World Systems: Diasporas, Colonies and Interaction in Uruk Mespotamia.* University of Arizona Press, Tucson.
2004 Structural Parameters and Sociocultural Factors in the Economic Organization of North Mesopotamian Urbanism in the Third Millennium B. C. In *Archaeological Perspectives on Political Economies*, edited by G. Feinman and L. M. Nicholas, pp. 61–78. University of Utah Press, Salt Lake City.
2012a The Development of Indigenous Social Complexity in Late Chalcolithic Upper Mesopotamia in the 5th–4th Millennia BC: An Initial Assessment. *Origini* 34:125–151.
2012b Food Preparation, Social Context, and Ethnicity in a Prehistoric Mesopotamian Colony. In *The Menial Art of Cooking: Archaeological Studies of Cooking and Food Preparation*, edited by S. R. Graff and E. Rodriguez-Alegria, pp. 47–63. University of Colorado Press, Boulder.
Stepien, M.
1996 *Animal Husbandry in the Ancient Near East: A Prosopographic Study of Third-Millennium Umma.* Capital Decisions Ltd, Bethesda.
Stiner, M. C.
2009 Prey Choice, Site Occupation Intensity and Economic Diversity in the Middle-Early Upper Palaeolithic at the Üçağızlı Caves, Turkey. *Before Farming* 2009:1–20.

Stiner, M. C., A. Gopher, and R. Barkai
2011 Hearth-Side Socioeconomics, Hunting and Paleoecology During the Late Lower Paleolithic at Qesem Cave, Israel. *Journal of Human Evolution* 60:213–233.
Stiner, M. C. and S. L. Kuhn
2016 Are We Missing the "Sweet Spot" between Optimality Theory and Niche Construction Theory in Archaeology? *Journal of Anthropological Archaeology* 44:177–184.
Stiner, M. C. and E. Tchernov
1998 Pleistocene Species Trends at Hayonim Cave: Changes in Climate versus Human Behavior. In *Neanderthals and Modern Humans in Western Asia*, edited by T. Akazawa, K. Aoki, and O. Bar-Yosef, pp. 241-262. Kluwer Academic, New York.
Studer, J.
2002 Dietary Differences at Ez Zantur Petra, Jordan (1st Century BC–AD 5th Century). In *Archaeozoology of the Near East*, Vol. V, edited by H. Buitenhuis, A. Choyke, M. Mashkour, and A. H. al-Shiyab, pp. 273–281. Rijksuniversitit, Groningen.
2007 Animal Exploitation in the Nabataean World. In *The World of the Nabataeans. Volume 2 of the International Conference The World of the Herods and the Nabataeans Held at the British Museum, 17–19 April 2001*, edited by K. D. Politis, pp. 251–272. Franz Steiner Verlag, Stuttgart.
Studnitz, M., M. B. Jensen, and L. J. Pedersen
2007 Why Do Pigs Root and in What Will They Root? A Review on the Exploratory Behaviour of Pigs in Relation to Environmental Enrichment. *Applied Animal Behaviour Science* 107:183–197.
Styring, A. K., M. Charles, F. Fantone, M. M. Hald, A. McMahon, R. H. Meadow, G. K. Nicholls, A. K. Patel, M. C. Pitre, A. Smith, A. Sołtysiak, G. Stein, J. A. Weber, H. Weiss, A. Bogaard
2017 Isotope Evidence for Agricultural Extensification Reveals How the World's First Cities Were Fed. *Nature Plants* 3:17076.
Sussman, L. J.
2005 The Myth of the Trefa Banquet: American Culinary Culture and the Radicalization of Food Policy in American Reform Judaism. *American Jewish Archives Journal* 57:29–52.
Sykes, N.
2015 *Beastly Questions: Animal Answers to Archaeological Issues*. Bloomsbury, London.
Talmage, F.
1972 *The Book of the Covenant of Joseph Kimḥi*. Pontifical Institute of Mediaeval Studies, Toronto.
Tambiah, S. J.
1969 Animals Are Good to Think and Good to Prohibit. *Ethnology* 8:423–459.
TAVO Map A IV 4 1984 Middle East. Mean Annual Rainfall and Variability. *Tübinger Atlas des Vorderen Orients*. Dr. Ludwig Reichert Verlag, Wiesbaden.
TAVO Map A VI 2 1990 Middle East. Early Holocene Vegetation (c. 8000 B.P.). *Tübinger Atlas des Vorderen Orients*. Dr. Ludwig Reichert Verlag, Wiesbaden.
Taxel, I., A. Glick, and M. Pines
2017 Majdal Yābā: More Insights on the Site in Medieval and Late Ottoman to Mandatory Times. *Journal of Islamic Archaeology* 4:49–86.
Taylor, R. B., E. C. Hellgren, T. M. Gabor, and L. M. Ilse
1998 Reproduction of Feral Pigs in Southern Texas. *Journal of Mammalogy* 79:1325–1331.

Tchernov, E.
1979 Quaternary Fauna. In *The Quaternary of Israel*, edited by A. Horowitz, pp. 257–290. Academic Press, New York.
1998 The Faunal Sequence of the Southwest Asian Middle Paleolithic in Relation to Hominid Dispersal Events. In *Neanderthals and Modern Humans in Western Asia*, edited by T. Akazawa, K. Aoki, and O. Bar-Yosef, pp. 77–90. Kluwer Academic, New York.
Tchernov, E. and F. F. Valla
1997 Two New Dogs, and Other Natufian Dogs, from the Southern Levant. *Journal of Archaeological Science* 24:65–95.
Thurston, B. J. and D. Attwater
1990 *Butler's Lives of The Saints, Complete Edition*. Christian Classics, Westminster.
Traubman, T.
2005 Israel's Pig Farms: A Picture of Inhumanity. *Haaretz*, 16 November. Tel Aviv.
Trella, P. A.
2010 Evaluating Collapse: The Disintegration of Urban-Base Societies in Third Millennium B.C. Upper Mesopotamia. PhD dissertation, Department of Anthropology, University of Virginia, Charlottesville.
Trentacoste, A.
2013 Faunal Remains from the Etruscan Sanctuary at Poggio Colla (Vicchio di Mugello). *Etruscan Studies* 16:75–105.
2016 Etruscan Foodways and Demographic Demands: Contextualizing Protohistoric Livestock Husbandry in Northern Italy. *European Journal of Archaeology* 19:279–315.
Tresset, A. and J.-D. Vigne
2007 Substitution of Species, Techniques and Symbols at the Mesolithic/Neolithic Transition in Western Europe. In *Going Over: The Mesolithic/Neolithic Transition in NW Europe*, edited by A. Whittle and V. Cummings, pp. 189–210. Proceedings of the British Academy 144. Oxford University Press, London.
Trut, L. N.
1999 Early Canid Domestication: The Farm-Fox Experiment. *American Scientist* 87:160–169.
Trut, L. N., I. N. Oskina, and A. V. Kharlamova
2009 Animal Evolution During Domestication: The Domesticated Fox as a Model. *Bioessays* 31:349–360.
Turnbull, P. F. and C. A. Reed
1974 The Fauna from the Terminal Pleistocene of Palegawra Cave, a Zarzian Occupation Site in Northeastern Iraq. *Fieldiana* 63:81–146.
Turner, V.
1967 *The Forest of Symbols: Aspects of Ndembu Ritual*. Cornell University Press, Ithaca.
Twiss, K. C.
2006 A Modified Boar Skull from Çatalhöyük. *Bulletin of the American Schools of Oriental Research* 342:1–12.
2008 Transformations in an Early Agricultural Society: Feasting in the Southern Levantine Pre-Pottery Neolithic. *Journal of Anthropological Archaeology* 27:418–442.
2012 The Archaeology of Food and Social Diversity. *Journal of Archaeological Research* 20:357–395.
2017 Animals of the Sealands: Ceremonial Activities in the Southern Mesopotamian "Dark Age." *Iraq* 79:257–267.

Uerpmann, H.-P.
2003 Environmental Aspects of Economic Changes in Troia. In *Troia and the Troad: Scientific Approaches*, edited by G. A. Wagner, E. Pernicka, and H.-P. Uerpmann, pp. 251–262. Springer-Verlag, Berlin.
Uerpmann, H.-P., M. Uerpmann, and S. A. Jasim
2000 Stone Age Nomadism in SE Arabia: Palaeo-Economic Considerations on the Neolithic Site of Al-Buhais 18 in the Emirate of Sharjah, U.A.E. *Proceedings of the Seminar for Arabian Studies* 20:229–234.
Uerpmann, M.
2017 Faunal Remains from the Islamic Fort at Luluyyah (Sharjah, UAE). *Arabian Archaeology and Epigraphy* 28:2–10.
Ur, J. A.
2010 Cycles of Civilization in Northern Mesopotamia, 4400–2000 BC. *Journal of Archaeological Research* 18:387–431.
Valeri, V.
2000 *The Forest of Taboos: Morality, Hunting, and Identity among the Huaulu of the Moluccas.* University of Wisconsin Press, Madison.
Valla, F. R.
1988 Aspects du sol de l'abri 131 de Mallaha (Eynan). *Paléorient* 14:283–296.
Van de Mieroop, M.
2007 *A History of the Ancient Near East.* 2nd ed. Blackwell, London.
van den Brink, E. C. M., L. K. Horwitz, R. Kool, N. Liphschitz, H. K. Mienis, and V. Zbenovich
2015 Excavations at Tel Lod: Remains from the Pottery Neolithic A, Chalcolithic, Early Bronze Age I, Middle Bronze Age I, and Byzantine Periods. *'Atiqot* 82:141–218.
van der Toorn, K.
1999 Magic at the Cradle. In *Mesopotamian Magic: Textual, Historical, and Interpretative Perspectives*, edited by T. Abusch and K. van den Toorn, pp. 139–147. Styx, Groningen.
van Gennep, A.
1960 *The Rites of Passage.* Routledge, London.
van Koppen, F.
2001 The Organisation of Institutional Agriculture in Mari. *Journal of the Economic and Social History of the Orient* 44:451–504.
2006 Pigs in Lower Mesopotamia During the Old Babylonian Period (2000–1600 BC). In *De la domestication au tabou*, edited by B. Lion and C. Michel, pp. 181–194. De Boccard, Paris.
Van Lerberghe, K.
1996 The Livestock. In *Administrative Documents from Tell Beydar*, edited by I. Farouk, W. Sallaberger, P. Talon, and K. Van Lerberghe, pp. 107–117. Brepols, Turnhout.
Van Neer, W.
1997 Archaeozoological Data on the Food Provisioning of Roman Settlements in the Eastern Desert of Egypt. *Archaeozoologia* 9:137–154.
Van Neer, W. and B. De Cupere
2000 Faunal Remains from Tell Beydar (Excavation Seasons 1992–1997). In *Tell Beydar: Environmental and Technical Studies*, edited by K. Van Lerberghe and G. Voet, pp. 69–115. Brepols, Turnhout.

van Zeist, W. and S. Bottema
1991 *Late Quaternary Vegetation of the Near East: Beihefte zum Tubinger Atlas des Vorderen Orients, Reihe A18*. Dr. L. Reichert Verlag, Wiesbaden.
Vanpoucke, S., B. De Cupere, and M. Waelkens
2007 Economic and Ecological Reconstruction at the Classical Site of Sagalassos, Turkey, Using Pig Teeth. In *Pigs and Humans: 10,000 Years of Interaction*, edited by U. Albarella, K. Dobney, A. Ervynck, and P. Rowley-Conwy, pp. 269–281. Oxford University Press, Oxford.
Vanpoucke, S., I. Mainland, B. De Cupere, and M. Waelkens
2009 Dental Microwear Study of Pigs from the Classical Site of Sagalassos (SW Turkey) as an Aid for the Reconstruction of Husbandry Practices in Ancient Times. *Environmental Archaeology* 14:137–154.
Vigne, J.-D.
2011a Les Suinés (Sus scrofa). In *Shillourokambos: Un établissement Néo-lithique Pré-céramique à Chypre—Les fouilles du Secteur 1*, edited by J. Guilaine, F. Briois, and J.-D. Vigne, pp. 919–969. École Française d'Athènes, Paris.
2011b The Origins of Animal Domestication and Husbandry: A Major Change in the History of Humanity and the Biosphere. *Comptes Rendus Biologies* 334:171–181.
2015 Early Domestication and Farming: What Should We Know or Do for a Better Understanding? *Anthropozoologica* 50:123–150.
Vigne, J.-D., F. Briois, A. Zazzo, G. Willcox, T. Cucchi, S. Thiébault, I. Carrère, Y. Franel, R. Touquet, C. Martin, C. Moreau, C. Comby, and J. Guilaine
2012 First Wave of Cultivators Spread to Cyprus at Least 10,600 y Ago. *Proceedings of the National Academy of Sciences* 109:8445–8449.
Vigne, J.-D., I. Carrère, F. Briois, and J. Guilaine
2009a The Early Process of Mammal Domestication in the Near East: New Evidence from the Pre-Neolithic and Pre-Pottery Neolithic in Cyprus. *Current Anthropology* 52:S255–S271.
Vigne, J.-D., A. Zazzoa, J.-F. Saliège, F. Poplin, J. Guilaine, and A. Simmons
2009b Pre-Neolithic Wild Boar Management and Introduction to Cyprus More than 11,400 Years Ago. *Proceedings of the National Academy of Sciences* 106: 16135–16138.
Vila, E.
1995 Analyse de la faune des secteurs nord et sud du Steinbau I. In *Ausgrabungen in Tell Chuēra in Nordost Syrien I: Vorbericht über die Grabungskampagnen 1986 bis 1992*, edited by W. Orthmann, pp. 267–279. Saarbrücker Druckerei, Saarbrücker.
1998 *L'Exploitation des animaux en Mésopotamie au IVème et IIIème millénaires avant J.-C. (Monographie du CRA 21)*. CNRS Editions, Paris.
2005 Analyse archaéozoologique de la faune de Tell Shiukh Fawqani. In *Tell Shiukh Fawqani: 1994–1998*, Vol. II, edited by L. Bachelot and F. M. Fales, pp. 1081–1108. S.A.R.G.O.N. Editrice e Libreria, Padova.
2006 Données archéozoologiques sur les suidés de la periode Halaf a l'Âge du fer. In *De la domestication au tabou*, edited by B. Lion and C. Michel, pp. 137–153. De Boccard, Paris.
2010 Étude de la faune mammalienne de Tell Chuera, Secteurs H et K (2000–2007) et de Kharab Sayyar, Secteur A (Bronze ancien, Syrie). In *Ausgrabungen auf dem Tell Chuēra in Nordost-Syrien: II. Vorbericht zu den Grabungsgampagnen 1998 bis 2005*,

edited by J.-W. Meyer, pp. 223–292. Vorderasiatische Forschungen der Max Freiherr von Oppenheim-Stiftung. Harrassowitz, Wiesbaden.

Vila, E. and A.-S. Dalix
2004 Alimentation et idéologie: La Place du sanglier et du porc à l'Âge du bronze sur la côte levantine. *Anthropozoologica* 39:219–236.

Vila, E. and D. Helmer
2014 The Expansion of Sheep Herding and the Development of Wool Production in the Ancient Near East: An Archaeozoological and Iconographical Approach. In *Wool Economy in the Ancient Near East and the Aegean: From the Beginnings of Sheep Husbandry to Institutional Textile Industry*, edited by C. Breniquet and C. Michel, pp. 22–40. Oxbow Books, Oxford.

von den Driesch, A.
1993 Faunal Remains from Habuba Kabira in Syria. In *Archaeozoology of the Near East*, edited by H. Buitenhuis and A. Clason, pp. 52–59. Universal Book Services, Leiden.
1997 Tierreste aus Buto im Nildelta. *Archaeofauna* 6:23–39.

von den Driesch, A. and J. Boessneck
1981 *Reste von Haus—und Jagdtieren aus der Unterstadt von Boğazköy-Hattuša: Grabungen 1958-1977*. Gebr. Mann Verlag, Berlin.

von den Driesch, A. and A. Dockner
2002 Animal Exploitation in Medieval Siraf, Iran, Based on the Faunal Remains from the Excavations at the Great Mosque (Seasons 1966–1973). *Bonner Zoologische Beiträge* 50:227–247.

Waetzoldt, H.
1972 *Untersuchungen zur neusumerischen Textilindustrie*. Studi Economici e Tecnologici. Centro per la Antichita e la Storia dell'Arte Vicino Oriente, Rome.

Wallace, S. L.
1938 *Taxation in Egypt from Augustus to Diocletian*. Princeton University Press, Princeton.

Wallace, W. C.
1844 Jewish Hygiene. *Boston Medical and Surgical Journal* 31:309–316.

Walmsley, A. G.
1988 Pella/Fiḥl after the Islamic Conquest (AD 635–c.900): A Convergence of Literary and Archaeological Evidence. *Mediterranean Archaeology* 1:142–159.

Walmsley, A. G., P. G. Macumber, P. C. Edwards, S. J. Bourke, P. M. Watson, R. V. S. Wright, B. Churcher, R. Sparks, K. Rielly, K. da Costa, and M. O'Hea
1993 The Eleventh and Twelfth Seasons of Excavations at Pella (Tabaqat Fahl), 1989–1990. *Annual of the Department of Antiquities of Jordan* 37:165–231.

Walton, J. T.
2015 The Regional Economy of the Southern Levant in the 8th–7th Centuries BCE. PhD dissertation, Department of Near Eastern Language and Civilizations, Harvard University, Cambridge.

Wapnish, P. and D. N. Fulton
2018 Animal Remains from Middle Bronze Age Ashkelon. In *Final Reports of the Leon Levy Expedition to Ashkelon: 6. The Middle Bronze Age Ramparts and Gates of the North Slope and Later Fortifications*, edited by L. E. Stager, J. D. Schloen, and R. J. Voss, pp. 489–508. Eisenbrauns, Winona Lake, Indiana.

2020 Exploring the Ritual Divide: Diverse Faunal Behaviors at Iron I Ashkelon. In *Final Reports of the Leon Levy Expedition to Ashkelon: 7. The Iron Age I,* edited by L. E. Stager, D. M. Master, and A. J. Aja, pp. 705–726. Eisenbrauns, Winona Lake, Indiana.

Wapnish, P. and B. Hesse

1988 Urbanization and the Organization of Animal Production at Tell Jemmeh in the Middle Bronze Age Levant. *Journal of Near Eastern Studies* 47:81–94.

1991 Faunal Remains from Tel Dan: Perspectives on Animal Production at a Village, Urban and Ritual Center. *Archaeozoologica* 4: 9–86.

Ward, D. and K. McKague

2007 *Water Requirements of Livestock.* Ontario Ministry of Agriculture, Food and Rural Affairs Fact Sheet. http://www.omafra.gov.on.ca/english/engineer/facts/07-023.pdf.

Wasserman, N.

2016 *Akkadian Love Literature of the Third and Second Millennium BCE.* Leipziger Altorientalistische Studien (LAOS) 4. Harrassowitz, Wiesbaden.

Watson, J. B.

1977 Pigs, Fodder, and the Jones Effect in Postipomoean New Guinea. *Ethnology* 16:57–70.

Watson, L.

2004 *The Whole Hog: Exploring the Extraordinary Potential of Pigs.* Smithsonian Books, Washington, DC.

Wattenmaker, P.

1987 Town and Village Economies in an Early State Society. *Paléorient* 13:113–122.

Wealleans, A. L.

2013 Such as Pigs Eat: The Rise and Fall of the Pannage Pig in the UK. *Journal of the Science of Food and Agriculture* 93:2076–2083.

Weber, J. A.

2001 A Preliminary Assessment of Akkadian and Post-Akkadian Animal Exploitation at Tell Brak. In *Excavations at Tell Brak: 2. Nagar in the Third Millennium BC,* edited by D. Oates, J. Oates, and H. McDonald, pp. 345–350. McDonald Institute for Archaeological Research and the British School of Archaeology in Iraq, Cambridge.

2006 Economic Developments of Urban Proportions: Evolving Systems of Animal-Product Consumption and Distribution in the Early and Middle Bronze Ages in Syro-Mesopotamia. PhD dissertation, Department of Anthropology, University of Pennsylvania, Philadelphia.

2008 Elite Equids: Redefining Equid Burials of the Mid- to Late 3rd Millennium BC from Umm el-Marra, Syria. In *Archaeozoology of the Near East,* Vol. VIII, edited by E. Vila, L. Gourichon, A. M. Choyke, and H. Buitenhuis, pp. 499–519. Maison de l'Orient et de la Mediterranée, Lyon.

Weber, M.

1992 [1930] *The Protestant Ethic and the Spirit of Capitalism.* Translated by T. Parsons. Routledge Classics, London.

Weber, S. L. and M. D. Price

2016 What the Pig Ate: A Plant Microfossil Study of Pig Dental Calculus from 10th–3rd Millennium BC Northern Mesopotamia. *Journal of Archaeological Science: Reports* 6:819–827.

Weingarten, S.
2007 Food in Roman Palestine: Ancient Sources and Modern Research. *Food and History* 5:41–67.
Weinstock, J.
1995 Some Bone Remains from Carthago, 1991 Excavation Season. In *Archaeozoology of the Near East*, Vol. II, edited by H. Buitenhuis and H.-P. Uerpmann, pp. 113–118. Backhuys Publishers, Leiden.
Weiss, E., W. Wetterstrom, D. Nadel, and O. Bar-Yosef
2004 The Broad Spectrum Revisited: Evidence from Plant Remains. *Proceedings of the National Academy of Sciences* 101:9551–9555.
Weiss, H.
2012 Quantifying Collapse: The Late Third Millennium Khabur Plains. In *Seven Generations Since the Fall of Akkad*, edited by H. Weiss, pp. 1–24. Harrassowitz, Wiesbaden.
Weiss, H., M.-A. Courty, W. Wetterstrom, F. Guichard, L. Senior, R. H. Meadow, and A. Curnow
1993 The Genesis and Collapse of Third Millennium North Mesopotamian Civilization. *Science* 261:995–1004.
Wex, M.
2005 *Born to Kvetch: Yiddish Language and Culture in All Its Moods*. St. Martin's Press, New York.
White, L. A.
1959 *The Evolution of Culture*. McGraw-Hill, New York.
White, S.
2011 From Globalized Pig Breeds to Capitalist Pigs: A Study of Animal Cultures and Evolutionary History. *Environmental History* 16:94–120.
Whitehead, H.
2000 *Food Rules: Hunting, Sharing, and Tabooing Game in Papua New Guinea*. University of Michigan Press, Ann Arbor.
Whitley, J.
2001 *The Archaeology of Ancient Greece*. Cambridge University Press, Cambridge.
Wiener, M. H.
2017 Causes of Complex Systems Collapse at the End of the Bronze Age. In *"Sea Peoples" Up-to-Date: New Research on Transformations in the Eastern Mediterranean in the 13th–11th Centuries BCE*, edited by P. M. Fischer and T. Bürge, pp. 43–74. Austrian Academy of Sciences, Vienna.
Wiggerman, F. A. M.
2010 Dogs, Pigs, Lamaštu, and the Breast-Feeding of Animals by Women. In *Cuneiform Monographs: 41. Von Göttern und Menschen*, edited by T. Abusch, M. J. Geller, S. M. Maul, and F. A. M. Wiggerman, pp. 407–413. Brill, Boston.
Wilcke, C.
1985 Liebesbeschwörungen aus Isin. *Zeitschrift für Assyriologie und Vorderasiatische Archäologie* 75:188–209.
Wilcox, J. T. and D. H. Van Vuren
2009 Wild Pigs as Predators in Oak Woodlands of California. *Journal of Mammalogy* 90:114–118.
Wilford, J. N.
1994 First Settlers Domesticated Pigs Before Crops. *New York Times*, 31 May. New York.

Wilken, R. L.
2012 *The First Thousand Years: A Global History of Christianity*. Yale University Press, New Haven.
Wilkens, B.
2000 Faunal Remains from Tell Afis (Syria). In *Archaeozoology of the Near East*, Vol IVB, edited by M. Mashkour, A. M. Choyke, H. Buitenhuis, and F. Poplin, pp. 29–39. ARC-Publicaties 32, Groningen.
Wilkins, A. S., R. Wrangham, and W. T. Fitch
2014 The "Domestication Syndrome" in Mammals: A Unified Explanation Based on Neural Crest Cell Behavior and Genetics. *Genetics* 197:795–808.
Wilkins, J. and R. Nadeau (editors)
2015 *A Companion to Food in the Ancient World*. Wiley-Blackwell, New York.
Wilkinson, T.
2003 *Archaeological Landscapes of the Near East*. University of Arizona Press, Tucson.
Wilkinson, T. A. H.
1999 *Early Dynastic Egypt*. Routledge, London.
Willcox, G., S. Fornite, and L. Herveux
2008 Early Holocene Cultivation Before Domestication in Northern Syria. *Vegetation History and Archaeobotany* 17:313–325.
Willcox, G. and D. Stordeur
2012 Large-Scale Cereal Processing Before Domestication During the Tenth Millennium cal BC in Northern Syria. *Antiquity* 86:99–114.
Willis, P.
1977 *Learning to Labour: How Working Class Kids Get Working Class Jobs*. Gower Publishing Company, Aldershot.
Wolf, C. A.
1988 Analysis of Faunal Remains from Early Upper Paleolithic Sites in the Levant. In *The Early Upper Paleolithic: Evidence from Europe and the Near East*, edited by J. F. Hoffecker and C. A. Wolf, pp. 73–95. British Archaeological Reports International Series 437, Oxford.
Wood, J. W., G. R. Milner, H. C. Harpending, and K. M. Weiss
1992 The Osteological Paradox: Problems of Inferring Prehistoric Health from Skeletal Samples. *Current Anthropology* 33:343–370.
Wossink, A.
2009 *Challenging Climate Change: Competition and Cooperation among Pastoralists and Agriculturalists in Northern Mesopotamia (c. 3000–1600 BC)*. Sidestone Press, Leiden.
Yasuda, Y., H. Kitagawa, and T. Nakagawa
2000 The Earliest Record of Major Anthropogenic Deforestation in the Ghab Valley, Northwest Syria: A Palynological Study. *Quaternary International* 73:127–136.
Yasur-Landau, A.
2010 *The Philistines and Aegean Migration at the End of the Late Bronze Age*. Cambridge University Press, Cambridge.
Yener, K. A., C. Edens, J. Casana, B. Diebold, H. Ekstrom, M. Loyet, and R. Özbal
2000 Tell Kurdu Excavations 1999. *Anatolica* 26:31–117.
Yeomans, L., T. Richter, and L. Martin
2017 Environment, Seasonality and Hunting Strategies as Influences on Natufian Food Procurement: The Faunal Remains from Shubayqa 1. *Levant* 49:85–104.

Yeshurun, R.
2013 Middle Paleolithic Prey Choice Inferred from a Natural Pitfall Trap: Rantis Cave, Israel. In *Zooarchaeology and Modern Human Origins: Human Hunting Behavior During the Later Pleistocene*, edited by J. L. Clark and J. D. Speth, pp. 45–58. Springer, New York.

Yeshurun, R., G. Bar-Oz, and M. Weinstein-Evron
2007 Modern Hunting Behavior in the Early Middle Paleolithic: Faunal Remains from Misliya Cave, Mount Carmel, Israel. *Journal of Human Evolution* 53:656–677.

Yoskowitz, J.
2010 In Israel, a Pork Cookbook Challenges a Taboo. *New York Times*, 28 September. New York.

Zagarell, A.
1986 Trade, Women, Class, and Society in Ancient Western Asia. *Current Anthropology* 27:415–430.

Zeder, M. A.
1988 Understanding Urban Process through the Study of Specialized Subsistence Economy in the Near East. *Journal of Anthropological Archaeology* 7:1–55.
1991 *Feeding Cities: Specialized Animal Economy in the Ancient Near East*. Smithsonian Institution Press, Washington, DC.
1994 After the Revolution: Post-Neolithic Subsistence in Northern Mesopotamia. *American Anthropologist* 96:97–126.
1996 The Role of Pigs in Near Eastern Subsistence: A View from the Southern Levant. In *Retrieving the Past: Essays on Archaeological Research and Methodology in Honor of Gus W. Van Beek*, edited by J. D. Seger, pp. 297–312. Cobb Institute of Archeology, Winona Lake, Indiana.
1998a Environment, Economy, and Subsistence in Northern Mesopotamia. *Bulletin of the Canadian Society for Mesopotamian Studies* 33:55–67.
1998b Pigs and Emergent Complexity in the Ancient Near East. In *Ancestors for the Pigs*, edited by S. Nelson, pp. 109–122. MASCA Research Papers in Science and Archaeology 15. University of Pennsylvania Museum of Archaeology and Anthropology, Philadelphia.
2003 Food Provisioning in Urban Societies: A View from Northern Mesopotamia. In *The Social Construction of Ancient Cities*, edited by M. L. Smith, pp. 156–183. Smithsonian Books, Washington, DC.
2008 Domestication and Early Agriculture in the Mediterranean Basin: Origins, Diffusion, and Impact. *Proceedings of the National Academy of Sciences* 105:11597–11604.
2009 The Neolithic Macro-(R)evolution: Macroevolutionary Theory and the Study of Culture Change. *Journal of Archaeological Research* 17:1–63.
2011 The Origins of Agriculture in the Near East. *Current Anthropology* 52:S221–S235.
2012a The Broad Spectrum Revolution at 40: Resource Diversity, Intensification, and an Alternative to Optimal Foraging Explanations. *Journal of Anthropological Archaeology* 31:241–264.
2012b Pathways to Animal Domestication. In *Biodiversity in Agriculture: Domestication, Evolution, and Sustainability*, edited by P. Gepts, T. R. Famula, and R. L. Bettinger, pp. 227–259. Cambridge University Press, Cambridge.
2015 Core Questions in Domestication Research. *Proceedings of the National Academy of Sciences* 112:3191–3198.

2017 Out of the Fertile Crescent: The Dispersal of Domestic Livestock through Europe and Africa. In *Human Dispersal and Species Movement: From Prehistory to the Present*, edited by N. Boivin, R. Crassard, and M. Petraglia, pp. 261–303. Cambridge University Press, Cambridge.

2018 Why Evolutionary Biology Needs Anthropology: Evaluating Core Assumptions of the Extended Evolutionary Synthesis. *Evolutionary Anthropology* 27:267–284.

Zeder, M. A. and S. R. Arter

1994 Changing Patterns of Animal Utilization at Ancient Gordion. *Paléorient* 20:105–118.

Zeder, M. A. and M. D. Spitzer

2016 New Insights into Broad Spectrum Communities of the Early Holocene Near East: The Birds of Hallan Çemi. *Quaternary Science Reviews* 151:140–159.

Zuckerman, S.

2007 "... Slaying Oxen and Killing Sheep, Eating Flesh and Drinking Wine ...": Feasting in Late Bronze Age Hazor. *Palestine Exploration Quarterly* 139:186–204.

Index

For the benefit of digital users, indexed terms that span two pages (e.g., 52–53) may, on occasion, appear on only one of those pages.

'Ain Ghazal, 55–56
'Ain Mallaha, 31–32
'Apiru, 119–20
4.2ka Event, 69, 76–77
Aelian, 135
Akitu festival, 134
Akrotiri-Aetokremnos, 37
Alevi Muslims, 189–90
Alexander, 142, 155
Amarna, 80–82
Amazigh, 178
Amsterdam, 190
Animal Rights, 21–22, 136, 182–83
Anthony, Saint, 170–71
Antiochus IV, 155
Arabs
  Early history, 172
  Origins of Islam, 172
Arslantepe, 57
Arsur, 177–78
Ashkelon, 125–26
Asiab, Teppeh, 39
Assyria, 134
Augustine of Hippo, 168
Ayios Konstantinos, 148

Baal, 87
Benedictions of Labarna, The, 84
Bet Yerah, Tel, 73–74
Beth Shean, 127
Beydar, Tell, 152
Bible
  Writing of, 117, 118–19, 122, 130–33
Bigotry
  Hate crimes involving pork, 6, 135–36, 160–61, 163, 168
  Against Jews, 99, 168, 186
    Judensau, 186
  Against Muslims, 163–64, 186, 188
    Trump, Donald, 163–64
  Against Christians, 185–89
Book of the Dead, 88–89
Brak, Tell, 69

Caesarea, 157, 158, 159
Caledonian Boar, 149–50
Carthage, 136
Çatalhöyük, 55–56
Cattle, 64–65, 77
Çayönü Tepesi, 43–44, 52
Chalcolithic, 56
Chassidic, 162
Choga Mami, 55–56
Christianity, 158–59, 164–68
  Pig-keeping and, 177
Chthonic rituals, 85–86, 101–2
Circumcision, 129, 164, 165, 173–74, 188
Class,
  Development of, 56–58
  Pig-keeping and, 69–70, 72, 75–76, 106–7, 124, 151–52, 166
  Record-keeping and, 77–78
  Archaeological bias and, 82
  Taboos and, 106–7, 112–13
Columella, 146–47
Comana, 135–36, 160–61
Commensalism, 18–20, 36
  As pathway to domestication, 41
Complex societies, 48–49, 56–57
Crusades, 172–73, 177–78
Cyprus, 37

Dendara, 72
Dental calculus, 53

Diaspora, 123–24, 131–32, 135
Diener, Paul, 106–7
Disgust, 92–93, 100, 109–10, 189–90
DNA, see "Genetics"
Dogs, 27–28, 32, 36, 84, 85, 87, 89–90, 169
Domestication
  Definition, 16
  Features of domestic animals, 17–18, 34
  Pig domestication, 18–19, 34–35, 41–42
    China, 19–20
  Zooarchaeological study of, 20, 43–44
  Appearance of first domestic
      pigs, 44–45
Domuztepe, 55–56
Douglas, Mary, 103–4

Ebla, 75
Egalitarianism, 48–49, 55, 120–21
Ekron (Tel Miqne), 125–26
Elephantine, 123–24
Elephants, 150–51
Eloquent Peasant, The, 78–79
Elunum (festival), 83
Ephesus, 151–52
Epistle of Barnabas, 167
Equids, 65
Eros, 148
Erotic symbolism, 84
Erymanthian Boar, 149–50
Ethnicity
  Archaeological detection of, 125
Ethnogenesis, 108, 119
  In Judah, 122, 128
Eumaeus, 148–49
Europe, 51

Feasting, 38–39, 55, 56–57, 83
Fertility rituals, 84, 148
  Thesmophoria, 85–86, 148, 152
  Frazer's theories of pigs
      and, 101–2
Finkelstein, Israel, 119
Foodways, 125–26, 129
Frazer, James, 92, 101, 104
Freud, Sigmund, 95

Game Management, 19, 36, 37
Gath (Tell es-Safi), 125–26
Genetics, 51–52, 137–39
Gerasene demoniac, 160, 169
Göbekli Tepe, 39–40
Gordion Tepe, 124, 152–53
Greek culture, 142
  Pig husbandry in, 145
  Pig sacrifice in, 148
  Relation to Judaism, 154–55, 156–57
Greek Magical Papyri, 135
Gritille Höyük, 177

Hallan Çemi, 35–37, 39
Hammurabi's Code, 66
Hamoukar, Tell, 69
Harris, Marvin, 76–78
Hasmonean dynasty, 156
Hatra, 152
Hazor, Tel, 127
Herod, 157–58
Herodotus, 98, 135
Hisn al-Bab, 177
  History of, 180–81
Homo erectus, 28
Hui Muslims, 190
Hunting of boar
  Elites and, 56, 157
  Masculinity and, 29, 50–51, 87,
      148–49, 183
  Modern-day, 183–84
Hyksos, 89
Hypoplasias, 44, 53, 153–54

Ibn Khaldun, 63
Iliad, The, 149–50
Informal economy, 71, 179–81
Institutions
  Definition of, 62–63
  Uses of livestock, 64–66
  Uses of pigs, 69–70
Instrumentalism, see Ethnogenesis
Isotopes, 154
Israel
  Pig husbandry, 182–83
  Controversies about pork, 181–82
  Settlements and wild boar, 184
Israelites, 116–17,
  Origins of, 119–21
  Pig taboo among, 126–27, 128–29

Jerusalem, 121–22, 123
Josephus, 154, 155, 157

Kagemni, Tomb of, 83
Kom al Ahmar, 151

Lagash, 78
Lamashtu, 87
Lard, 79–80
Larson, Greger, 137
Late Bronze Age
  Pig husbandry in, 80–82
  Collapse, 116
Leilan, Tell, 69
Lidar Höyük, 137–38
Lower Paleolithic, 28

Maccabees, 156, 160–61
Maimonides, 99–100
Majdal Yaba, 178
Manichaeism, 142–43
Marj Rabba, 73–74
Meat, symbolic importance of, 38–39, 94–95
Megiddo, 127
Meiri, Meirav, 138–39
Methen, Tomb of, 78
Mezraa-Teleilat, 53–54
Middle Paleolithic, 28–29
Milgrom, Jacob, 102
Mishnah, see Talmud
Mozan, Tell, 69, 85–86, 102, 152
Muslim-Christian Relations, 175–76, 185–86
Muslim-Jewish Relations, 173–74; 175–76, 185

Natufian, 31–32
Naukratis, 124
Nazareth, 178
Neanderthals, 27
Neolithic, traits of, 33
New Guinea, 22, 35, 45–46, 93, 100

Odyssey, *The*, 29, 148–49
Ohalo II, 31
Orthodoxy, 143, 165
Orthopraxy, 141, 143, 173

Palestinians, 178, 184
Parthian Empire, 142–43, 152
Pastoralism, 109–10
Paul of Tarsus, 165–66
Pella, 177
Pergamon, 151
Persia, 117–18, 123–24, 155
Petronius, 99
Philistines, 125
  Pig husbandry and, 125–26
  Genetic turnover in pigs and, 138–39
Philo, 99, 143, 161, 165
Phoenicians, 136
Pig Husbandry
  Industrial, 21–22
  Intensive, 21, 45–46, 52–53, 152–54
  Extensive, 22–23, 52, 78
  Environmental conditions and, 50, 51, 66, 76–77, 105
Pig principles, 24–25, 96, 126
Pliny, 150–51
Poggio Colla, 145
Pollan, Michael, 191
Pollution, 89–90
Porphyry of Tyre, 136
Prey pathway, 41
Primordialism, see Ethnogenesis
Prodigal Son, Parable of the, 169

Qasile, 125–26
Qiryat Ata, 73–74
Quran, 173–76

Rabbinic Judaism, 164
Redding, Richard, 110–11
Reform Movement (in Judaism), 192
Robkin, Eugene, 106–7
Roman culture, 142–43
  Pig husbandry in, 145, 146–47
  Pork as delicacy in, 145–46
  Pigs as wealth in, 147
Roman-Jewish relations, 98–99, 154–59
  Pigs and, 159–64
Rosenblum, Jordan, 161, 163

Sacrifices of pigs, 83–84
Sagalassos, 153
Saqqara, 83–84

Sassanian Empire, 142–43, 152, 172
Scott, James, 66
Sea Peoples, 116
Secondary products
  Definition of, 49
  Revolution, 59
  Wool, 59, 65, 75
  Traction, 59, 64
Schwartz, Seth, 88
Sheep and goats, 50, 59, 65
Silberman, Neil Asher, 119
Simoons, Frederick, 107–8, 109–10
Skhul, 28–29
Soap, 79–80
Soler, Jean, 107–8
Strabo, 135–36
Substitution (in ritual), 86–87, 148, 169
Suidae, 13
Suovetaurilia, 148
Sus scrofa
  Evolution, 13–14
  Intelligence, 14
  Reproductive Behavior, 14–15
  Diet, 15
  Water Requirements, 15
Sus strozzi, 13–14, 28
Swans, 114–15
Swine Flu, H1N1, 1
Sykes, Naomi, 114–15

Taboo
  Etymology, 92
  Definition, 92–93
  Radioactivity metaphor for, 93
  Gender and, 93
  Liminality and, 93
  Meat and, 94–95
  Composition of biblical, 122
  Archaeological detection of, 126–27
Tacitus, 98–99, 104, 107–8
Talmud, 156–57, 160, 164
Thesmophoria, see Fertility rituals
Traction, see Secondary products
Tradition, 71, 73, 113
Transgression, 95, 189–93
Tribal ideal, 120, 130–31
Trichinosis, 100
Troy, 72
Tsaf, Tel, 55

Üçağızlı Cave, 29
Ugarit, 87
Upper Paleolithic, 30–31
Urbanism, 62, 69, 82
  Pig husbandry and, 69, 127, 145, 177–78
Uruk, 78

Varro, 146, 148

Warfare, 65, 150–51
Wisdom literature, Babylonian, 124, 127, 130, 134
Wool, see "Secondary products"

Yiftahel, 73–74
Younger Dryas, 27–28

Zabaleen
  2009 pig cull, 1–2, 180–81
  Pig husbandry, 180, 181
  History of, 180–81
Zeder, Melinda, 33, 41
Ziyaret Tepe, 124
Zooarchaeology, 4–5
  Feasting, 39
  Domestication at Çayönü Tepesi, 43–44
  Pig husbandry strategies, 52–54
Zoroastrianism, 142–43